Shyamasree Ghosh
Nanomaterials Safety

Also of interest

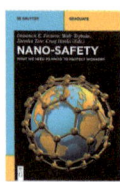

Nano-Safety
What We Need to Know to Protect Workers
Fazarro, Trybula, Tate, Hanks (Eds.), 2017
ISBN 978-3-11-037375-2, e-ISBN 978-3-11-037376-9

Nanoparticles
Jelinek, 2015
ISBN 978-3-11-033002-1, e-ISBN 978-3-11-033003-8

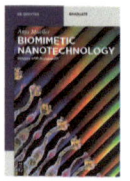

Biomimetic Nanotechnology
Senses and Movement
Mueller, 2018
ISBN 978-3-11-037914-3, e-ISBN 978-3-11-037916-7

Nanoscience and Nanotechnology
Advances and Developments in Nano-sized Materials
Van de Voorde (Ed.), 2018
ISBN 978-3-11-054720-7, e-ISBN 978-3-11-054722-1

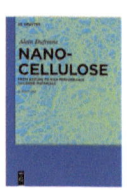

Nanocellulose
From Nature to High Performance Tailored Materials
Dufresne, 2017
ISBN 978-3-11-047848-8, e-ISBN 978-3-11-048041-2

Shyamasree Ghosh

Nanomaterials Safety

Toxicity and Health Hazards

DE GRUYTER

Author
Dr. Shyamasree Ghosh
National Institute of Science Education and Research (NISER), Bhubaneswar
School of Biological Sciences
Jatni Campus
PO: Bhimpur-Padanpur
Via-Jatni, District: Khurda 752050
Orissa, India
shyamasree_b@yahoo.com

ISBN 978-3-11-057808-9
e-ISBN (PDF) 978-3-11-057909-3
e-ISBN (EPUB) 978-3-11-057836-2

Library of Congress Control Number: 2018956856

Bibliographic information published by the Deutsche Nationalbibliothek
The Deutsche Nationalbibliothek lists this publication in the Deutsche Nationalbibliografie; detailed
bibliographic data are available on the Internet at http://dnb.dnb.de.

© 2019 Walter de Gruyter GmbH, Berlin/Boston
Typesetting: Integra Software Services Pvt. Ltd.
Printing and binding: CPI books GmbH, Leck
Cover image: eugenesergeev / iStock / Getty Images Plus

www.degruyter.com

Dedicated to my parents.

Acknowledgment

I deeply and sincerely acknowledge untiring efforts of all the scientist, researchers, technologists and policy makers working in the field of nanotechnology, School of Biological Sciences, National Insititute of Science Education and Research (NISER) at Bhubaneswar; the Department of Atomic Energy, Government of India; my alma mater, Presidency College, Kolkata (now Presidency University); Ballygunge Science College, the University of Calcutta; Indian Institute of Chemical Biology (IICB), Kolkata; the Council of Scientific and Industrial Research (CSIR), India; and Jadavpur University are acknowledged.

I sincerely thank Professor Dhrubajyoti Chattopadhyay, Vice Chancellor, Amity Univeristy, Kolkata, Professor Ashoke Ranjan Thakur, Ex VC West Bengal University of Technology, India and my guide Dr. Chitra Mandal, Former Director, IICB, India for their guidance.

I thankfully acknowledge all the staff and the esteemed members of DeGruyter for their immense support in the making of the book.

Shyamasree Ghosh, PhD

https://doi.org/10.1515/9783110579093-201

Contents

1 Nanomaterials safety and health hazard

Abstract: Nanotechnology involves matters of size in a scale of 1/1,000,000,000 of a meter and nanomaterial devices range in size from few to about 100 nm. Since the original inception of the concept, considerable progress has been made in the field of nanotechnology and its diverse applications have touched almost every aspect of the human life from daily use personal products to medicines and to nanodevices for diagnosis, therapy, food, chemicals and energy.

Despite the tremendous progress of nanotechnology and their applications in different products, the nature of engineered nanoparticles (NPs), environmental NPs and ultrafine particles in air from sources, such as diesel, remains uncharacterized, and their effects on the health and environment remain largely unknown. They also pose to be an occupational health hazard to workers involved in their manufacture and handling and other workers who are unintentionally exposed to NPs at workplaces. In the recent years, nanotoxicity and safety have attracted worldwide attention, with more involvement from the industry, academia and legislations to understand the effects of NPs, to design NPs that are biocompatible and less or nontoxic, to implement regulatory measures so that workers are safe while handling NPs and legislations to control unsafe use of NPs.

In this chapter, we discuss (i) a brief history of nanotechnology, (ii) its advances and applications in human benefit, (iii) hazards associated with NPs, (iv) nanosafety practices from a global perspective, (vi) regulations in NP research and (vii) challenges in NP research and the road ahead.

Keywords: Nanotechnology, bionanotechnology, stem cell (SC), biosensors, adjuvants, World Health Organization (WHO), tuberculosis, biodegradable, biocompatible, malaria, *Plasmodium falciparum,* antimalarials, artemisinin, artemisinin-based combination therapies, plasmid DNA, quantum dots, blood–brain barrier, nanocrystalline silicon, chemiluminiscence or bioluminescence, carbon nanotube (CNT), magnetic random access memory, high definition, graphene, toxicokinetic, inhalation, skin, dermal, ultrafine nanoparticles (UNPs), polymethylmethacrylate, high-aspect ratio nanoparticles, zeta potential, material safety data sheet, environmental health and safety, engineered NPs, gastrointestinal tract, in silico

1.1 Introduction

Nanotechnology is a branch of science and technology encompassing matter that functions on a molecular or an atomic scale. The term "nano" has originated from the Greek word "nanos" referring to dwarfs. Nanotechnology involves items, matter or functional operations at a nanometer scale that is 1/1,000,000,000, that is, one billionth of a meter, and nanomaterial devices range in size from a few to about 100 nm.

https://doi.org/10.1515/9783110579093-001

Nanotechnology has a wide range of applications in the human life, ranging from the creating new materials to devices and sensors, and it also finds applications in plants, medical, healthcare, agriculture, energy, communication, computation, electronics and chemical industries.

Bionanotechnology is an application of biotechnology and nanotechnology towards human benefits. It has touched our day-to-day lives, starting from personal healthcare products such as fluoride crystals in toothpaste to making of drugs and applications in targeting deadly diseases like infectious diseases and cancer and to applications in devices to control functions of stem cell (SC) that are undifferentiated progenitor pluripotent cells capable of self-renewal and differentiation into progeny cells. Thus, they play an important role in regenerative medicine.

Applications of nanomaterials in the development of medical appliances, diagnostics, preventive, therapeutics, pharmaceuticals, personal healthcare products, environment, and industries (Figure 1.1) are an active area of global research and are also a step towards modern medicine and treatment. Although nanotechnology finds applications in almost every aspect of our lives, their usages raise concerns regarding the potential harmful effects to the health and environment.

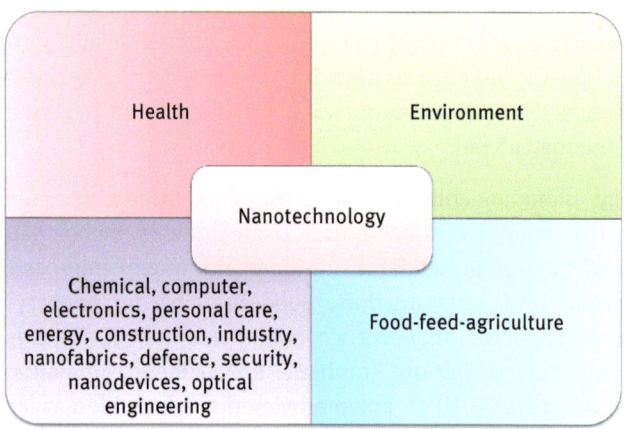

Figure 1.1: Applications of nanotechnology for human benefit.

1.2 History of nanotechnology

Nanotechnology has found applications in the chemical sector such as steel manufacturing, paint and rubber industry even before its current-day understanding. James Clerk Maxwell, a pioneering physicist, in the nineteenth century from Scotland, studied electricity and magnetism in the light of optics through his eponymous equations and found that electromagnetism had a microscopic basis. This is now being associated with a nanometer scale and forms an interesting domain of modern day research. Science of nanotechnology further progressed with the study of colloids, which included gels, sols and emulsions where one chemical remains dispersed in the other and it cannot be separated out from the other.

In the nineteenth century, colloids and gold sols were studied by the Vienna-born chemist Richard Adolf Zsigmondy (1865–1929). Zsigmondy was able to explain the heterogeneous nature of colloidal solutions, and in 1925, he was awarded the Noble Prize in chemistry. The study of nanoparticles (NPs) was pioneered by the renowned physical chemists, Irvin Langmuir (1881–1957) and Katherine B. Blodgett (1898–1910), in the twentieth century who developed the modern day science known as plasma physics.

Irvin Langmuir, known for his pioneering work on surface chemistry, elucidated the role of chemical forces at the interfacial surfaces between substances discovering the properties of adsorption, a phenomenon by which atoms and molecules show adhesion to a surface forming a layer of adsorbate. In 1932, he was awarded Nobel Prize in chemistry for his observation of adsorbate with monolayer thickness and his work in surface chemistry of matter.

The renowned American theoretical physicist Richard Feynman (1918–1988) in 1959 highlighted the importance of operating matter on a small scale in his speech "There's Plenty of Room at the Bottom." He envisaged the development of a functional nanomotor, and and writing of Encyclopedia Britannica on a pin head with the concept of controlling and manipulation of atoms in the size range 10^{-9} m or a nanometer. His lectures led to the revolution in modern day physics. Richard P. Feynman together with Sin-Itiro Tomonaga and Julian Schwinger in 1965 were awarded Noble Prize in physics for the research in elementary particles and quantum electrodynamics with deep-ploughing phenomena. His ideas and research led to the revolution and development of modern day concepts and their applications in nanotechnology, which has touched almost every sphere of our lives.

A decade later in 1974, Norio Taniguchi (1912–1999), professor at Tokyo Science University in Japan, coined the term *nanotechnology*. A few years later, K. Eric Drexler, an engineer, worked in the area of molecular nanotechnology, and in 1991 he worked on his thesis "Nanosystems: Molecular Machinery, Manufacturing, and Computation" in MIT. Born in 1943 in Finland, Tuomo Suntola working in the field of materials science, developed a thin film growth technique called atomic layer deposition and patented this technology that led to the development of nanotechnology a step further.

Klaus Schulten's work on computational biophysics in 2000 led to the development of the field of bionanotechnology. Schulten around 2005 coined the term "computational microscope" to the process of imaging technique encompassing applications of nanosystems, with a vision of applications in a wide range of biological processes including photosynthesis to signaling by ion channels, neural network organization and brain function to mechanics of muscle protein. Schulten developed the concept of *molecular dynamics simulation* with peptides and gold particles, thus developing the concept of bionanoengineering and biology of protein in solution with a nanosystem of gold particles.

Since the 1980s the study of nanostructures and their properties has been developing in a completely new dimension. The discovery of scanning tunnel microscope and atomic force microscope (AFM) by IBM Zurich enabled atomic level visualization of materials. The year 1985 marked the discovery of buckyball structure of carbon consisting of 60 carbon atoms (Figure 1.2), which now has a wide range of biological applications.

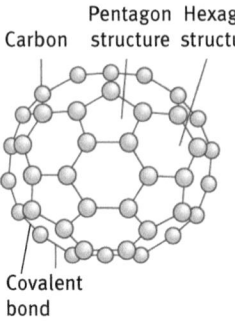

Figure 1.2: A structural representation of buckminster fullerene or buckyball wherein the carbon pentagon is linked to five carbon hexagon resulting in the curved structure.

In the 1990s research from IBM using 35 individual atoms of xenon (Xe) contributed to further development in the understanding of nanostructures. In 1991, carbon nanotubes (CNTs) were discovered. Semiconductor nanocrystals led to the development of quantum dots (QDs), which have properties in between semiconductors and discrete molecules.

In the late 1990s and early 2000s, nanotechnology developed across the world and both industries and academia were working towards its development and applications in various fields for human benefit. In the Unites States, the Office of Science and Technology Policy was established and an Interagency Working Group on Nanotechnology was developed. Different global organizations monitor the activities and developments in nanotechnology and take active part with the major focus on nanotechnology research encompassing nanoscale properties, synthesis, characterization, applications to human benefit, toxicity and economic benefits. The need of the hour is also the education and public awareness in this developing field.

1.3 Applications of NPs

Nanotechnology has developed considerably since its inception and its rapid expansion is touching our daily lives. Designed matter with desired and improved physical and chemical properties of materials by virtue of their strength, durability, reactivity, electrical conduction, biocompatibility and surface properties have enabled designing of tailor-made materials for the human benefit. These desired properties designed in the NPs have enabled their large-scale applications in different industrial sectors of information technology, security, medical devices, food, pharmaceuticals, medicine, diagnostics, therapeutics, electronics, environment, transportation and energy.

Nanotechnology is enabling improvisation of conventional methods and revolutionizing and generating new technologies and applications in the field of information technology, defense, security, medicine, transportation, information, computers, electronics, energy, food safety, environmental science, health and medicine (Figure 1.3). Beneficial applications of nanotechnology are discussed in subsequent sections.

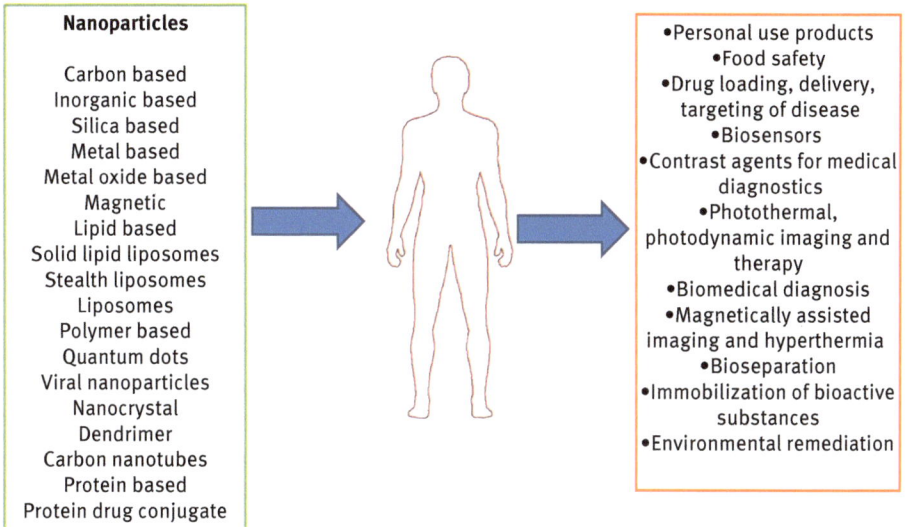

Nanoparticles

Carbon based
Inorganic based
Silica based
Metal based
Metal oxide based
Magnetic
Lipid based
Solid lipid liposomes
Stealth liposomes
Liposomes
Polymer based
Quantum dots
Viral nanoparticles
Nanocrystal
Dendrimer
Carbon nanotubes
Protein based
Protein drug conjugate

•Personal use products
•Food safety
•Drug loading, delivery, targeting of disease
•Biosensors
•Contrast agents for medical diagnostics
•Photothermal, photodynamic imaging and therapy
•Biomedical diagnosis
•Magnetically assisted imaging and hyperthermia
•Bioseparation
•Immobilization of bioactive substances
•Environmental remediation

Figure 1.3: Nanotechnology and NP in biomedical applications.

1.3.1 Applications in human health

Nanomedicine finds applications in generating improved properties of easy drug administration, targeted localized delivery and protection of the drug from degradation till it reaches the target site, intracellular delivery, prolonged exposure to drug, stealth function, efficient drug activity with reduced dosage and side effects, improved patient comfort and effective management of treatment.

Nanoscale additives find applications in fabric surface treatments, thereby leading to properties for resisting bacterial growth in addition to resistance to wrinkling and staining. Clear nanoscale films are being developed with antimicrobial properties. They find applications as better treatment strategies. Solid-state nanopore materials are being employed in the development of novel gene sequencing technologies, with property of single-molecule detection at low cost and high speed and user-friendly approaches.

Nanoscale biosensors and electronics find applications in health monitoring in diagnostics, preventive and prognostic therapy. Gold nanoparticles (AuNps) due to their relatively inert nature find applications in the treatment of diseases including cancer. Nanobiotechnology finds applications in diagnosis and therapy of atherosclerosis and in regenerative medicine including bone, teeth and neural tissue modeling and engineering. It also finds applications in improvisation of adjuvants, delivery devices and design of vaccines.

An infectious disease such as tuberculosis (Tb) is a result of infection by *Mycobacterium tuberculosis* (*Mtb*) bacteria that affects the lungs. Tb is curable and preventable but is globally ranked as one of the first ten life-claiming diseases. In 2016, 1.7 million people died of this disease, with an estimate of 250,000 children and over 95% of deaths from Tb were recorded from low- and middle-income countries. Multidrug-resistant Tb (MDR-Tb) remains a major threat to public health in 2016 [1]. Drug resistance emerges due to inappropriate use of anti-Tb medicines, poor quality drugs and patients abandoning the treatment prior to completion. MDR-Tb does not respond to isoniazid and rifampicin, and requires extensive chemotherapy; in cases of severe drug resistance, often no treatments are available. The World Health Organization (WHO) estimated 600,000 new cases with resistance to rifampicin with 490,000 MDR-Tb [1]. Latent infection by *Mtb* can lead to a condition where individuals do not transmit the disease and show no symptoms of acute infection, but later may develop acute symptoms. It is envisaged by WHO to eradicate Tb by 2050 with effective technologies. The treatment of Tb through conventional approaches (Table 1.1) suffers from drugs with side effects, toxicity, low permeability, early degradation before reaching site and drug resistance and therefore effective delivery systems need to be developed, with prolonged effects and slow release and alternative routes of delivery. Nanotechnology holds great promise in designing biodegradable, stealth, biocompatible polymers in drug delivery with encapsulated drugs to enable slow and controlled release of drugs [2] and in diagnosis, treatment and prevention of Tb; however, their safety is being considered prior to being used by people. Low-cost NPs-based Tb diagnosis and therapy (Figure 1.4) by using diagnostic kits and NP-based drug carriers with stability and carrier capacity and controlled release of drugs, with increased drug availability, are being tested. Aerosol vaccine to prevent Tb infection is also being tested [1–3].

The development of nanodelivery systems through inhalation routes is being exploited for delivery of anti-Tb drugs directly to the lungs, which are the sites of infection; thus, local administration can reduce systemic effects of the disease [3].

Malaria is an infectious disease caused by *Plasmodium falciparum* (*P. falciparum*), claiming lives globally, and is of global concern [5]. Conventional therapy by antimalarials [6–7] suffers from the generation of increased drug resistance in *P. falciparum* parasites. Artemisinin-based combination therapies with increased efficiency and lowered resistance generation, although are the best available therapies in the current times, suffer from being costly and therefore are unaffordable, and due to

Table 1.1: Treatment and drawbacks in the treatment of Tb by conventional methods [4].

Drugs	Treatment	Drawbacks	Need
Isoniazid, pyrazinamide, rifampicin and ethambutol	First-line drugs	1. Fight drug-resistant Tb 2. Severe adverse effects 3. Appearance of MDR strains 4. Limited ability to penetrate granulomas 5. Reduced effects on dormant bacilli 6. The oral route associated with severe systemic side effects	1. Shorten Tb treatment duration 2. Prevent resistance 3. Reduce lung injury 4. Routes other than oral
Streptomycin, kanamycin, amikacin, capreomycin and viomycin, fluoroquinolones (ofloxacin, levofloxacin, gatifloxacin and moxifloxacin) and other oral agents (ethionamide, prothionamide, cycloserine, terizidone and paraamino salicylic acid)	The second-line drugs used when first-line drugs fail or MDR-Tb	1. Less effective 2. More toxic 3. Unavailable in many countries 4. High costs	
Bedaquiline and delamanid, newly approved	MDR-Tb treatment, when other alternatives are not available		

their poor water solubility they need to be administered at high doses to increase their bioavailability [6–7]. Thus, there lies a need for better delivery systems with improved and effective bioavailability. Combination therapy, together with targeted delivery with nanomediated drug delivery vehicles, is promising in malaria treatment [8] (Figure 1.5). Nanotechnology with improved drug delivery formulations offer promises, but issues pertaining to safety and applicability to human remain largely a modern-day research focus.

NPs find applications as vehicles for gene transfer and delivery, administration of DNA using the coding regions of viral or bacterial genome as vaccination strategies [9–13], thus enabling activation of humoral and cellular immune pathways especially in infectious diseases, viral diseases and cancer [14–15]. NPs also find uses

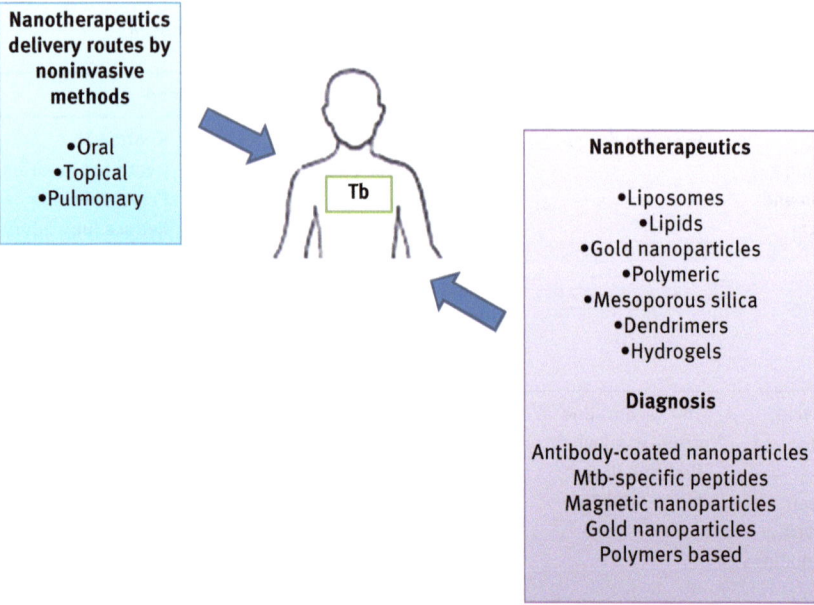

Figure 1.4: Applications of NPs in Tb research [1–4, 169].

in diagnosis and prognosis of infectious diseases (Figure 1.6). Nanodelivery of DNA saves it from degradation by cell enzyme as compared to naked DNA delivery and bears the advantage of enabling targeted and controlled delivery to antigen presenting cells including B cells, dendritic cells and macrophages [13].

Acquired immunodeficiency syndrome (AIDS) caused by the human immunodeficiency virus (HIV) is a deadly disease, claiming 36 million lives globally [16]. Although effective antiretroviral drugs are used in disease control and they prevent mass transmission, there is no cure for HIV; DNA vaccines are being researched to evoke humoral and cellular immunity against HIV. But delivery of naked DNA suffers from problems of degradation by enzymes including DNases and lysosomes, and therefore delivery of DNA by nanodevices or generation of nanovaccines using polymers, liposomes, lipid, peptide and inorganic nanomaterials is being tested for their efficacy as effective HIV DNA vaccines [17].

Delivery of RNA molecules together with inorganic NP is being researched to overcome the problems with DNA vaccines and traditional vaccines [18].

Silica NPs (42 nm), shows promise in transfer of DNA, gene and transfer and targeted molecular imaging of cancer due to their low cell toxicity [19] and is being tested in disease like cystic fibrosis, by designing of a immunosensor with detection system of laser-induced fluorescence to screen the immunoreactive trypsin (IRT) in cystic fibrosis [20] and treatment of skin diseases due to their controlled and sustained release of drugs to skin, efficient skin penetration of encapsulated drug, and therefore finds application in transcutaneous vaccination, transdermal gene

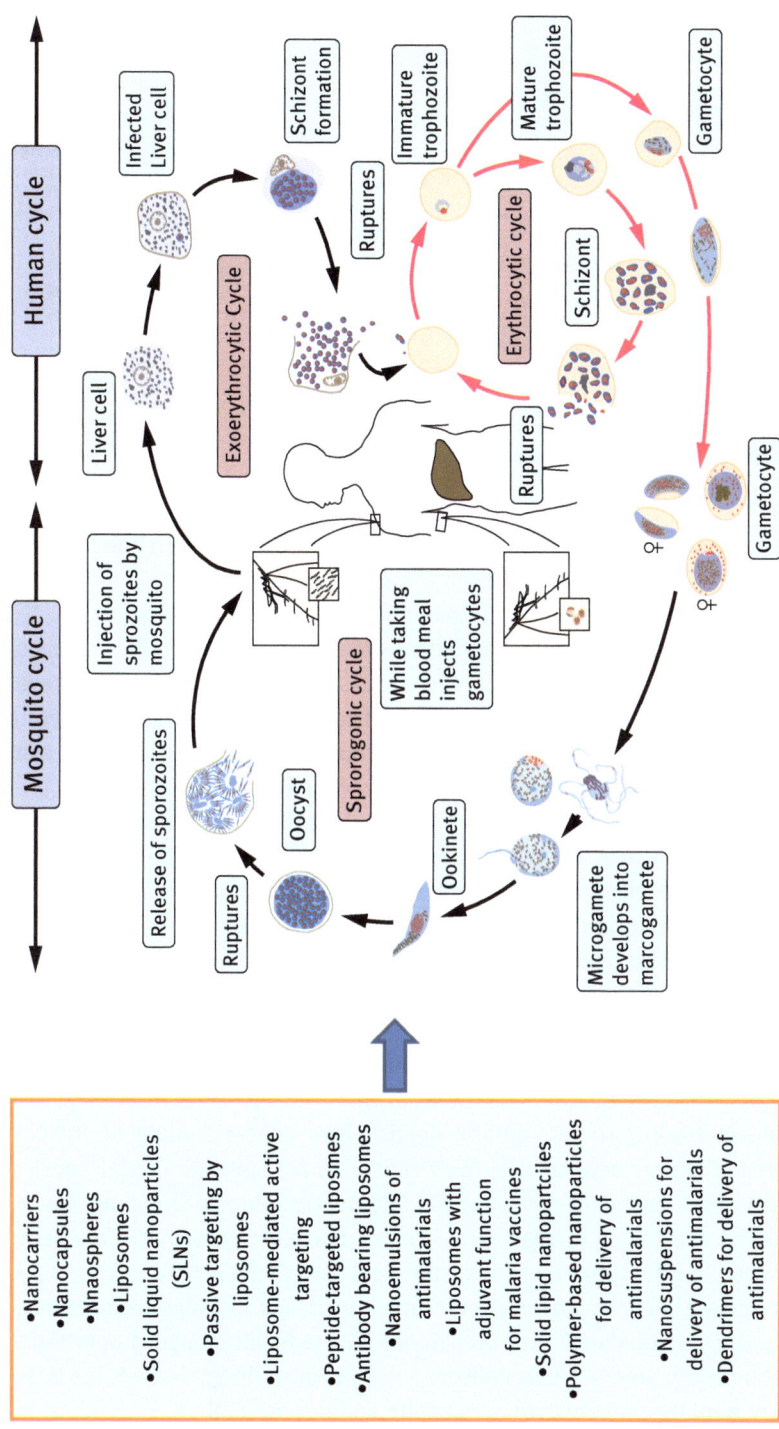

Figure 1.5: Life cycle of malarial parasite and application of NPs in malaria research [5–8, 170–171].

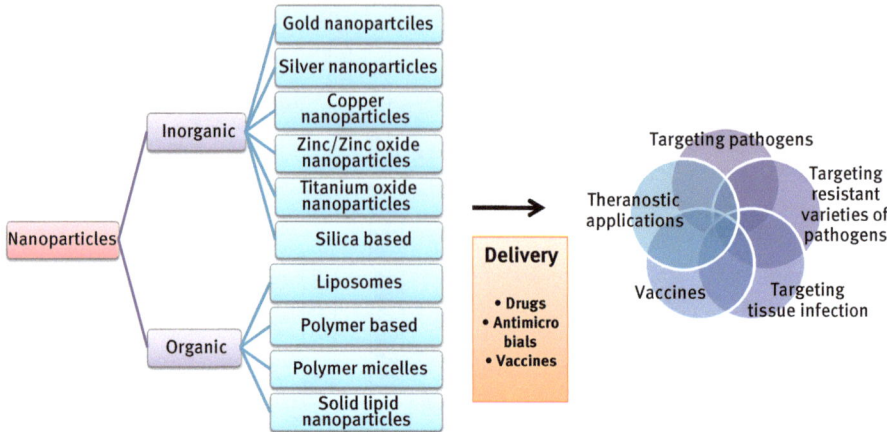

Figure 1.6: Applications of NPs in research of infectious disease.

therapy [21], but questions pertaining to their safety to the environment and health is under research [21–22].

NPs are being tested for specifically targeting eye retinal epithelium cells [23]. Lipid-based nanodelivery systems are reported to be biocompatible, with efficient muco-adhesion, increased precorneal retention and increased uptake by corneal epithelial cells for ocular delivery. Gene delivery by lipid-based nanodelivery systems to the retina in targeting retinal disorders is reported to have promising effects [24]. Multiple synthetic polymer NPs have shown promises in delivery of medicine to retinal cells [25].

Polygamma glutamic acid coated silver NPs conjugated to cyclic arginylglycylaspartic acid peptides that play an important role in cell adhesion to the extracellular matrix (ECM) are reported to cause controlled release of arginylglycylaspartic acid peptides at neovascularization sites or retinal angiogenesis cells and also cause apoptosis without affecting the normal retinal cells, thus revealing the potential of therapy of pathological retinal angiogenesis that might lead to visual impairment and complete blindness due to drug delivery by silver NPs (AgNPs 25).

According to the report submitted by WHO, 8.8 million deaths in 2015 were reported for people suffering from cancer. The highest rate of mortality is reported from lung cancer leading to 1.69 million deaths, liver cancer leading to 788,000 deaths, colorectal cancer contributing to 774,000 deaths, stomach cancer causing 754,000 deaths and breast cancer causing 571,000 deaths globally [26]. Lung cancer is being recorded alarmingly increasing in women. Recently nanodevices are being exploited in treatment for lung cancer detection and drug delivery [27].

QDs (<10 nm) are being tested for specific targeting of lung vasculature and tumors [28,29]. Design of "theranostic" multifunctional NPs with integrated properties of diagnosis, monitoring, targeted drug delivery and controlled drug release is the latest contribution of nanotechnology to human health and disease [30].

Indomethacin-loaded nanospheres of size <200 nm designed by surface modification with reduced toxicity has been proved to be better and more effective for targeted drug delivery and different formulations including Indomethacin-loaded methoxy poly (ethylene glycol)/poly(d,l-lactide) amphiphilic diblock copolymeric nanospheres have been tested in rodents [31]. Methoxy poly(ethylene glycol)/poly(epsilon-caprolactone) nanospheres have been proved effective in in vitro and in vivo delivery of hydrophobic drugs indomethacin and paclitaxel in normal mice [32] and development of spray-dried indomethacin-loaded polyester nanocapsules and nanospheres [33] is also being tested.

Carbohydrate binding ligands on polylactic-*co*-glycolic acid (PLGA) nanospheres could improve drug delivery and activity [34]. The biocompatible and biodegradable nature of PLGA with more versatility and controlled degradation rates over polylactic acid and polyglycolic acid finds potential applications in bone regeneration and has been approved by the American Food and Drug Administration (FDA) [35].

The blood–brain barrier forms a barrier to the central nervous system (CNS) that allows small lipid-soluble molecules to pass selectively through the capillary endothelial membrane but restricts the entry of pathogens or toxins and prevents the drug delivery, thus posing a major problem to the treatment of neurological and brain disorders. NPs offer advantages in designing of pharmaceutical that can cross the blood–brain barrier and offer effective and noninvasive way of treatment of cerebral and neurodegenerative diseases such as Alzheimer's diseases (AD) and Parkinson's disease (PD). Most commonly used nanocarriers include poly (butylcyanoacrylate), PLGA, poly (lactic acid) NPs, liposomes, inorganic metals and carbon QDs drug delivery system [36]. Magnetic NPs are being exploited as agents to monitor diagnosis, prognosis and outcomes of treatment in diseases of the CNS [37], but their toxicity and effects on the human tissues remain largely to be explored before their administration in human [38–42].

Polymer NP formulations of polyisobutylcyanoacrylate [43] and poly (ε-caprolactone) NP loaded with 5-fluorouracil have been reported for their potential applications in the treatment of colon cancer [44]. Anti-Tb drug-loaded gelatin and polyisobutyl-cyanoacrylate NPs have been reported to combat mycobacterial infection [45].

Restenosis is the recurrence of stenosis – a physiological condition involving narrowing of a blood vessel, like an artery, or large blood vessel, leading to restricted blood flow. It can also occur after treatment for clearing blockage and subsequently the blood vessel becomes renarrowed as an adverse event of endovascular procedures. Bare metal stents, although have been applied in the treatment of percutaneous coronary intervention for the last three decades, suffer from the problem of in-stent restenosis, requiring repeat revascularization, are associated with increased morbidity and mortality and pose a challenge to therapy. Fluorescent QDs [46–48] have shown promising applications in the detection of lymph nodes and their uptake by endothelial cells and find applications in the diagnosis and therapy of cardiovascular restenosis [49–51].

Biocompatible fluorescent carbon QDs have proved to be a promising platform for chemosensing and biosensing [52]. Attempts are being made to design paper-based, low-cost, easily available, sensitive, user-friendly and laboratory-free devices as diagnostics in the detection of biomolecules using fluorescent dyes and QDs carbon dots [53]. Conjugation of QDs with carbohydrates and lectins is thought to be generating very sensitive and specific probes to investigate disease diagnosis and prognosis [54].

The toxicity of NPs can be affected by surface area, number and concentration [55–57], and therefore research in nanotechnology is a step toward personalized medicine with better life expectancy rates.

From References [58–65].

1.3.2 Biosensors

Biosensor (Table 1.2) comprises of a biological macromolecule, including tissue, bacteria, yeast, antigens, receptors, antibodies, DNA and cell organelles, that reacts with the biological macromolecules and leads to generation of signal from their concentrations. Biosensors function by detecting the chemicals and estimate the product in the biochemical reaction. They find large-scale

Table 1.2: Nanotechnology in major biological applications.

Field	Applications
SC biology	NPs used in the detection of postimplantation or monitoring their migration to sites of lesion of SCs by optical imaging [58] find importance in the development of SC-based therapies Imaging of signaling pathways in SCs [58].
Detection of specific molecule	Development of optical tags Development of imaging techniques including surface-enhanced Raman scattering and surface-enhanced resonance Raman scattering [58,59] Development of nanoplex biotags and their detection by Raman microscopy [58, 60]
Medical applications	Impants Stents Biosensors [58, 61] and electrodes nanoarrays in the detection of toxic molecules and infectious agents, early diagnosis of genetic and infectious diseases [58, 62] Molecular imaging of targeted sites in diseases [58] Targeted delivery of NPs Cardiac diagnostics [58] Transdermal delivery improving bioavailability to target sites [58] Drug delivery in infectious diseases including malaria and Tb

Table 1.2 (continued)

Field	Applications
Food industry	Targeting food-related disorders such as obesity, diabetes and so on
	Applications in food-processing steps, sensors in food-processing detecting allergens, residues, unwanted matters and detection of microbes and pathogens
	Drug delivery through envelope of nutrients [58, 63]
Personal care products	Color management
	Improving brightness and whiteness of teeth [58, 64]
	Skin color management and protection from UV lights [58]
Vaccine adjuvants	Proteoliposome from *Neisseria meningitidis* bacterium [58, 65]
	Development of new adjuvant based on the interaction between pathogen recognition receptor on host cells and pathogen-associated molecular patterns

applications as tools for better imaging of supermolecular structures and for detecting their quantity of production and also find applications in early diagnosis and prognosis of a disease. The signal is most often detected by luminous radiation and methods including chemiluminiscence or bioluminescence are applied as signals. However, sensitivity or ability to detect minute quantities of signal is another area of research. Recently highly sensitive photodetecting systems are being developed, with a detection spectrum ranging from 300 to 500 nm. Thin film nanocrystalline silicon alloys connected with dielectrical layers are being applied for increased photosensitivity and detecting ability [58]. The conjugation of AuNPs [66] and metal nanoparticles has been reported to detect enzymes, antibodies and aptamers, which are short single-stranded oligonucleotides that can bind to various molecules with high affinity and specificity, DNA sequences and whole cells [67]. Nanobased biosensors find different applications (Figure 1.7). Biosensors enable detection of prostate-specific antigen [68]. Nanobiosensors offer advantages to suit specific applications, being flexible, fast and efficient sensing systems with increased selectivity and sensitivity. In the recent years, they are being designed as cheap and user-friendly devices and find many applications.

Food needs to be tested for chemical and microbiological agents including pathogenic bacteria that can spoil the food. Recently optical electrochemical colorimetric, fluorescent and immune bionanosensors are being developed [69] to detect even traces of food contaminants. Mycotoxin including aflatoxin (AF) and AFs B1 known for carcinogenic effects are released by some fungi in food and feed is known to spoil them [70]; therefore, their detection in trace amounts in food using nanotechnological biosensors holds great promise. Fast and highly sensitive, selective nanobiosensors capable of sensing insecticides, pesticides, herbicides, phenolics, dioxins, hormones, microbes, allergens and genetically modified foods are being developed [71].

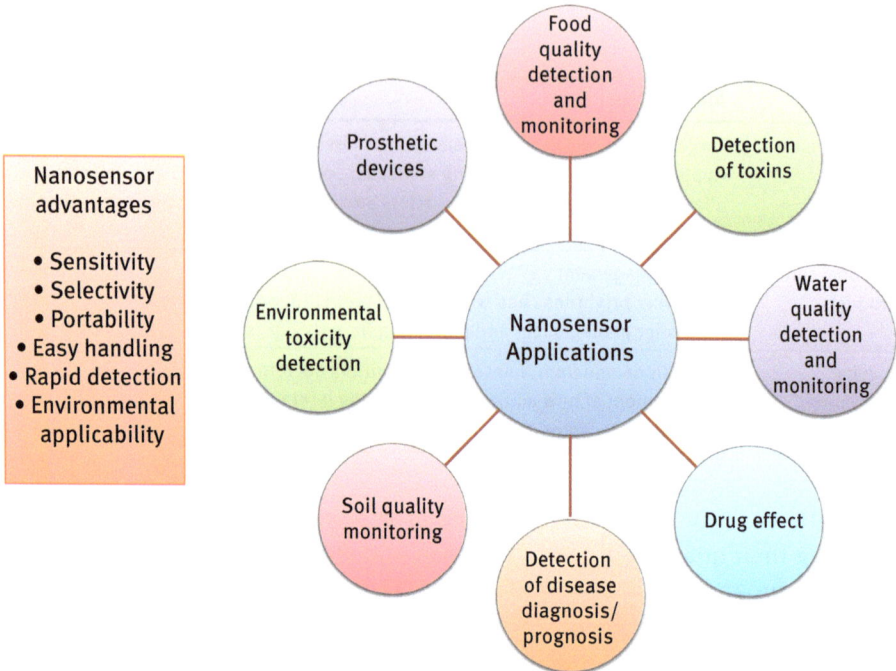

Figure 1.7: Nanosensor applications and advantages.

1.3.3 Applications in energy

Generation of renewable alternative, environment-friendly, low-cost energy resources, with less consumption and improved fuel production efficiency to meet global demands, is the need of the hour. Nanotechnology provides the potential of efficient energy utilization from conventional sources of energy including fossil and nuclear fuels and leverage the production of renewable energy sources like geothermal energy, sun, wind, water, tides or biomass through new technologies and promising inventions. Solar panels for efficient conversion of sunlight to electricity in generating cheap solar power are another aim of nanotechnology. In the field of primary energy sources, nanocoated, wear-resistant drill probes have developed low-cost efficient systems for oil, natural gas or geothermal energy. High-duty nanomaterials are utilized in the development of light, rugged, corrosion-resistant rotor blades and other components including gear boxes of wind and tide- power plants. Carbon nanotubes (CNTs) with epoxy coating find applications in designing of long, strong, but lightweight windmill blades. Nanotechnologies show promises of effective use of solar energy through photovoltaic systems, development of energy-efficient crystalline silicon solar cells, development of thin-layer solar cells using silicon, copper, indium, selenium, development of dye solar cells or polymer solar cells with high potential, optimization of the layer design of component structures of organic semiconductor mixtures and utilization of QDs and wires in generating over 60% efficient solar cell.

The efficient conversion of primary energy sources into electricity, heat and kinetic energy has been shown to be markedly improved through nanomaterial-mediated development of heat and corrosion protection layers for turbine blades in power plants or aircraft engines or by the use of lightweight materials like titanium aluminides. Nano-optimized membranes have potential applications in the separation of carbon dioxide (CO_2) and its utilization in environment-friendly power generation in coal power plants.

Scrubbers designed with CNT and membranes are being searched for their efficiency to separate CO_2 from power plant exhaust. Nanodevices, electrodes, catalysts and membranes have enabled efficient conversion of chemical energy and are being utilized in buildings and electronics. Thermoelectric energy conversion seems to be comparatively promising. Nanomaterial semiconductors have promising applications in efficient thermoelectric conversion and utilization of waste heat generated in automobiles, computers and human body for portable electronics in textiles. Nanodevices and thin-film solar electric panels are being designed to convert the waste heat to usable electrical power, thus enabling energy conservation and their effective utilization. Applications of CNTs in generating low-resistance wires are being exploited in the development of electric grid. It is finding applications in the development of efficient, lightweight, high-power density, with the use of long-lasting charge and quick charging batteries. Nanoporous metal hydrides or metalorganic compounds find applications in hydrogen storage. Nanosensory devices and electronic components find applications in control and monitoring of such grids.

Nanotechnology finds applications in the development of energy storage devices such as batteries and super capacitors. Development of lithium-ion batteries has enabled better storage of energy. Nanoporous zeolites find applications as heat stores in heating grids or in industries [72–75]. Nanoporous hydrogen storage materials are being designed for efficient fuel storage devices.

Nanotechnology promises efficient energy usage, thereby preventing unnecessary energy consumption both in industry and individual establishments. Nanocomposite based lightweight construction materials find applications in automobiles, transportation vehicles. NPs also find application in fuel additives, generation of low rolling resistance tires. Nanoporous materials find application in thermal insulation material, nanomaterials in control of light and heat flux like switchable glasses. They offer promising application of reduced energy consumption in buildings. Nanofoams find applications in the construction of superinsulation systems in building that minimize the convective heat transport. Gas tight polymer nanocomposites find applications in reducing hydrocarbon emissions from vehicle tanks.

Development of nanomaterials for the design of light, safe, efficient vehicles of transport with reduced fuel consumption like aircraft, spacecraft and ships is being applied. Low-cost nanomaterials achieved by nanoengineering of materials like aluminum, steel, asphalt, concrete and other cementitious materials, with better performance, resiliency and longevity, find applications as components of transportation devices. Efficient energy generating and transmitting qualities and applications of nanoscale sensors and devices with low cost along with increasing structural integrity and performance of bridges, tunnels, rails and parking structures find major applications.

1.3.4 Applications in environmental remediation

Nanotechnology seems to hold promise in the management of environmental pollution and sustainable environment over and above the conventional methods. Nanomaterials provide high surface areas and can concentrate in small volumes, and this property enables them to remove toxic contaminating agent in water like industrial effluents, polluted river water and so on, both in a cost effective and efficient manner, thereby playing a role in environmental remediation (Figure 1.8). Generating clean drinking water through rapid, low-cost detection and treatment of impurities in water is a major domain in which nanotechnology finds its application.

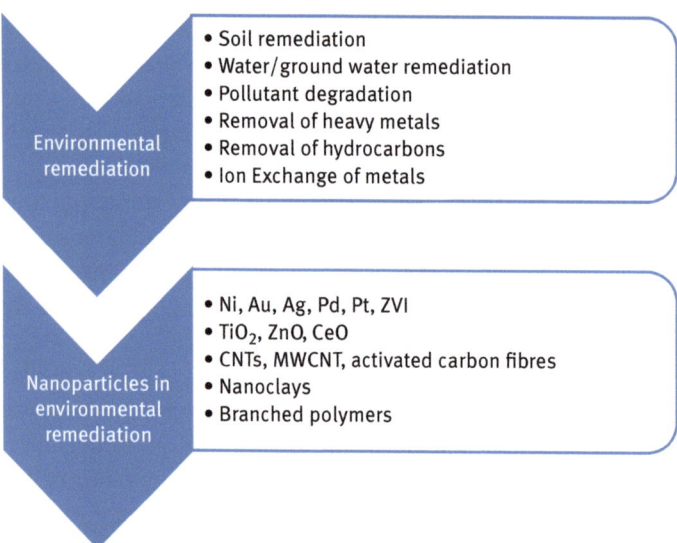

Figure 1.8: Nanoparticle (NP) applications in environmental remediation.

Nanopores in thin film membrane have enabled the energy-efficient desalination of water. Industrial water has been rendered clean and free of pollutants in ground water by applying nanotechnology. Spherical-shaped NPs, cylindrical nanorods, hexagonal pore-shaped NPs and diverse compositions of silicon dioxide (SiO_2), zirconium dioxide (ZrO_2), cadmium sulfide (CdS) and functional engineered NPs find applications in effective removal of contaminations including organophosphonate pesticides, heavy metals, cesium, mercury, radionuclides, polycyclic aromatic hydrocarbons (PAHs), dense nonaqueous phase liquids, carbon contaminants, oxometallelate anions and other contaminants in water source. Studies have revealed that 25 nm aggregates of spherical calcium oxide (CaO) nanoformulate aerogels can effectively destroy chlorocarbon solvents including carbon tetrachloride (CCl_4), chloroform ($CHCl_3$), tri-chloroethylene and tetra-chloroethylene [76]. The aerogel of 4.7 nm magnesium oxide (MgO) has been reported to react with 1-chlorobutane in water at higher temperatures. Doped with small amount of chlorine (Cl_2) or bromine (Br_2), MgO has revealed microcidal activity against Gram-positive and -negative bacteria and bacterial spores. When coated with surfactants, they were reported to remove pesticides from water [76].

Al_2O_3-MgO mixed phase has been reported to be effective in the destruction of CCl_4, SO_2 and paraoxan [77], and nanocrystals of copper oxide (CuO) and nickel oxide (NiO) have been reported to be effective in the destruction of SO_2 and paraoxan [78]. Bimetallic NPs find applications in the removal of dense nonaqueous phase liquids and toxic heavy metals, 10–30 nm ZVI NPs on polyflo resins has been proved effective against Cr(VI) and Pb(II) contamination [79], 17–97 nm diameter amphiphilic polyurethane (APU), NPs have been reported to remove PAH contamination, PEG modified urethane acrylate can remove PAH contamination sorbed into soil particles and nonaqueous phase liquids [80]. Nanostructured transition metal phosphonates including diverse chemistry of molybdenyl phenyl-phosphonate, zinc phenylphosphonate, zinc biphenylenebis(phosphonate) and Zr(IV) biphenylenebis(phosphonate) find applications as fine ion-exchange materials [81–85, 76].

Generation of self-assembled monolayers on mesoporous supports finds applications as sorbents to environmental contaminants like mercury (Hg) from ground water and act as a sensor material. A thiol-self-assembled monolayers on mesoporous support has been reported to remove Hg from contaminated oil [86], cadmium (Cd), gold (Au) and silver (Ag) [76]. CNTs find applications in selective binding and removal of radionuclides and CO_2 [86] from contaminated sources.

Nanofabrics, magnetic water-repellant NPs with applications in oil cleanup and spills have been developed. Nanotechnology-based sensors and solutions are being developed for sensing chemical or biological molecules and pathogens in air and soil. Polymer-based nanomembranes, less than 100 nm thickness, are known to act as molecular filters. Molecularly imprinted materials with strong membrane permeability and natural aquaporins, in proteins, are being used to design industrial membranes.

Powerful adsorbents and catalytic nanoscale titanium dioxide (TiO_2) find applications in removing contaminants. Nanosilver ceramic filters find applications in antibacterial and antiviral activities.

1.3.5 Textiles

Nanotechnology finds applications in designing antibacterial textiles, preventing cross contaminations, and methicillin-resistant *Staphylococcus aureus* resistant bandages with medical applications. Fungi has been reported to find application in replacing dangerous colorants thereby is being used in development of self-cleaning nanofrabrics. UV blocking, flame retardant applications, applications related to wireless biomonitoring of vital functions are few applications of nanotechnology. Stain repellant, wrinkle free and electrical conductivity are properties that NPs can confer to textiles and fibers. Garments using nanomaterials capable of sensing and responding to external signals of electricity, color or physiological agents are being designed [87].

1.3.6 Electronic applications

The field of computing and electronics such as transistors, magnetic random access memory, which are of small size, efficient and fast, has been gifted as nanotechnology can be applied well to this field. QDs are being used in high-definition displays and televisions with more energy efficiency and display of vibrant colors. Materials are being designed or engineered toward generation of flexible and stretchable electronics with medical, aerospace applications, smart phone and e-reader semiconductor nanomembrane. Graphene and cellulosic nanomaterials find applications in sensors and photovoltaics. Smart phone flash memory chips and thumb drives, ultraresponsive hearing aids, antipathogen coatings on keyboards and cell phone casings, smart cards/smart packaging and e-book reader display units are other applications where nanotechnology has revolutionized. In terms of cost and efficiency, NP copper suspensions offer advantages over lead-based solders and other hazardous materials commonly used to fuse electronics.

1.3.7 Food industry

Nanotechnology finds applications in the food industry (Figure 1.9), and engineered NPs capable of reducing fat, salt or sugar in food without altering the taste also find applications. Development of new packaging system to keep food fresh for longer

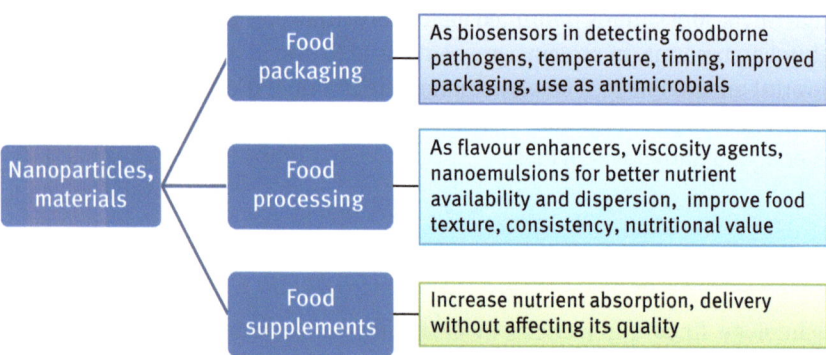

Figure 1.9: Applications of nanomaterials in the food industry.

periods of time even without having to store these in fridges to save energy consumption is another area where nanotechnology is being employed.

Identification of freshness of food and contamination mediated by microbes in packs is important in the interest of consumers across the globe. Nanotechnology finds applications in producing agents to improve color, texture and flavor of food. TiO_2 NPs, SiO_2 NPs and amorphous silica find applications as food additives. TiO_2 is being applied as an agent for coloring the powdered sugar coating on food. Nanomaterials including nanocomposites, polymers, zinc oxide (ZnO), magnesium oxide (MgO) NPs and amorphous silica, used in packing food are employed for their potential to act as mechanical barriers to contaminants. Aerosol of engineered water nanostructures (EWNS) is reported to effectively remove foodborne pathogens including *Escherichia coli (E. coli)*, *Listeria sp* and *Salmonella sp.* on containers for food preparation [88–96]. Nanomaterials enable enhanced absorption and bioavailability of nutrients.

Nanosensors find applications in checking food storage and during food transport in refrigerated devices. Biosensors are being used to detect microbial and other contamination and regulate the food environment, and they are also being applied to check nutrient deficiency in edible plants [88–96].

While companies across the globe are working on such applications of nanotechnology and their uses in the food industry, there is a global concern on the effects of NPs in food on the human gastrointestinal (GI) tract and overall health [97]. Silver hydrosol, a nanoscale form of silver due to its lack of information on its effects on human health, is recommended to be banned by the UKs Food Standard Agency. In March 2009, the European Food Safety Agency published opinion on applications of nanotechnology in the field of food and animal feed safety [98]. Assessing the potential risk involved in this process and regulations of use of engineered nanomaterials in food additives, enzymes, flavorings, food contact materials, novel foods, food supplements, feed additives and pesticides was published in May 2011 [98]. The Food and Drug Administration (FDA) in USA has issued a draft guidance for industrial use of nanomaterials in animal feed [99]. However, no internationally accepted standard protocols for testing of NP toxicity in food or feed are currently available, and this area calls in for more research in the coming years. A need for set up of databases on the research and development sector of nanotechnology and their applications in food industry has been realized. There is a growing need of risk assessment of NPs in food safety prior to their applications in foods. Research involving efforts from the academia, nongovernmental organizations and the industry is important to understand the food safety of NPs. Strict controls on the regulations and government legislations need to be worked out in this field of applications of NPs in food [97–99].

1.4 Hazards of NPs

Despite the potential advantages and promises that nanotechnology offers in applications, they also have harmful effects and therefore their use is restricted to testing their safety prior to human applications. The major hazards of toxicity of NPs lie in its chemistry, size, shape, composition, surface properties, routes of inhalation, interaction with biological macromolecules, solubility, nondegradable properties, bioavailability, tissue distribution and property to accumulate in routes of entry, tissues and cells.

Majority of the NPs remain uncharacterized in composition and properties and biological effects. Both in their nascent form and when engineered, they pose a major threat to the human health due to their uncharacterized nature and unknown interactions with the biological systems and unexplored effect on cells and tissues. Thus, the characterization by their size, shape, surface properties, chemical composition, fate of translocation, degradation processes and their toxicokinetic properties remain a major domain of modern-day research.

Most NP uptake leads to generation of toxic responses, irritation to the tissues, inflammatory response predominant in lungs and liver with release of chemokines, proinflammatory cytokines, inflammatory mediator molecules, free radical, reactive oxygen species (ROS) production, overall inflammation physiology, tissue fibrosis, necrosis, apoptosis and cancer.

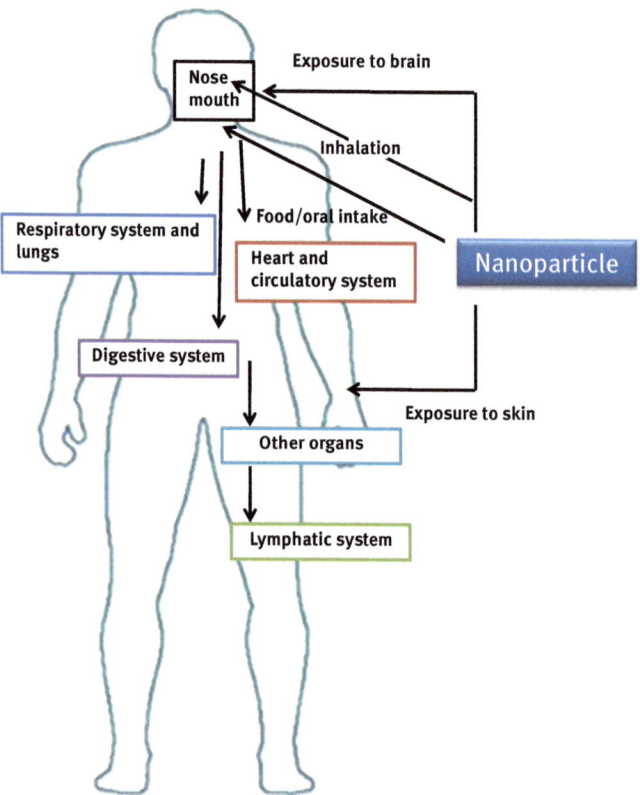

Figure 1.10: Routes of exposure of NPs and the human body.

Routes of NP exposure (Figure 1.10), their uptake and transportation and their effects on the tissues and cells and the way these nanomaterials are taken up into the body largely remain a major domain modern-day research.

1.4.1 Routes of NP exposure

NPs could be generated in the environment from three different sources including the natural source like forest fire and volcanic eruptions, anthropogenic sources like pollution, industrial activities of chemicals, polishing and so on, producing byproducts on combustion of diesel and manmade sources including manufactures of surface modified or engineered NPs with commercial applications. Multiple primary and secondary exposure pathways to nanomaterials are adversely affecting the health of workers and common man.

The routes of exposure to nanoparticles involves intake through inhalation by which nanoparticles move through the respiratory tract with three different parts including nose, airways and alveoli, oral and move through the intestinal tract, or dermal absorption through the skin or intentional exposure by injection by intravenous, intraperitoneal or intramuscular routes for treatment depending on the nanomaterial and the specific application [100–102]. Therefore, it is important to understand the routes of exposure and the systemic effects imposed. Although estimation, prediction or assessing occupational, consumer and environmental exposures are being studied currently, there is a dearth of data on the effects on pediatric individuals across the globe [102–104] as they are more sensitive than the adult population [99–104].

1.4.1.1 Exposure through inhalation

Inhalation is the most frequent route of entry of NPs in human and therefore their accumulation in the respiratory system is largely controlled by their physical characteristics, size, shape and weight of the particle, architecture of the respiratory tract including the anatomy and diameter of airways; these pose a major health hazard [105–114]. Inhaled NPs, either intentionally or unintentionally from air, affect mostly the lungs, alveoli, alveolar and bronchial tissue. Inhaled nascent or engineered NPs lead to inflammatory processes, oxidative damage, cytotoxicity, granuloma formation and sometimes even cancer. Smaller NPs and ultrafine nanoparticles (UNPs) and fibers can penetrate deeper and can deposit in the lung alveoli by diffusion. Larger fibers can block the airways. However, the factors determining severity are not completely understood. This finds particular importance as occupational health hazard and hazard from air pollution. The phagocytosis of the fibers leads to activation of macrophages, with release of chemokines, cytokines, ROS, oxidative stress and mediator of inflammation, causing pulmonary inflammation and fibrosis that could even lead to cancer. Inhaled NPs are known to cause damage to the cardiovascular system (CVS), leading to arrthymias, myocardial infarction and rupture of atherosclerotic plaques in the coronary artery with the formation of thrombus and blockage of blood vessels [76]. NPs are reported to cause damage to the neurological system, olfactory nerves, in addition to respiratory system. Particulate matter in air pollution is known to affect human health, causing pulmonary and cardiac disorders, but the components are not well understood and the downstream effects are not well studied [115–119]. Therefore, understanding their effects (Figure 1.11) and determining their toxic effects find importance in the biology and health of man.

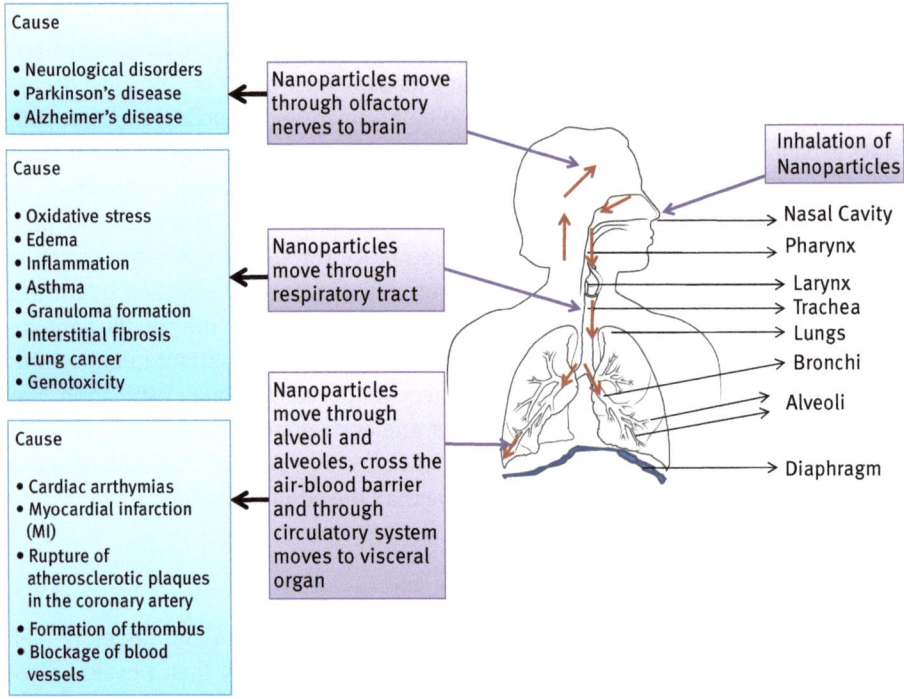

Figure 1.11: Exposure through inhalation and the different effects.

1.4.1.2 Exposure through skin/dermal routes

NPs can gain access to the body through being contact with the skin and GI tract. Skin is composed of layers including epidermis, dermis and subcutaneous layers. The outer layer of the epidermis is called the stratum corneum layer that contains a layer of keratin-containing cells. The skin is covered with hair. The skin acts as the first line of defense barrier to entry of pathogens and foreign particles, being a component of the innate immune system, thus conferring protection against infection. NPs can penetrate the skin through the pathways of (i) hair follicles, sweat glands and skin furrows; (ii) intracellular route through epidermal layers and (iii) intercellular route in between epidermal cells. The entry of NPs depends on its size, charge, morphology and material. TiO_2 NPs can penetrate the stratum corneum layer and can reach the epidermis and dermis [58, 102, 113]. Uptake of NPs through dermal exposures has been reported as exposures in workplace. Penetration of NPs into deeper layers of the skin is controlled by their smaller size. Glass and rockwool fibers can lead to contact dermatitis when exposed to the skin [58, 102, 113]. Single-walled (SW) CNTs can lead to apoptosis of skin cells in time and dose-dependent manner and multi-walled (MW) CNTs have been reported to be uptaken in vacuoles of skin cells but not in the nucleus with the release of IL-8, indicative of irritation to the skin cells generated as a consequence of NP uptake [58, 76, 113]. The effects of the NPs on the human physiology that gain access through skin are highly complex and are not yet well understood (Figure 1.12).

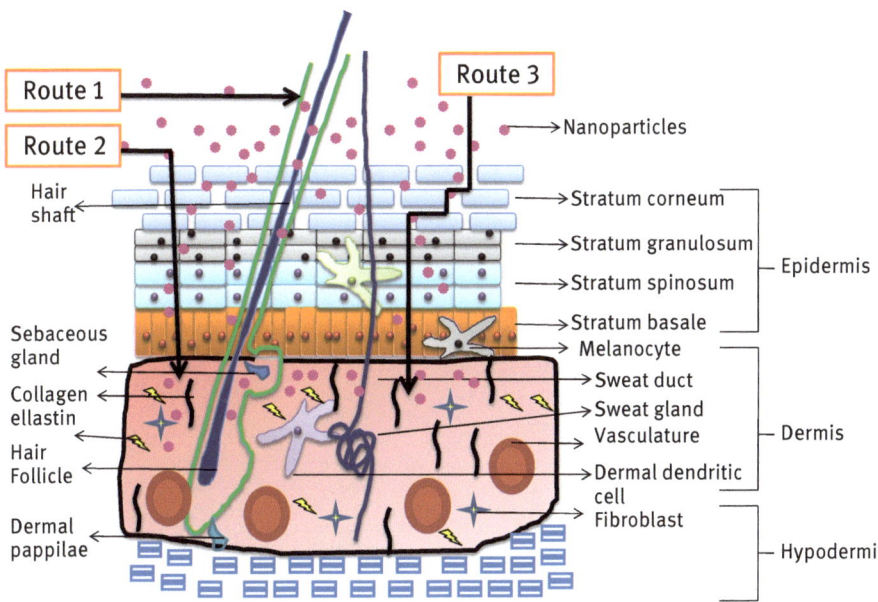

Figure 1.12: Exposure of NPs through skin: diagrammatic representation of NP penetration pathways through skin by one of the three different ways: through the (1) hair follicles, sweat glands and skin furrows; (2) intracellular route through epidermal layers and (3) intercellular route in between epidermal cells. The pathways taken by NP depend on its size, charge, morphology and material.

1.4.1.3 Entry through GI tract

Through the GI tract, all NPs enter intentionally or unintentionally by the oral route (Figure 1.13) through food, drugs, medicines or by secondary translocation of inhaled NPs. Exposures to contaminated environments may lead to their intake by food, water or aquatic sea food such as fish, molluscs, crabs, prawns, squids and so on. The M cells, which are specialized cells located in follicle-associated epithelium of intestinal

Peyer's patches of gut-associated lymphoid tissue in the GI tract, play an important role in transporting NPs from the lumen of the small intestine to mucosal lymphoid tissues, leading to various disorders including Crohn's disease. NPs pass through the intestinal epithelium by either of the two routes including paracellular transport involving transfer of NPs through intercellular space between the epithelial cells and transcellular transport where NPs move across the cell across both the apical and basolateral membranes.

The gut microbiome controls the biology of the gastric system. NPs may affect the absorption, cause accumulation and potentially may affect or alter the gut microbes and their normal functions. Nanometals and metal oxides, carbon-based NPs, and polymer/dendrimers have been reported to exert the toxic effects. Overgrowth of Gram negative bacteria and adherences of NPs to lipopolysaccharide can affect NP absorption, leading to their enhanced delivery. NP entry in the lumen may affect gut microbial metabolism by affecting nutrient absorption or xenobiotic metabolism [120–122]. Research reveals that engineered NPs, including CNTs, TiO_2, cerium dioxide, ZnO NPs and nano-(silica and silver), may affect the gut microbiome, leading to diseases such as colitis, obesity and immunological dysfunctions [123].

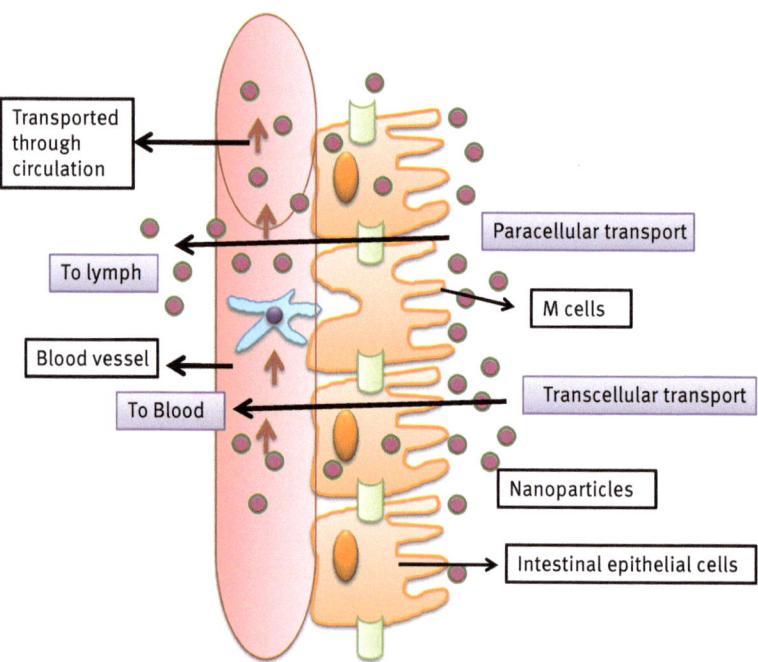

Figure 1.13: Transport of NPs through intestinal epithelium.

1.4.1.4 Work place exposure

The effect of NPs on the biological system is poorly understood as a result of workplace exposure. Exposures to TiO_2, carbon black, nickel powder, silica fumes or byproducts like welding fumes, metal fumes, beryllium or diesel exhaust [124–125] are being reported to have toxic effects on human health. Although TiO_2 and carbon black are studied for decades, little information is available for the others [126]. SW nanotubes [127], carbon nanofibers [128], engineered NPs [129], exposure to amorphous SiO_2 fume in smelters and CNT-doped concrete [129] have been reported to have toxic effects. Different studies have shown the effects of NPs on workers' health exposed to nanomaterials [130–131]. Recent researches have revealed that air pollution can affect the SCs [132] and is extremely detrimental to health.

1.4.2 Hazardous nature

The situation is far more complex when we consider the huge diversity and properties of nanomaterials and their different forms along with their different toxicological properties. Thus, a complete understanding of toxicity of NPs is of utmost importance to assess their hazardous nature.

The hazards arise due to their properties as discussed in the following sections.

1.4.2.1 The surface effects

Surface characteristics of NPs control the distribution of materials in tissues. NPs undergo adsorption on macromolecules on exposure to body tissues and fluids at the routes of entry into the host body. Adsorption is a property influenced by particle surface properties including its chemistry and energy. Engineered NPs with designed modification or functionalization of the NP surfaces [133] for conferring desired properties in the NP [134], like increasing the affinity of the NPs, designing of targeted delivery systems [135], analytical purposes [136] and designing of optical labeled biomolecules [137] pose another level of alteration in the surface chemistry of the NP. Thus, both passive surface and active agents on surface of NPs and their chemical composition contribute to potential hazard nature of NP. While metallic NPs are reported to react chemically, ionic crystal NPs are observed to be accumulated in cytoplasm or body fluid through circulation. Particle surface chemistry with properties of solubility equilibrium, catalytic properties, surface charge, surface adsorption and desorption of molecules that are functions of the atomic or molecular composition, the physical surface structure, chemical purity, functionalization and surface coating are also important aspects that could affect surface chemistry. Modification of surfaces has been shown to affect the toxicity of CNTs and TiO_2. Functionalization and surface modification of NPs add up to another level of complexity associated with NPs. Surface coating with surfactants of polymethyl methacrylate affects their distribution [76]. While polycationic macromolecules show strong association with cell membranes, globular polymers show difficulty in membrane attachment [76] and their effect on human health (Figure 1.14) and are important issues that need further research.

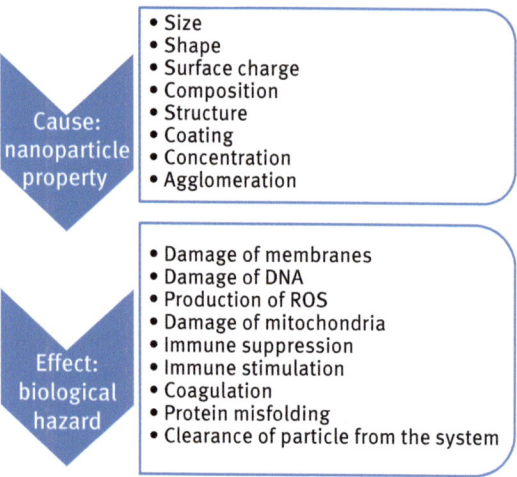

Figure 1.14: Cause and effect of NP toxicity.

1.4.2.2 Size, shape and composition

NPs' biological system interaction is largely governed by their size, shape and chemistry. Size controls the distribution and penetration of tissues by NPs. Reduction in size to the nanoscale level leads to an increase of surface-to-volume ratio, thereby increasing the number of chemical molecules on the surface, leading to an increase in intrinsic toxicity [113]. Thus, the particle size can control the dose–response relationship in relation to its solubility and toxicity [107]. The chemical formulation and its intrinsic toxic nature determine the particle toxicity [138] and govern its downstream effects on cells and tissues. While both carbon black and TiO_2 could cause lung inflammation in rats, carbon black was reported to exhibit more toxic effects when compared to TiO_2 NPs [139], indicating that the chemical nature of the particles governs its cytotoxicity. UNPs from diesel exhaust when inhaled were reported of their toxic effects [140]. Among different NPs of polyvinyl chloride, TiO_2, SiO_2, cobalt (Co) and nickel (Ni), only Co was reported of their toxic effects on endothelial cells [141], wherein the chemical properties and size governed their toxicity [141]. Composition could affect the in vivo distribution of NPs. Polymethylmethacrylate NPs with two polymers, administered intraperitoneally, were reported to travel to spleen when compared to polystyrene [142], which were mostly retained in the adipose tissue of the peritoneal cavity.

1.4.2.3 Shape

The shape of NPs has been reported to play an important role in determining the toxic nature of NPs. High aspect ratio indicative of smaller one or two of the three dimensions as compared to other dimension of a particle is associated with toxicity of NPs. Fibers are high-aspect ratio materials and fibers with a dimension of length greater than 5 μm, less than 3 μm width and length-to-width ratio greater than 3:1 are defined as a respirable fiber by the WHO[143]. Such particles with dimensions in nanoscale, with the ratio greater than 3:1, would be considered a high-aspect ratio NP. High-aspect ratio NPs with the following characteristics of (i) thinner than 3 μm, (ii) longer than 10–20 μm, (iii) biopersistent nature and (iv) not dissolving or breaking into shorter fibers may remain for long time in the pleural cavity, leading to inflammation and several diseases such as lung cancer. Inhaled NP fibers and nanotubes lead to inflammation in the respiratory physiology governed by physical parameters of thinness and length.

Asbestos fibers are known for their carcinogenic potentials. Intratracheal administration of SWCNTs has been reported to cause lung granulomas [144–145]. CNTs are reported to be more toxic on respiratory system when inhaled when compared to quartz particles on a dose per mass basis.

1.4.2.4 Solubility and persistence

Their bioavailability, solubility and degradation are governed by their solubility and persistence in biological systems [146]. While many NPs show solubility and biodegradability, many of them show accumulation in cells and tissues. The composition, chemical properties and degraded end products of the former category of NPs control their effect on host macromolecules, while for the latter category, with less to nil solubility and nonbiodegradable nature, the tendency is to accumulate in cells and tissues for a considerable period, posing a major threat to health [147–152]. Biodegradability and availability property of NPs are a major domain of modern-day research.

1.4.2.5 Aggregation/agglomeration state

NPs tend to form aggregates and agglomerates and the size of aggregates/agglomerates influences the residence time and reduces the potential for a NP to be inhaled. Their aggregation/agglomeration is controlled by external environment like air and dispersion media. They may also undergo disaggregation and disagglomeration within the respiratory system, thereby penetrating the lungs.

1.4.2.6 Surface area

Properties of particles are governed by surface area-to-mass ratio and therefore the effects of NPs are more when compared to those by larger particles. Thus, even for materials with similar compositions, dose responses may vary with the surface area. The relationship of surface area of NPs in determining exposures and doses is not completely studied. Thus, there is a need to study toxicity of NPs expressed in terms of surface area as well as mass.

1.4.2.7 Surface charge

The surface charge expressed as zeta potential of a particle affects the adsorption, cellular uptake, interaction with biomolecules and their reactions. Metal and metal oxide NPs with high zeta potential are reported to cause inflammation and therefore are directly related to their toxicity property. However, the zeta potential and toxicological effect relationship remains a topic for future research and for assessing risk of NP toxicity.

1.4.2.8 Other properties

Photocatalytic activities of NPs may cause inflammation as they become more reactive when exposed to light. NPs with a property of high acidity or alkalinity could lead to localized irritation in lungs, skin or GI tract.

1.4.3 Effects on health

Concerns over the hazardous nature of CNTs date back to the last two decades due to their physical similarities with asbestos fibers; experimental studies have confirmed their hazardous nature [151]. The United States National Institute for Occupational Safety and Health (NIOSH [152]) has recommended an exposure level of 1 $\mu g/m^3$ of element carbon as respirable mass 8-h time-weighted average concentration [152]. For TiO_2 NPs, 20–30 nm particles have been reported to be more toxic to respiratory health than their microparticle (>100 nm) forms [153]. Inhaled NPs have been reported to be deposited in the alveoli and can translocate throughout the body [105–106]. For TiO_2 NPs, NIOSH has proposed an occupational exposure limit of 0.3 mg/m^3, while for fine TiO_2 particles it is 2.4 mg/m^3 [153]. Transfer and systemic distribution of NPs after entry by oral routes has been reported with systemic toxic effects, affecting the liver and other organs of the reticuloendothelial system [102–104]. NPs can affect cardiovascular by inducing oxidative stress, inflammation, release of ROS, cell death or DNA damage leading to the formation of atherosclerotic lesions and thrombosis (Figure 1.15) [102–104].

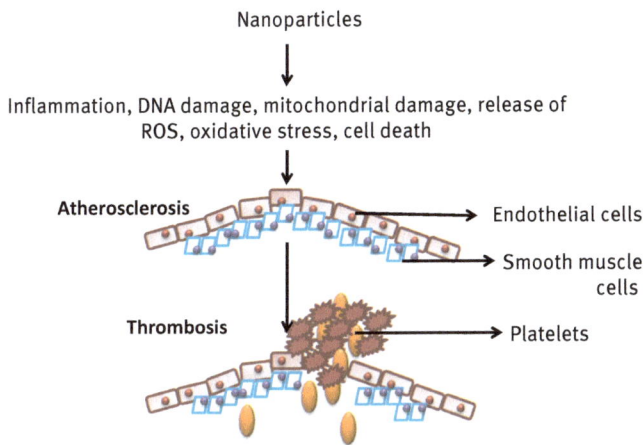

Nanoparticles

Inflammation, DNA damage, mitochondrial damage, release of ROS, oxidative stress, cell death

Atherosclerosis

Endothelial cells

Smooth muscle cells

Thrombosis

Platelets

Figure 1.15: NPs mediating thrombosis.

1.5 Nanosafety practices: global perspective

Acute toxicity and mortality and chronic toxicity have been associated with the toxicity of NPs globally and thus the concerns over the hazards caused by NPs on human health and environment have been a global cause of concern since the last two decades. Various safety guidelines have been drafted toward the safe use and handling of NPs. Nanosafety practices find importance largely in an industry in which workers are exposed to NPs and are an occupational hazard. These guidelines focus on the usage of laboratory protection equipment like using high-efficiency particulate air filter, which is a type of mechanical air filter that forces air through a fine mesh, trapping harmful particles including pollen, dust, mites and smoke and act as a barrier to exposure of airborne hazardous material from reaching the human body. To ensure safety of the workers from exposure to NPs hazards, emphasis on the use of personal protective equipment such as respirators, masks, gloves and face shields toward is highly recommended. Considering safety measures is as important as considering the hazardous nature of the material recorded in the material safety data sheet (MSDS) of the material used, the process adopted, NPs generated during the process of synthesis and waste disposal. The toxicity in NPs is generated from a number of components including its size, shape, aggregation, chemical composition, complex surface chemistry, engineering and charge and solubility. The awareness of the environmental health and safety practice measures and strict compliance of guidelines of handling of NPs need to be implemented both in the industry and for the benefit of the health and survival of the common man.

Although globally attempts have been made in managing the risks of NPs, there are still no clear guidelines on the threshold exposure limit of different NPs and their consequence in human health – neither from environmental nor anthropogenic sources. No clear guideline exists as mandatory protective measures from NP exposure during their synthesis. Although instruments to detect the concentration of airborne particles are being manufactured, devices to understand and separate NPs from the environment and those generated from the industry remain a bottleneck issue. Their uncharacterized nature and unknown effects on human health are other safety issues with regard to NP use, applications and exposure. More than 60 countries globally are taking initiative for drafting regulations and policies with regard to nanotechnology and their impact on health and environment and safety; however, more international coordination is required toward achieving sustainable economic, social and environmental benefits for all. Regulations related to manufacturing, use, disposal of nanowaste across countries need to be accessed [102–104, 154–156] and developed.

1.6 Regulations

Industries, academia and nongovernment organizations are making efforts to draft policies and perspectives on environmental health and safety for the applications of engineered NPs with environmental benefits and minimized negative adverse health effects. Worldwide efforts have enabled understanding and awareness on the need of safety from NP exposure and their applications to control and prevent environmental pollution when compared to conventional methods. However, globally there a need for more awareness. The Royal Society and the Royal Academy of Engineering in UK in 2004 has published reports on the risks and benefits of nanotechnology; the first European Union action plan on nanosciences and nanotechnologies in the years between 2005 and 2009 was adopted in 2005, and new action plans for 2011–2015 have also been prepared. The Nanotechnology White paper published by Environmental Protection Agency (EPA) in 2007 documented the role of NPs on the environment and in 2009 and the Nanomaterial Research Strategy (NRS) was drafted. In US both the FDA and National Institute for Occupational Safety and Health (NIOSH) have documented research strategies in this domain [157].Thus, further research in these areas seems pertinent to develop guidelines toward practicing of safety in the use, applications and exposure of NPs. The European legislation on Registration, Evaluation, Authorization and Restriction of Chemicals includes nanomaterials [158–159]. Several activities have been initiated by the ECHA (European Chemical Agency), European Commission and the Organization for Economic Cooperation and Development on NP characterization and risk assessment and providing substantial feedback [158–159] and various nanomaterials have been evaluated [158–159]. Among the European Union (EU), France has initiated a protocol of declaration of nanomaterials by industries into production, import, distribution or formulation in quantities more than 100 g/year, while Belgium,

Denmark, Germany and Italy emphasize on the design of database for nanomaterial production and usage. Industry relies on the technical guidance provided by ECHA for chemicals in general. A number of key nanospecific guidance updates in the domain of sample preparation, documenting key physicochemical properties, exposure and hazard assessment, [149]. Technical guidance provided by ECHA on chemical risks and the Organization for Economic Cooperation and Development's (OECD) approximately 150 test guidelines on physicochemical properties, human toxicity, ecotoxicity, and so on need to be followed by the industry [102–104, 161].

1.7 Problems in NP research and the road ahead

Hazardous properties of NPs and their effects on the human health need to be understood to know the toxicological properties. Their diversity in size, charge, chemical composition, surface properties, size, shape, surface area, zeta potentials, aggregation, particle distribution, surface functionalization or engineering govern their biological effects and interaction with biological molecules. The interaction of NPs with biomolecules are to be understood to study the toxicological effects. There is little information on their uptake and translocation after exposure through different routes such as inhalation, neuronal uptake, translocation across lung epithelium and ingestion translocating through GI tract after oral exposure and exposure by skin and is a major domain of research. In vitro methods and in silico approaches are potential alternative approaches in the study of nanotechnology in the recent years.

1.8 Nanotechnology and computational study

Computational approaches enable the study of science and technology in the field of nanotechnology, and their everyday expanding domain of applications and hazards leads to the development of a new branch of science termed as nanotechnology informatics.

The DaNa project funded by the German Federal Ministry for Education and Research has devised an interdisciplinary approach to process information of research results of nanomaterials and their biological interactions and effects on humans and the environment, with critical evaluation for the knowledge of the common man and creation of a knowledge base [162].

Biocorona formation is a consequence of NP and biological surface interaction and modulates cellular uptake, transport, biodistribution and biodisposition. Therefore, understanding of NP's surface adsorption forces that control biocorona formation finds importance in risk assessment of engineered nanomaterial. Computational approaches find importance in devising algorithms in understanding the formation of biocorna across different environments, where the NP interacts with the human body and environment [163].

Curation and integration of data from research on nanomaterials is being attempted by computational scientists using the concepts from nanoinformatics to generate a digital nanotechnology data repository to understand and share information by all researchers, users and stake holders. Nanomaterial registry data curation tool, designed based on informatics concepts of minimal information sets and controlled vocabulary, is an efficient one for curating data from huge information sources [164].

The toxicity studies of engineered nanomaterials are generating huge data for reliable processing and analyses and require high throughput screening (HTS). A web-based platform for HTS data analyses tools utilizing statistical methods and concepts from informatics, including plate normalization method, self-organizing map-based clustering analysis and visualization of raw and processed data using heat maps and self-organizing map, has been developed for integration, screening, analysis and study of toxicity of engineered NPs [165]. Computational methods of data mining, modeling and simulation enable the extraction, management and storage data from different NPs, their characteristics, effects on human, animals and environment, but this domain of science faces the challenge of vast heterogeneity of NPs and their effects on the biological system and environment as well as need of integrated inputs of scientists, from different diverse disciplines including medical professionals, computational scientists, chemists, physicists, industry and pharmacologist [166]. Theoretical mechanistic models are being designed for profiling the biological and toxicological effects of oxide nanomaterials including their reactivity, protein adsorption and membrane adhesion from experimental data or from statistical regression methods. They find importance in understanding the toxic effects and fate of the NPs, and enable assessment of hazard and safety of nanomaterials and high-throughput screening as well as enable safe designing of engineered NPs [167].

The National Cancer Institute (NCI) Center for Biomedical Informatics and Information Technology, together with the NCI Alliance for Nanotechnology in Cancer (Alliance), has enabled the development data curation, repository and a data-sharing portal known as caNanoLab, enabling access to data from experimental and literature sources from the NCI Nanotechnology Characterization Laboratory, the Alliance and the cancer nanotechnology laboratory [168].

1.9 Discussion

NPs hold promise in diverse day-to-day applications and human health and environment. However, none of their surface characteristics pose a threat toward toxicity, which warrant their large-scale applications. Research is being conducted to detect the safe use, designing of NPs and their applications in different diseases from infectious to neurological to cancer. There is also a global need for appropriate international regulations for the manufacturing units involved in manufacture and trading of NPs. Prior to their use in food and as agents in the human body, their toxicity should be appropriately tested. Computational nanotechnology is a newly developing domain, which efficiently aids in profiling of toxic effects of NPs. In the subsequent chapters, we will discuss the effects of different nanoparticles and their applications in human life and health.

References

[1]   http://www.who.int/mediacentre/factsheets/fs104/en/.
[2]   M.L Hans, A.M Lowman Biodegradable nanoparticles for drug delivery and targeting Curr. Opin. Solid State Mater. Sci. 2002;6(4): 319–327.
[3]   Costa A, Pinheiro M, Magalhães J, Ribeiro R, Seabra V, Reis S, Sarmento B. The formulation of nanomedicines for treating tuberculosis. Adv Drug Deliv Rev. 2016;102:102–115.
[4]   Jnawali HN and Ryoo S First – and Second–Line Drugs and Drug Resistance "Tuberculosis – current issues in diagnosis and management" Chapter 10. 2013.
[5]   http://www.who.int/malaria/en/.
[6]   Nicholas J. White. Antimalarial drug resistance. J Clin Invest. 2004 Apr 15;113(8):1084–1092.
[7]   Robert A, Benoit-Vical F, Dechy-Cabaret O, Meunier B. From classical antimalarial drugs to new compounds based on the mechanism of action of artemisinin. Pure Appl. Chem. 2001;73(7):1173–1188.
[8]   Thakkar M, S B. Combating malaria with nanotechnology-based targeted and combinatorial drug delivery strategies. Drug Deliv Transl Res. 2016 Aug;6(4):414–425.
[9]   Salem AK, Searson PC, Leong KW. Multifunctional nanorods for gene delivery. Nat Mater. 2003;2:668–671. 2003.
[10]   Cui Z, Mumper RJ. Microparticles and nanoparticles as delivery systems for DNA vaccines. Crit Rev Ther Drug Carrier Syst. 2003;20:103–137.
[11]   Cui Z, Mumper RJ. Topical immunization using nanoengineered genetic vaccines. J Controlled Release, 2002;81:173–184. 2002.
[12]   Zhang W, Yang H, Kong X, Mohapatra S, San Juan-Vergara H, Hellermann G, Behera S, Singam R, Lockey RF, Mohapatra SS. Inhibition of respiratory syncytial virus infection with intranasal siRNA nanoparticles targeting the viral NS1 gene. Nat Med 2005; 11:56–62.
[13]   Shah MA, He N, Li Z, Ali Z, Zhang L. Nanoparticles for DNA vaccine delivery. J Biomed Nanotechnol. 2014 Sep;10(9):2332–2349.
[14]   Ramesh R, Ito I, Saito Y, Wu Z, Mhashikar AM, Wilson DR, Branch CD, Roth JA, Chada S. Local and systemic inhibition of lung tumor growth after nanoparticle-mediated mda-7/IL-24 gene delivery. DNA Cell Biol. 2004;23:850–857.
[15]   Gordon EM, Hall FL. Nanotechnology blooms, at last. Oncol Rep. 2005;13:1003–1007.
[16]   http://www.who.int/mediacentre/factsheets/fs360/en/.
[17]   Dong Y, Yang J, Zhang J, Zhang X. Nano-delivery vehicles/adjuvants for DNA vaccination against HIV. J Nanosci Nanotechnol. 2016, Mar;16(3):2126–2133.
[18]   DeLong RK, Curtis CB. Toward RNA nanoparticle vaccines: Synergizing RNA and inorganic nanoparticles to achieve immunopotentiation. Wiley Interdiscip Rev Nanomed Nanobiotechnol. 2017 Mar;9(2). doi: 10.1002/wnan.1415.

[19] Ravi Kumar MN, Sameti M, Mohapatra SS, Kong X, Lockey RF, Bakowsky U, Lindenblatt G, Schmidt H, Lehr CM. Cationic silica nanoparticles as gene carriers: Synthesis, characterization and transfection efficiency in vitro and in vivo. J Nanosci Nanotechnol. 2004;4:876–881.

[20] Seia MA, Stege PW, Pereira SV, De Vito IE, Raba J, Messina GA. Silica nanoparticle-based microfluidic immunosensor with laser-induced fluorescence detection for the quantification of immunoreactive trypsin. Anal Biochem. 2014 Oct 15;463:31–37.

[21] Nafisi S, Schäfer-Korting M, Maibach HI. Perspectives on percutaneous penetration: Silica nanoparticles. Nanotoxicol. 2015;9(5):643–657.

[22] Konstan MW, Davis PB, Wagener JS, Hilliard KA, Stern RC, Milgram LJ, Kowalczyk TH, Hyatt SL, Fink TL, Gedeon CR, Oette SM, Payne JM, Muhammad O, Ziady AG, Moen RC, Cooper MJ. Hum Gene Ther. Compacted DNA nanoparticles administered to the nasal mucosa of cystic fibrosis subjects are safe and demonstrate partial to complete cystic fibrosis transmembrane reconstitution. 2004, 15: 1255–1269.

[23] Bourges JL, Gautier SE, Delie F, Bejjani RA, Jeanny JC, BenEzra D, Behar-Cohen FF. Ocular drug delivery targeting the retina and retinal pigment epithelium using polylactide nanoparticles. Invest Ophthalmol Vis Sci, 2003;44:3562–3569.

[24] Battaglia L, Serpe L, Foglietta F, Muntoni E, Gallarate M, Del Pozo Rodriguez A, Solinis MA. Application of lipid nanoparticles to ocular drug delivery. Expert Opin Drug Deliv. 2016 Dec;13(12):1743–1757.

[25] Lin H, Yue Y, Maidana DE, Bouzika P, Atik A, Matsumoto H, Miller JW.Vavvas. DG drug delivery nanoparticles: Toxicity comparison in retinal pigment epithelium and retinal vascular endothelial cells. Semin Ophthalmol. 2016;31(1–2):1–9.

[26] http://www.who.int/mediacentre/factsheets/fs297/en/.

[27] Badrzadeh F, Rahmati-Yamchi M, Badrzadeh K, Valizadeh A, Zarghami N, Farkhani SM, Akbarzadeh A. Drug delivery and nanodetection in lung cancer. Artif Cells Nanomed Biotechnol. 2016;44(2):618–634.

[28] Åkerman ME, Chan WCW, Laakkonen P, Bhatia SN, Ruoslahti E. Nanocrystal targeting in vivo. PNAS. 2002;99:12617–12621.

[29] Ciarlo M, Russo P, Cesario A, Ramella S, Baio G, Neumaier CE, Paleari L. Use of the semiconductor nanotechnologies "quantum dots" for in vivo cancer imaging. Recent Pat Anticancer Drug Discov. 2009 Nov;4(3):207–215.

[30] Hofferberth SC, Grinstaff MW, Colson YL. Nanotechnology applications in thoracic surgery. Eur J Cardiothorac Surg. 2016 Jul;50(1):6–16.

[31] Kim SY, Lee YM, Kang JS. Indomethacin-loaded methoxy poly(ethylene glycol)/poly(D,L-lactide) amphiphilic diblock copolymeric nanospheres: pharmacokinetic and toxicity studies in rodents. J Biomed Mater Res A. 2005 Sep 15;74(4):581–590.

[32] Kim SY, Lee YM, Baik DJ, Kang JS. Toxic characteristics of methoxy poly(ethylene glycol)/ poly(epsilon-caprolactone) nanospheres; in vitro and in vivo studies in the normal mice. Biomaterials. 2003 Jan;24(1):55–63.

[33] Raffin Pohlmann A, Weiss V, Mertins O, Pesce da Silveira N, Stanisçuaski Guterres S. Spray-dried indomethacin-loaded polyester nanocapsules and nanospheres: development, stability evaluation and nanostructure models. Eur J Pharm Sci. 2002 Sep;16(4–5):305–312.

[34] Weissenböck A, Wirth M, Gabor F. WGA-grafted PLGA- nanospheres: Preparation and association with Caco-2 single cells. J Control Release. 2004;99:383–392.

[35] Félix Lanao RP, Jonker AM, Wolke JG, Jansen JA, van Hest JC, Leeuwenburgh SC. Physico-chemical properties and applications of poly(lactic-co-glycolic acid) for use in bone regeneration. Tissue Eng Part B Rev. 2013 Aug;19(4):380–390.

[36] Zhou Y, Peng Z, Seven ES, Leblanc RM. Crossing the blood-brain barrier with nanoparticles. J Control Release. 2017 Dec 19;270:290–303.

[37] D'Agata F, Ruffinatti FA, Boschi S, Stura I, Rainero I, Abollino O, Cavalli R, Guiot C. Magnetic nanoparticles in the central nervous system: Targeting principles, applications and safety issues. Molecules. 2017 Dec 21;23(1).

[38] Lockman PR, Koziara JM, Mumper RJ, Allen DD. Nanoparticle surface charges alter bloodbrain barrier integrity and permeability. J Drug Target. 2004;12:635–641.

[39] Lockman PR, Koziara JM, Roder KE, Paulson J, Abbruscato TJ, Mumper RJ, Allen DD. In vivo and in vitro assessment of baseline blood-brain-barrier parameters in the presence of novel nanoparticles. Pharm Res. 2003;20:705–713.

[40] Kreuter J. Nanoparticulate systems for brain delivery of drugs. Adv Drug Deliv Rev. 2001;47:65–81.

[41] Kreuter J. Influence of the surface properties on nanoparticle-mediated transport of drugs to the brain. J Nanosci Nanotechnol. 2001;4:484–488.

[42] Kreuter J, Ramge P, Petrov V, Hamm S, Gelperina SE, Engelhardt B, Alyautdin R, Briesen H, Begley DJ. Direct evidence that polysorbate-80-coated poly(butylcyanoacrylate) nanoparticles deliver drugs of the CNS via specific mechanisms requiring prior binding of drug to the nanoparticles. Pharm Res. 2003;20:409–416.

[43] Nishioka Y, Yoshino H. Lymphatic targeting with nanoparticulate system. Adv Drug Del Rev. 2001;47:55–64.

[44] Ortiz R, Cabeza L, Arias JL, Melguizo C, Álvarez PJ, Vélez C, Clares B, Áranega A, Prados J. Poly(butylcyanoacrylate) and Poly(ε-caprolactone) Nanoparticles Loaded with 5-Fluorouracil Increase the Cytotoxic Effect of the Drug in Experimental Colon Cancer. AAPS J. 2015 Jul;17(4):918–929.

[45] Sarfraz M, Shi W, Gao Y, Clas SD, Roa W, Bou-Chacra N, Löbenberg R. Immune response to antituberculosis drug-loaded gelatin and polyisobutyl-cyanoacrylate nanoparticles in macrophages. Ther Deliv. 2016;7(4):213–228.

[46] Kim SY, Lee YM, Baik DJ, Kang JS. Toxic characteristics of methoxy poly(ethylene glycol)/poly(ε-caprolactone) nanospheres; in vitro and in vivo studies in the normal mice. Biomaterials. 2003;24: 55–63.

[47] Kim S, Lim YT, Soltesz EG, DeGrand AM, Lee J, Nakayama A, Parker JA, Mihaljev T, Laurence RG, Dor, DM, Cohn LH, Bawendi MG, Frangioni JV. Near-infrared fluorescent type II quantum dots for sentinel lymph node mapping. Nature Biotechn. 2004. 22:93–97.

[48] Soltesz EG, Kim S, Laurence RG, DeGrand AM, Parungo CP, Dor DM, Cohn LH, Bawendi MG, Frangioni JV, Mihaljevic T. Intraoperative sentinel lymph node mapping of the lung using near-infrared fluorescent quantum dots. Ann Thorac Surg. 2005;79:269–277.

[49] Davda J, Labhasetwar V. Characterization of nanoparticle uptake by endothelial cells. Int J Pharmac. 2002233:51–59.

[50] Uwatoku T, Shimokawa H, Abe K, Matsumoto Y, Hattori T, Oi K, Matsuda T, Kataoka K, Takeshita A. Application of nanoparticle technology for the prevention of restenosis after balloon injury in rats. Circ Res. 2003;92: e62–e69.

[51] Westedt U, Barbu-Tudoran L, Schaper AK, Kalinowski M, Alfke H, Kissel T. Effects of different application parameters on penetration characteristics and arterial vessel wall integrity after local nanoparticle delivery using a porous balloon catheter. Eur J Pharmac Biopharmac. 2004;58:161–168.

[52] Feng H, Qian Z. Functional carbon quantum dots: A versatile platform for chemosensing and biosensing. Chem Rec. 2017 Nov 24.

[53] Wu M, Lai Q, Ju Q, Li L, Yu HD, Huang W. Paper-based fluorogenic devices for in vitro diagnostics. Biosens Bioelectron. 2018 Apr 15;102:256–266. doi: 10.1016/j.bios.2017.11.006. Epub 2017 Nov 16.

[54] Cunha CRA, Oliveira ADPR, Firmino TVC, Tenório DPLA, Pereira G, Carvalho LB Jr., Santos BS, Correia MTS, Fontes A. Biomedical applications of glyconanoparticles based on quantum dots. Biochim Biophys Acta. 2017 Nov 7;1862(3):427–439.

[55] Brown DM, Wilson MR, MacNee W, Stone V, Donaldson K. Size dependent proinflammatory effects of ultrafine polystyrene particles: a role for surface area and oxidative stress in the enhanced activity of ultrafines. Toxicol Appl Pharmacol. 2001;175:191–199.

[56] Oberdörster G, Finkelstein JN, Johnston C, Gelein R, Cox C, Baggs R, Elder ACP. HEI research report; Acute pulmonary effects of ultrafine particles in rats and mice. Research Report 96. August 2000;Health Effects Institute, www.healtheffects.org/pubs-research.htm.

[57] Höhr D, Steinfartz Y, Schins RP, Knaapen AM, Martra G, Fubini B, Borm PJ. The surface area rather than the surface coating determines the acute inflammatory response after instillation of fine and ultrafine TiO2 in the rat. Int J Hyg Environm Health. 2002;205:239–244.

[58] Reisner DE. Biotechnology Global Perspectives. 2009;CRC Press.

[59] Doering, Williams E, Mechanisms and application of Single-Nanoparticle surface-enhanced Raman Scattering, PhD Thesis, University of Indiana, 2003.

[60] Holland NB et al., Biomimetic engineering of non-adhesive glycocalyx-like surfaces using oligosaccharide surfactant polymers, Nature. 1998;392(799).

[61] Mirzabekov AD, Biochips in biology and medicine of XXI century, Bulletin of Russian Academy of Sciences, 2003;73(5):412–416.

[62] DietrichHRS et al., Nanoarrays, A method for performing enzymatic assays. Anal. Chem. 2004;76:4112–4117.

[63] Colvin, Vicky L. Nature Biotechnology. 2003;21:1166.

[64] Patent, Unilever, WO 2006/097332 A2, Colourant compositions and their use.

[65] Rodriguez T et al., Interactions of proteoliposomes from serogroup B Neisseria meningitides with bone marrow-derived dendritic cells and macrophages: Adjuvants effects and antigen delivery. Vaccine. 2005;23(10):1312–1321.

[66] Miao Z, Gao Z, Chen R, Yu X, Su Z, Wei G. Surface-bioengineered gold nanoparticles for biomedical applications. Curr Med Chem. 2018 Jan 16.

[67] Malekzad H, Zangabad PS, Mirshekari H, Karimi M, Hamblin MR. Noble metal nanoparticles in biosensors: recent studies and applications. Nanotechnol Rev. 2017 Jun 27;6(3):301–329.

[68] Damborska D, Bertok T, Dosekova E, Holazova A, Lorencova L, Kasak P, Tkac J. Nanomaterial-based biosensors for detection of prostate specific antigen. Mikrochim Acta. 2017 Jul 14;184(9):3049–3067.

[69] Lv M, Liu Y, Geng J, Kou X, Xin Z, Yang D. Engineering nanomaterials-based biosensors for food safety detection. Biosens Bioelectron. 2018 May 30;106:122–128.

[70] NRai M, Jogee PS, Ingle AP.Emerging nanotechnology for detection of mycotoxins in food and feed. Int J Food Sci Nutr. 2015;66(4):363–370.

[71] Nikoleli GP, Nikolelis DP, Siontorou CG, Karapetis S, Varzakas T. Novel biosensors for the rapid detection of toxicants in foods. Adv Food Nutr Res. 2018;84:57–102.

[72] Cientifica, Nanotechnologies and Energy", whitepaper, Cientifica, London, 2/2007;www.cientifica.eu.

[73] CLSA, Solar Power Sector outlook", July 2004, www.clsa.com.

[74] EPIA, Greenpeace "Solar Generation IV – 2007". Bericht der European Photovoltaics Industry Association und Greenpeace International. 2007.

[75] Application of Nano- technologies in the Energy Sector, 2008, Hessian Ministry of Economy, Transport, Urban and Regional Development https://www.hessen-nanotech.de/mm/NanoEnergy_web.pdf.

[76] Remediation Nanomaterials-Toxicity Health Environmental Issues Edited by Challa SSR Kumar, Wiley, 2006.

[77] Corrie L. Carnes, Jennifer Stipp, and Kenneth J. Klabunde Synthesis, Characterization, and adsorption studies of nanocrystalline copper oxide and nickel oxide. Langmuir. 2002; *18* (4):1352–1359.

[78] Corrie L. Carnes, Pramesh N. Kapoor, and Kenneth J. Klabunde Synthesis, Characterization, and adsorption studies of nanocrystalline aluminum oxide and a bimetallic nanocrystalline aluminum oxide/magnesium. OxideChem. Mater. 2002;14(7):2922–2929.

[79] Sherman M. Ponder, John G. Darab, and Thomas E.Mallouk Remediation of Cr(VI) and Pb(II) aqueous solutions using supported, nanoscale zero-valent Iron. Environ Sci. Technol. 2000;34(12): 2564–2569.

[80] Warapong Tungittiplakorn, Claude Cohen, and Leonard W. Lion engineered polymeric nanoparticles for bioremediation of hydrophobic contaminants. Environ.Sci. Technol. 2005;39(5):1354–1358.

[81] Damodara M. Poojary, Yiping Zhang, Baolong Zhang, and A. ClearfieldSynthesis, X-ray Powder Structure, and Intercalation behavior of molybdenyl phenylphosphonate, MoO2(O3PC6H5). cntdot.H2O Chem.Mater. 1995;7(5): 822–827.

[82] Poojary DM, and Clearfield A Coordinative intercalation of alkylamines into layered zinc phenylphosphonate. crystal structures from X-ray powder diffraction data.J.Am. Chem. Soc. 1995;117 (45):11278–11284.

[83] Poojary DM, Zhang B, and Clearfield A. Pillared layered metal phosphonates. syntheses and X-ray powder structures of copper and zinc Alkylenebis(phosphonates). J. Am. Chem. Soc. 1997;119(51):12550–12559.

[84] Zhang B, Poojary DM, and Clearfield A Synthesis and characterization of layered zinc biphenylylenebis(phosphonate) and three mixed-component Arylenebis(phosphonate)/phosphates. Inorg. Chem. 1998;*37*(8):1844–1852.

[85] Wang Z, Heising JM, and Clearfield A Sulfonated microporous organic–inorganic hybrids as strong bronsted acids. J. Am. Chem. Soc. 2003. 125(34):10375–10383.

[86] Klasson KT, Treatment of mercury contaminated oil from mound site. Topical Report. Oak Ridge National Laboratory, Oak Ridge TN 37831, 2000.

[87] Yetisen AK, Qu H, Manbachi A, Butt H, Dokmeci MR, Hinestroza JP, Skorobogatiy M, Khadem-hosseini A, Yun SH. Nanotechnology in textiles. ACS Nano. 2016 Mar 22;10(3):3042–3068.

[88] Uboldi, C et al. 2012. "Amorphous silica nanoparticles do not induce cytotoxicity, cell transformation or genotoxicity in Balb/3T3 mouse fibroblasts." Mutat Res. 745(1–2):11–20.

[89] Oberdorster, G et al. 2005. "Nanotoxicology; An emerging discipline evolving from studies of ultrafine particles." Environ Health Perspect. 113:823–839.

[90] Bradley, EL et al. 2011. "Applications of nanomaterial in food packaging with a consideration of opportunities for developing countries." Trends Food Sci Technol. 22:604–610.

[91] Llorens, A et al. 2012. "Metallic-based micro- and nanocomposites in food contact materials and active food packaging." Trends Food Sci Technol. 24:19–20.

[92] Pyrgiotakis, G et al. 2015. "Inactivation of foodborne microorganisms using engineered water nanostructures (EWNS)." Environ Sci Technol. 49(6):3737–3745.

[93] Chaudhry, Q et al. 2008. "Applications and implications of nanotechnologies for the food sector." Food Addit Contam. 25(3):241–258.

[94] Bouwmeester, H et al. 2009. "Review of health safety aspects of nanotechnologies in food production." Regul Toxicol Pharmacol. 53:52–62.

[95] Buzby, JC. 2010. "Nanotechnology for food applications: More questions than answers." J Consumer Affairs. 44(3):528–545.

[96]   Moraru, CI et al. 2003. "Nanotechnology: A new frontier in food science." Food Technol 57:24–29.
[97]   Nature Nanotechnology. 2010. "Nanofood for thought." Nature Nanotechnol. 5:89.
[98]   www.efsa.europa.eu/en/topics/topic/nanotechnology.
[99]   www.regulations.gov.
[100]   Chaudhry Q Current and projected applications of nanomaterials. WHO workshop on nanotechnology and human health: Scientific Evidence and Risk Governance. Bonn, Germany, 10–11 December 2012.
[101]   Hansen SF Exposure pathways of nanomaterials. WHO workshop on nanotechnology and human health: Scientific Evidence and Risk Governance. Bonn, Germany, 10–11 December 2012.
[102]   Poland C Nanoparticles: Possible routes of intake. WHO workshop on nanotechnology and human health: Scientific Evidence and Risk Governance. Bonn, Germany, 10–11 December 2012.
[103]   Howard V. General toxicity of NM. WHO workshop on nanotechnology and human health: Scientific evidence and risk governance. Bonn, Germany, 10–11 December 2012.
[104]   http://apps.who.int/iris/bitstream/10665/108626/1/e96927.pdf.
[105]   Oberdörster G. Significance of particle parameters in the evaluation of exposure-dose-response relationships of inhaled particles. Inhal Toxicol. 1996;8(suppl):73–89.
[106]   Oberdörster G. Toxicology of ultrafine particles: in vivo studies. Phil Trans R Soc Lond A. 2000;358:2719–2740.
[107]   Oberdörster G, Oberdörster E, Oberdörster J. Nanotoxicology: an emerging discipline evolving from studies of ultrafine particles. Environ Health Perspect. 2005;113:823–839.
[108]   Donaldson K, Stone V.Current hypotheses on the mechanisms of toxicity of ultrafine particles Ann Ist Super Sanita. 2003;39(3):405–410.
[109]   Donaldson K, Stone V, Gilmour PS, Brown DM, MacNee W. Ultrafine particles: mechanisms of lung injury. Phil Trans R Soc Lond A. 2000;358:2741–2749.
[110]   Borm PJ. Particle toxicology: from coal mining to nanotechnology. Inhal Toxicol. 2002;14:311–324.
[111]   Donaldson K, Stone V, Clouter A, et al. Ultrafine particles. Occup Environ Med. 2001;58:211–216.
[112]   Donaldson K, Stone V, Tran CL, et al. Nanotoxicology. Occup Environ Med. 2004;61:727–728.
[113]   Dreher KL. Toxicological highlight.Health and environmental impact of nanotechnology: Toxicological assessment of manufactured nanoparticles. Toxicol Sc. 2004;77:3–5.
[114]   Kreyling WG, Semmler M, Möller W. Dosimetry and toxicology of ultrafine particles. J Aerosol Med. 2004;17:140–152.
[115]   Englert N. Fine particles and human health–a review of epidemiological studies. Toxicol Lett. 2004 Apr 1;149(1–3):235–242.
[116]   Pope C, Ziebland S, Mays N. Qualitative research in health care. Analysing qualitative data. BMJ. 2000 Jan 8;320(7227):114–116.
[117]   Samet JM, Zeger SL, Dominici F, Curriero F, Coursac I, Dockery DW, et al. The National morbidity, mortality, and air pollution study.Part II: morbidity and mortality from air pollution in the United States. Res Rep Health Eff Inst. 2000;94:5–70.
[118]   Peters A, Dockery DW, Muller JE, Mittleman MA. Increased particulate air pollution and the triggering of myocardial infarction. Circulation. 2001a;103:2810–2815.
[119]   Peters A, Frohlich M, Doring A, Immervoll T, Wichmann HE, Hutchinson WL, et al. Particulate air pollution is associated with an acute phase response in men; results from the MONICA-Augsburg Study. Eur Heart J. 2001b;22:1198–1120.

[120] Bu Q, Yan G, Deng P, Peng F, Lin H, Xu Y, Cao Z, Zhou T, Xue A, Wang YY, Cen X, Zhao YL. NMR-based metabonomic study of the sub-acute toxicity of titanium dioxide nanoparticles in rats after oral administration. Nanotechnol. 2010;21(12):125105.

[121] Cattani VB, Fiel LA, et al. Lipid-core nanocapsules restrained the indomethacin ethyl ester hydrolysis in the gastrointestinal lumen and wall acting as mucoadhesive reservoirs. EurJPharm. Sci. 2010;39(1–3):116–124.

[122] Bergin IL and Witzmann FA Nanoparticle toxicity by the gastrointestinal route: evidence and knowledge gaps. Int J Biomed Nanosci Nanotechnol. 2013; 3(1–2):doi:10.1504/IJBNN.2013.054515.

[123] Pietroiusti A, Magrini A, Campagnolo L. New frontiers in nanotoxicology: Gut microbiota/microbiome-mediated effects of engineered nanomaterials. Toxicol Appl Pharmacol. 2016 May 15;299:90–95.

[124] HSE NanoAlert Service, 2006.

[125] Möhlmann, C., 'Vorkommen ultrafeiner Aerosole an Arbeitsplätzen', Gefahrstoffe – Reinhaltung der Luft. 2005;65:469–471.

[126] Aitken, R.J., Chaudhry, M.C., Boxall, A.B.A., Hull, M., 'Manufacture and use of nanomaterials: current status in the UK and global trends'. Occup Med. 2006;56:300–306.

[127] Maynard, A.D., Baron, P. A., Foley, M., Shvedova, A.A., Kisin, E.R., Castranova, V., 'Exposure to carbon nanotube material: aerosol release during the handling of unrefined singlewalled carbon nanotube material', J. Toxicol. and Environ. Health, 2004;67:87–107.

[128] Mazzuckelli, J.F., Methner, M.M., Birch, M.E., Evans, D.E., Ku, B.-K., Crouch, K., Hoover, M.D. 'Case Study. Identification and characterization of potential sources of worker exposure to carbon nanofibers during polymer composite laboratory operations'. J.Occ.Environ.Hyg. 2007;4:125–130.

[129] Schneider, T., Evaluation and control of occupational health risks from nanoparticles, 2007, accessed on 13 November 2008. www.norden.org/pub/sk/showpub.asp?pubnr=2007:581.

[130] Brun, E., Op de Beeck, R., Van Herpe, S., Isotalo, L., Laamanen, I., Blotière, C., Guimon, M., Mur, J.-M., Orthen, B., Wagner, E., Flaspöler, E., Reinert, D., Galwas, M., Poszniak, M., Carreras, E., Guardino, X., Solans, X., European Risk Observatory Report - Expert forecast on emerging chemical risks related to occupational safety and health, Luxembourg, Office for Official Publications of the European Communities, 2008.

[131] Simon Kaluza, Judith kleine Balderhaar, Bruno Orthen, Bundesanstalt für Arbeitsschutz et al., Joanna Kosk-Bienko, European Agency for Safety and Health at Work (EU-OSHA), SpainWorkplace exposure to nanoparticles https://osha.europa.eu/en/tools-and-publications/publications/literature_reviews/workplace_exposure_to_nanoparticles.

[132] Ghosh S, Ansar W. Indoor air pollution: impact on health and stem cells. J Stem Cells. 2014;9(4):269–81. doi: jsc.2015.9.4.269.

[133] Schellenberger EA, Reynolds F, Weissleder R and Josephson L. Surface-functionalized nanoparticle library yields probes for apoptotic cells. Chembiochem. 2004;5:275–279.

[134] Nardin C, Thoeni S, Widmer J, Winterhalter M and Meier W. Nanoreactors based on (polymerized) ABA-triblock copolymer vesicles. Chem Comm. 2000;15:1433–1434.

[135] Hood JD, Bednarski M, Frausto R, Guccione S, Reisfeld RA, Xiang R and Cheresh DA. Tumor regression by targeted gene delivery to the neovasculature, Science. 2002;296(5577):2404–2407.

[136] Elghanian R, Storhoff JJ, Mucic RC, Letsinger RL and Mirkin CA. Selective colorimetric detection of polynucleotides based on the distance-dependent optical properties of gold nanoparticles. Science. 1977;277:1078–1081.

[137] Han MY, Gao XH, Su JZ and Nie S. Quantum-dot-tagged microbeads for multiplexed optical coding of biomolecules. Nature Biotech. 2001;19:631–635.

[138] Donaldson K, Stone V, Tran CL, Kreyling W and Borm PJA. Nanotoxicology. Occup Environ Med. 2004;61:727–728.

[139] Renwick LC, Brown D, Clouter A and Donaldson K. Increased inflammation and altered macrophage chemotactic responses caused by two ultrafine particle types. Occup Environ Med. 2004;61:442–446.

[140] Xia T, Korge P, Weiss JN, Li N, Venkatesen MI, Sioutas C and Nel A. Quinones and aromatic chemical compounds in particulate matter induce mitochondrial dysfunction: implications for ultrafine particle toxicity. Environ Health Perspect. 2004;112:1347–1358.

[141] Peters K, Unger RE, Kirkpatrick CJ, Gatti AM and Monari E. Effects of nanoscaled particles on the endothelial cell function in vitro: studies on viability, proliferation and inflammation. J Mat Sc: Mat in Med. 2004;15:321–325.

[142] Tomazic-Jezic VJ, Merritt K and Umbreit TH. Significance of the type and size of biomaterial particles on phagocytosis and tissue distribution. J Biomed Mater Res. 2001;55:523–529.

[143] Nanotechnology and human health: Scientific evidence and risk governance. Report of the WHO expert meeting 10–11 December 2012, Bonn, Germany http://apps.who.int/iris/bitstream/10665/108626/1/e96927.pdf.

[144] Lam C-W, James JT, McCluskey R, Hunter RL. Pulmonary toxicity of single-wall carbon nanotubes in mice 7 and 90 days after intratracheal instillation. Toxicol Sci. 2004;77:126–134.

[145] Warheit DB, Laurence BR, Reed KL, et al. Comparative pulmonary toxicity assessment of single-wall carbon nanotubes in rats. Toxicol Sci. 2004;77:117–125.

[146] http://apps.who.int/iris/bitstream/10665/108626/1/e96927.pdf.

[147] Baran ET, Özer N, Hasirci V. In vivo half life of nanoencapsulated L-asparaginase. J Mat Sc: Mat in Med. 2002;13:1113–1121.

[148] Cascone MG, Lazzeri L, Carmignani C, et al. Gelatin nanoparticles produced by a simple W/O emulsion as delivery system for methotrexate. J Mat Sc: Mat in Med. 2002;13:523–526.

[149] Duncan R. The dawning era of polymer therapeutics. Nat Rev Drug Disc. 2003;2:347–360.

[150] Kipp JE. The role of solid nanoparticle technonogy in the parental delivery of poorly water-soluble drugs. Int J Pharm. 2004;284:109–122.

[151] Poland CA et al. Carbon nanotubes introduced into the abdominal cavity of mice show asbestos-like pathogenicity in a pilot study. Nat.Nanotechnol. 3:423–428.

[152] NIOSH (2011). Occupational exposure to carbon nanotubes and nanofibers. Washington, DC, Department of Health and Human Services, Centers for Disease Control and Prevention, National Institute for Occupational Safety and Health (Current Intelligence Bulletin 65).

[153] Vogel U. Pulmonary and reproductive effects of nanoparticles. WHO workshop on nanotechnology and human health: Scientific Evidence and Risk Governance. Bonn, Germany, 10–11 December 2012.

[154] IRGC (2006). Nanotechnology risk governance (White Paper). Geneva, International Risk Governance Council.

[155] IRGC (2007). Nanotechnology risk governance: Recommendations for a global, coordinated approach to the governance of potential risks. Geneva, International Risk Governance Council.

[156] IRGC (2008). Risk governance of nanotechnology applications in food and cosmetics. Geneva, International Risk Governance Council.

[157] https://www.fda.gov/ScienceResearch/SpecialTopics/Nanotechnology/ucm309677.htm.

[158] Fabrega Climent J (2012). European Chemicals Agency activities on nanomaterials. WHO Workshop on Nanotechnology and Human Health: Scientific Evidence and Risk Governance. Bonn, Germany, 10–11 December 2012.

[159] Kobe A (2012). Nanomaterials in European Union regulation. WHO Workshop on Nanotechnology and Human Health: Scientific Evidence and Risk Governance. Bonn, Germany, 10–11 December 2012.

[160] Hankin S REACH implementation projects on nanomaterials: Outcomes and implementation. WHO Workshop on Nanotechnology and Human Health: Scientific Evidence and Risk Governance. Bonn, Germany, 10–11 December 2012.

[161] Kearns P OECD programme on nanomaterials. WHO Workshop on Nanotechnology and Human Health: Scientific Evidence and Risk Governance. Bonn, Germany, 10–11 December 2012.

[162] Kimmig D, Marquardt C, Nau K, Schmidt A, Dickerhof D. Considerations about the implementation of a public knowledge base regarding nanotechnology. Comput. Sci. Disc. 7:014001.

[163] Riviere JE, Scoglio C, Sahneh FD, and Monteiro-Riviere NA. Computational approaches and metrics required for formulating biologically realistic nanomaterial pharmacokinetic models. Comput. Sci. Disc.6. 2013;014005(15pp).

[164] Guzan KA, Mills KC, Gupta V, Murry D, Scheier CN, Willis DA and Ostraat ML. Integration of data: The nanomaterial registry project and data curation. Comput. Sci. Disc 6. 2013;014007 (8pp).

[165] Liu R, Hassan T, Rallo R and Cohen Y. HDAT: web-based high-throughput screening data analysis tools. Comput. Sci. Disc 6. 2013;014006(11).

[166] Iglesia D de La, Cachau RE, García-Remesal M and Maojo V. Nanoinformatics knowledge infrastructures: bringing efficient information management to nanomedical research. Computational science & discovery. 2013;6(1):014011. doi: 10.1088/1749-4699/6/1/014011.

[167] Burello E Profiling the biological activity of oxide nanomaterials with mechanistic models Comput. Sci. Disc. 2013; 014009 (6):1–16.

[168] Gaheen S *et al* caNanoLab: data sharing to expedite the use of nanotechnology in biomedicine. Comput. Sci. Disc. Volume 6(Number 1).

[169] Chang Liu, Zhen Zhao, Jia Fan, Christopher J. Lyon, Hung-Jen Wu, Dobrin Nedelkov, Adrian M. Zelazny, Kenneth N.Olivier, Lisa H. Cazares, Steven M. Holland, Edward A. Graviss, and Ye Hu. Quantification of circulating *Mycobacterium tuberculosis* antigen peptides allows rapid diagnosis of active disease and treatment monitoring PNAS. 2017 April 11;114 (15):3969–3974.

[170] Banerjee D, Ghosh S and Ansar W, Medical and Veterinary Entomology: The good and bad flies that affect human and animal life. Sch J Agric Vet Sci. 2015;2(3B):220–239.

[171] Santos-Magalhães NS, Furtado Mosqueira VC. Nanotechnology applied to the treatment of malaria, Adv.Drug.Deliv.Rev. 62(2010):560–575.

2 Nanotoxicity and the immune system

Abstract: Diseases that claim lives are a major concern to humans. Therefore, the major aims of research across the world are newer effective treatment agents and protocols for early detection of disease, designing of drugs with lesser side effects, better targeting and effective in less dose and less harmful nature to cure diseases with increased chances of disease-free survival and health management. Nanoparticles (NPs) find their importance in applications with nanomedicine and nanoimaging devices as better diagnostic and therapeutic strategies circumventing the problems in the conventional modes of diagnosis or treatment. There are several drugs approved by the FDA (Food and Drug Administration) in the United States with diverse formulations and applications. However, nanomaterials have been reported of their toxicity on the human immune system, affecting adversely. Exposure to NPs has been reported to cause acute and chronic toxicity affecting biological molecules with generation of reactive oxygen species, oxidative stress, hypersensitive responses, inflammation, cell death and autophagy. In this chapter we discuss the (i) innate and adaptive immune system, (ii) effect of NPs on the immune system and (iii) FDA-approved drugs and imaging devices.

Keywords: Nanomedicine, nanoimaging, nanotherapeutics, nanomaterials, immune system

2.1 Introduction

Nanoparticles (NPs) find their biomedical importance due to their increased bioavailability, sustained release, prolonged exposure, targeted delivery and biocompatible nature, and also act as carriers of drugs, antigens and imaging devices. Several nanomedicine formulations [1, 2] with human applications have been approved by the FDA (Food and Drug Administration) in the United States in the form of drugs and imaging applications against a host of disorders including cancer and many are in the stages of clinical trials. Doxil is an anthracycline topoisomerase inhibitor that indicates ovarian cancer, AIDS-related Kaposi's sarcoma and multiple myeloma, and Abraxane is a microtubule inhibitor that finds application in the treatment of metastatic breast cancer, advanced or metastatic nonsmall cell lung cancer and metastatic adenocarcinoma of the pancreas in combination with gemcitabine. These drugs have been approved by the FDA [1].

However, many NPs and formulations may interact with the cells of the immune system after entry into the body through different routes of oral, dermal and inhalation either intentionally or by chance as an environmental or workplace exposure, thereby affecting both the innate and adaptive immune responses. Metallic and carbon-based NPs tend to display toxicity [1].

https://doi.org/10.1515/9783110579093-002

It has been observed that the NP immune system interaction may lead to immune stimulation or immune suppression leading to one or more effects of immediate or delayed removal of the NP, hypertension, allergy and hypersensitive reaction. Oxidative damage, apoptosis and recently autophagy are reported to be the main events involved in NP and immune system interaction [3–6]. However, the signaling events involved are not yet well studied.

Although tremendous progress has been made in the last few decades in designing and targeting NPs in biological system, and still more are in preparation, considerable efforts are being taken globally by scientists and researchers to detect their effects on human health and test their efficacy and safety prior to human applications. Strategies for improving the safety of NPs by surface modification and pretreatment with immunomodulators are being searched for and nanosafety regulations pertaining to their clinical application are necessitated. Safety evaluation assays are being designed to be more appropriate for engineered NPs [1, 2]. In this chapter, we discuss the NP interaction in the light of (i) affecting the innate immunity, complement system, adaptive immunity; (ii) immune modulation showing either immune activation and immune suppression; (iii) downstream effects of oxidative stress, apoptosis, autophagy, immediate or delayed removal of the affected cell, hypersensitivity reaction, allergy responses and autoimmune responses; (iv) effect of different NPs on the immune system; and the (v) different nanomedicines approved by the FDA.

2.1.1 Immune system overview

2.1.1.1 Innate immune system

Innate immune system is the nonspecific, in-born immunity or natural immunity and forms the first-line of defense against infections and invading pathogens and agents and provides the physical and chemical barriers to infectious agents. Innate immune system recruits cells to sites of entry of foreign agents or sites of infection by the release of chemoattractants such as chemokines, cytokines, complement cascade activation, causing removal of lysed, dead cells and activates the adaptive immune system.

The first line of defense (Figure 2.1 and Table 2.1) includes epithelial cell barriers, sentinel cells in tissues including macrophages, dendritic cells, neutrophils, natural killer (NK) cells and plasma protein molecules. Microbial pathogens are known to activate complement that leads to lysis of the pathogen and causes its elimination. Phagocytic cells, including macrophages and neutrophils (Figure 2.1), lead to phagocytosis using degradative enzymes, signaling molecules, antimicrobial peptides (AMPs) and release of reactive oxygen species (ROS) which in turn triggers the inflammatory responses.

The entry of pathogens or foreign agents such as NPs through skin, respiratory tract and oral or gastrointestinal tract is guarded by epithelial barriers.

The innate immune system generating short-lived immune responses depends on pattern recognition receptors such as Toll-like receptors (TLRs), Nod-like receptors (NLRs), N-formyl peptide receptors and C-type lectin receptors (Figure 2.2) that

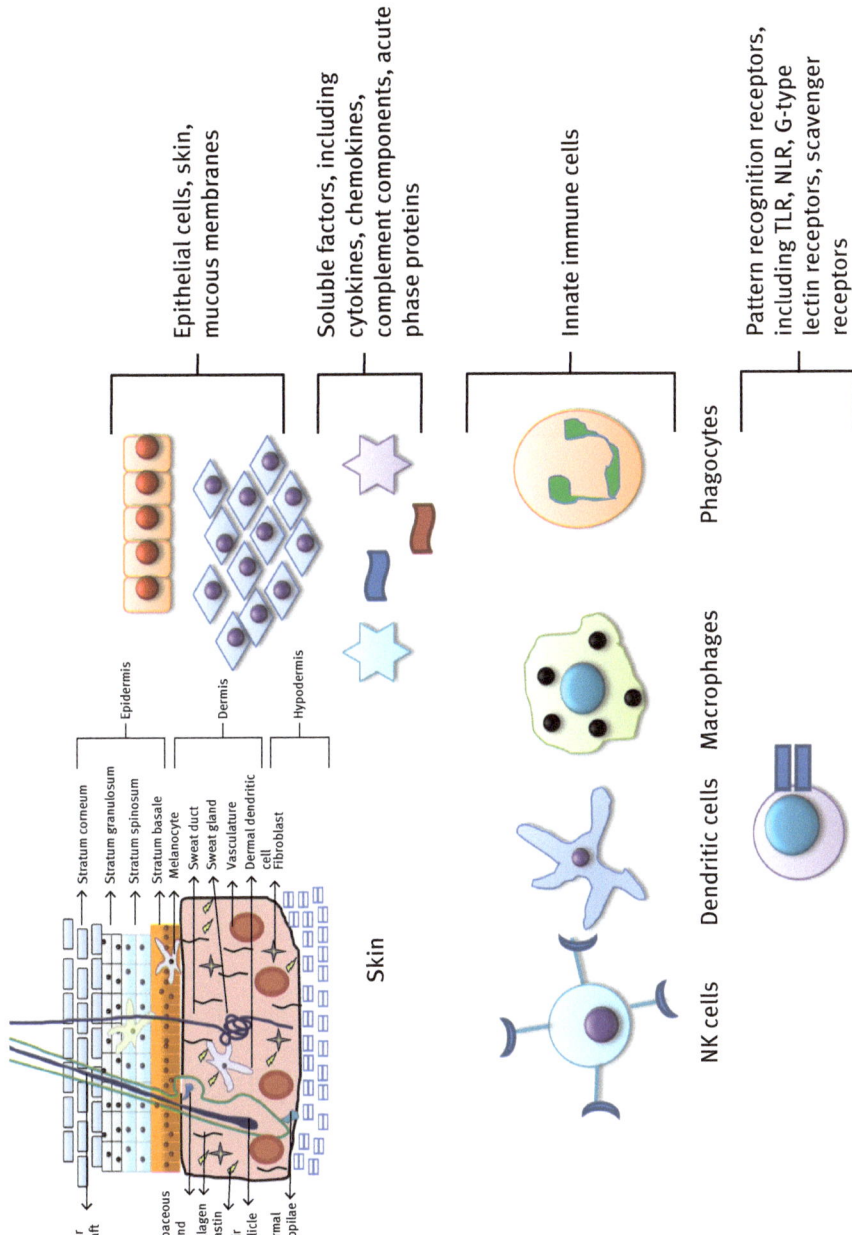

Figure 2.1: Innate immune system: first line of defense.

Table 2.1: Components of innate immune system.

First line of defense		*Mechanical barrier* including skin, hair, mucous membrane, digestive enzymes, gastric acidic pH
		Chemical inhibitor
		Normal microbial flora
Second line of defense	Cellular components	Cells including
		(i) *Phagocytic cells* including macrophages, dendritic cells and neutrophils
		(ii) *Cytotoxic cells* including γδ-T cells and NK cells.
		(iii) *Granulocyte cells* including mast cells, eosinophils, basophils and neutrophils
	Humoral components	*Soluble factors* including cytokines, chemokines, defensins, cathelicidins, pentraxins and complement components
		Inflammatory agents
		Acute phase proteins a, C reactive proteins and mannose-binding lectin

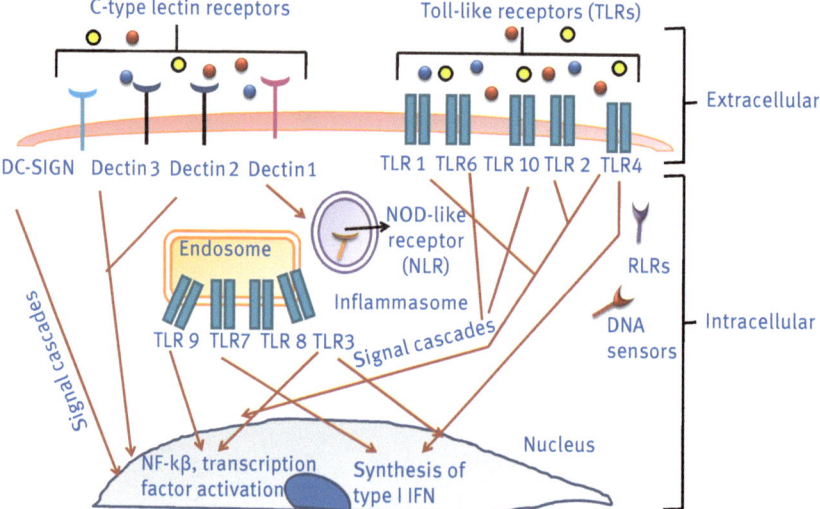

Figure 2.2: Different receptors of innate immune system: C-type lectins, TLRs, NLRs, IFN (interferon) and NFκB represents transcription factor nuclear factor kappa B.

detect distinct evolutionarily conserved sequences on pathogens called as pathogen-associated molecular patterns (PAMPS) or molecules released from damaged or necrotic cells called as damage-associated molecular patterns (DAMPS) that trigger intracellular signaling cascades resulting in the release of proinflammatory molecules, which mediate host responses to infection and eliminate the foreign agent and damaged cells to initiate the processes of tissue repair and lead to subsequent activation of adaptive immunity.

Epithelial cells are capable of producing AMPs including defensins and cathelicidins that can lyse bacteria. γδ-T cells in epithelial layer react against foreign infectious agents, but their role is not completely understood. Neutrophils (Figure 2.3) can act on bacteria and virus, and macrophages produce cytokines including interferon (IFN)-γ, interleukin (IL)-4 and IL-13 leading to inflammation reaction. Macrophages are important effector cells in both cell-mediated and humoral immune responses. Dendritic cells (Figure 2.3) act on microbes to cause inflammation and stimulate adaptive immune responses. Mast cells (Figure 2.3) are activated by microbial breakdown products that are recognized by TLRs. They release vasoactive amines such as histamine that cause vasodilatation with increased permeability in capillaries and cause lysis of bacteria by

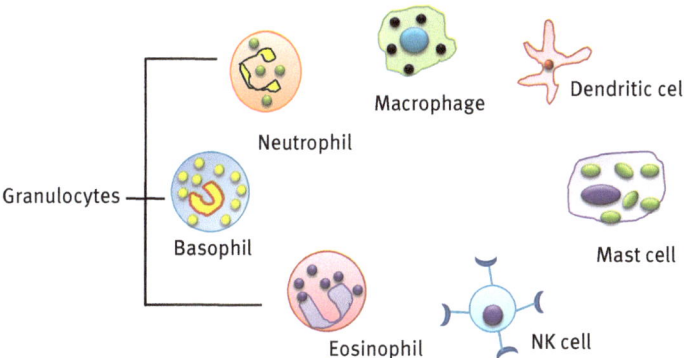

Figure 2.3: Innate immune system: cells.

release of degradative enzymes, release of lipid mediators such as prostaglandins and cytokines such as tumor necrosis factor (TNF) causing inflammation. Mast cells are known to generate defense against helminths and cause allergic disorders.

NK cells (Figure 2.3) are activated by IL-15, type I IFNs and IL-12 and can recognize infected cells and those under stress and lyse these cells by secreting IFN-γ that can activate macrophages and kill phagocytosed microbes. The NK cell receptor "NKG2D" recognizes the major histocompatibility complex (MHC) class I antigens expressed by stressed cells and another receptor CD16 that recognizes immunoglobulin (Ig)G-bound cells. This recognition leads to cell lysis by a process called antibody-dependent cellular cytotoxicity (ADCC) and is predominantly mediated by NK cells. These activating receptors on NK cells have immunoreceptor tyrosine-based activation motifs (ITAMs) in their cytoplasmic ends that undergo phosphorylation in the tyrosine residues when receptors bind to their activating ligands. The phosphorylated ITAMs bind cytoplasmic protein tyrosine kinases leading to downstream phosphorylation and activation of different signaling cascades causing exocytosis of cytoplasmic granules and production of IFN-γ.

NKT cells present on the epithelial layer and lymphoid organs recognize microbial lipids bound to CD1. B-1 cells, a subpopulation of B cells, predominant in mucosal cavities and mucosal tissues produce antibodies in response to microbial pathogens and their released toxins during their passage through intestinal walls. The marginal zone B cells present at the edges of lymphoid follicles act against polysaccharide-rich microbes in blood.

2.1.1.2 The complement system

The complement system (Figure 2.4) is composed of cascade of plasma proteins that gets activated directly by pathogens or indirectly by pathogen-bound antibody, leading to a chain of biochemical reactions on the pathogen surface and releases active components with various effecter functions. Complement activation can take place in one of the three different ways: (i) the classical pathway, in which the pathogen is bound by antibody. This antibody-bound pathogen acts as a triggering agent. (ii) The mannose-binding lectin (MBL) pathway is initiated by MBL or ficolin binding to certain sugars and is mediated by the activity of C4 and C2 complement components leading to the production of activated complement proteins. (iii) The alternative pathway is triggered by microbe or foreign materials, or damaged tissue-bound C3b complement protein. All three pathways proceed by the cleavage of a larger protein, leading to the activation of the next component. Initial activation of the complement cascade leads to formation of a C3 convertase enzyme, which cleaves C3 to produce C3b which bind to pathogen leading to complement activation. Pathogens coated with C3b are taken up by phagocytic cells. C3, C4 and C5 cleavage proteins cause infiltration of phagocytes at infection sites and activate them by binding to G protein–coupled receptors promoting phagocytosis. C3b bound to the C3 convertase leads to binding C5 leading to form C5b that leads to assembly of a membrane attack complex (MAC), which causes lysis on cell surface of pathogens. Regulatory proteins control complement activation and function and prevent tissue damage.

Complement system acts as an effective innate immune system in host defense to eradicate microbial pathogens, and can lead to acute and chronic inflammatory diseases such as sepsis and asthma. C3b coats microbes and promotes their binding to phagocytes by receptors for C3b expressed on phagocytes that lead to rapid ingestion and destruction by phagocytes. This process is also called opsonization. Complement components C3a and C5a chemoattract leukocytes including neutrophils and monocytes that cause leukocyte infiltration at complement-activated sites. When the

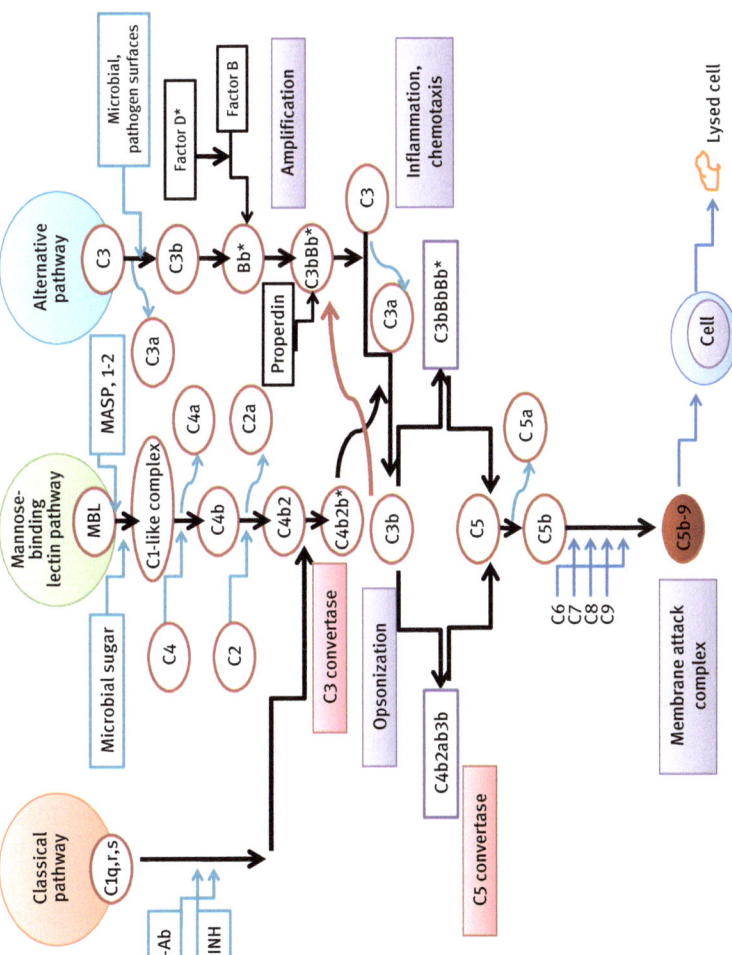

Figure 2.4: Complement system, complement components and reaction cascades. Three different pathways of initiation of complements including the classical pathway, mannose-binding lectin (MBL) pathway, mannan-binding protein (MBP) and mannose-associated serine protease 1/2 (MASP-1/2) are involved in activation of the lectin pathway. C1qrs, C2, C3 and C4 are the activation proteins of the classical pathway; C1-INH and C4-BP (C4-binding protein) are the control proteins. C3, factors B & D* and properdin are activation proteins for the alternative pathway and factors I* & H, decay accelerating factor and CR1 are the control proteins. These pathways lead to formation of membrane attack complex that causes cell lysis.

complement component C5a is activated, it can lead to inflammatory responses by attracting phagocytes, monocytes, neutrophils, eosinophils and T lymphocytes releasing enzymes and oxidants causing tissue damage [7–12]. As anaphylatoxins, C5a can mediate inflammation [7]. The MAC can form pores on bacterial cell walls and cause bacterial cell lysis.

2.1.1.3 Regulatory proteins
Complement function is regulated by different proteins [12]:
(i)   Factor H that targets C3/C5 convertase and regulates the complement system by cleavage of C3b
(ii)   Factor I also targets C3/C5 convertase and regulates complement system by cleavage of C3b and C4b
(iii)   Clusterin affects the complement complexes C5b–7 and keeps the components in solution thereby regulating complement activity
(iv)   C3 nephritic factor is an autoantibody to C3 convertase C3bBb
(v)   Decay accelerating factor (or CD55) affects the C3/C5 convertase and causes the decay of C3 and C5
(vi)   C1 inhibitor targets C1 and regulates by removing C1r and s
(vii)   CR1 targets the C3 or C5 convertase and causes the decay of these convertases
(viii) C4-binding protein targets the C3 convertase
(ix)   Membrane cofactor protein (or CD46) causes inactivation of C3b/C4b and targets the C3/C5 convertase
(x)   Protectin or CD59 inhibits the MAC by targeting C5b–8 complex and/or C5b-9 complex
(xi)   Vitronectin affects the C5b–7 component and keeps the components in solution

These regulatory proteins maintain the balance between acceleration and inhibition of complement activation that controls the host defense or tissue injury. Deficiency or malfunctioning of complement regulatory proteins can lead to diseases [12].

2.1.1.4 The adaptive immune system
The adaptive immune system (AIS) (Table 2.2) generates highly specific responses with specific self- and nonself-recognition providing long-term immunity. Any substance foreign to the host body and capable of eliciting an immune response is called an antigen. White blood cells or lymphocytes comprising B and T lymphocytes play a major role in adaptive immunity. B lymphocytes are synthesized in the bone marrow by the hematopoietic stem cells. On maturity, they express antigen-specific receptors with diverse specificities called as B-cell receptors that are membrane-bound monomeric forms of IgD and IgM [12]. Each antibody has two identical heavy chains and two light chains linked by disulfide bonds to form a "Y"-shaped molecule. The

Table 2.2: Components of the adaptive immune system from Janeway et al. [12].

Types of immunity	Cells	Types of cells
Cell mediated	T cells	CD8$^+$ T cells
		CD4$^+$ T cells
		Yδ T cells
		NK cells
Humoral immunity	B cells	Antibodies

two antigen-binding sites (Fab) of an antibody molecule is expressed on the exterior surface of B cell and are capable of binding specific pathogen epitopes to initiate the B-cell activation that produces plasma cells and memory B cells (Figure 2.5) of which the plasma cells secrete antibodies that enhance the reaction leading to elimination of pathogen. Immunological response mediated by B cells through production of antibodies is termed as humoral responses. The five types of Ig, including IgA, IgG, IgM, IgD and IgE, circulate in the bloodstream where they are exposed to and bind to antigen which leads to removal of the infected cell.

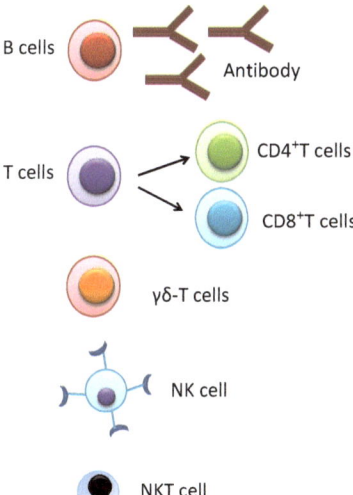

Figure 2.5: Cells of the adaptive immune system.

Cell-mediated immune (CMI) responses are mediated by T lymphocytes. They are produced in the bone marrow but under maturation in the thymus. They encounter specific antigen [12] in the blood. Activated T cells express T-cell receptors (TCRs) on the cell surface and can identify a foreign antigen presented to them on the surface of a host cell expressing MHC. Antigen is presented by antigen-presenting cells (APCs) including professional APCs such as macrophages, B cell and dendritic cells expressing class I/class II MHC complexes to TCRs on T cells and by nonprofessional APCs including

Table 2.3: Th1 and Th2 cytokines from Janeway et al. [12].

Cell	Cytokines	Function
Th1	Interferon (IFN)-γ, interleukin (IL)-2 and tumor necrosis factor (TNF)-β	Cell-mediated immunity and phagocyte-dependent inflammation, proinflammatory cytokines
Th2	IL-4, IL-5, IL-6, IL-9, IL-10, IL-13 and IL-25	Antibody responses (including those of the IgE class) and eosinophil accumulation, but inhibit several functions of phagocytic cells (phagocyte-independent inflammation), anti-inflammatory cytokines

all nucleated cells using MHC class I molecule coupled to beta-2 microglobulin that flags endogenous peptides of antigens on the cell membrane. CD4 and CD8$^+$ are T-cell surface coreceptors that play a role in T-cell recognition and activation by binding to class II and class I MHC on APC. The antigenic peptides from intracellular pathogens expressed by MHC class I molecules and presented to CD8$^+$ T cells undergo differentiation into cytotoxic T cells that kill infected cells. Peptide antigens from pathogens derived from ingested extracellular bacteria and toxins after processing in the cell and being presented by cell surface MHC class II molecules bind to CD4$^+$ Th (T helper) cells, which differentiate into two types of effector T cells called Th1 and Th2 cells (Table 2.3). The activation of TCR generates signal transduction events, and the secondary signals generated through coreceptors on the T cells led to activation of downstream pathways and release of cytokines and eventual removal of the infected cell.

2.1.2 Cytokines

Cytokines and chemokines [12, 13] are proteins with pleiotropic functions that act in signal transduction and play a role in the immune system, influencing growth and development, hematopoiesis, lymphocyte recruitment, T-cell subset differentiation and inflammation. Cytokines can exert autocrine effect, and the effect on the same cell or other cells is termed as the paracrine effect. The term cytokine was named by Cohen et al. in 1974, and the different ILs were also proposed. Mature CD4 helper and CD8 T cytotoxic T cells produce a variety of cytokines. After leaving the thymus they encounter APCs displaying either MHC class I molecules presenting antigenic peptides to CD8 T cells or MHC class II molecules presenting antigenic peptides to CD4 T cells. CD4 Th cells divide into two major subpopulations: Th1 cells secrete mainly IL-2, IFN-γ and TNF-β, whereas Th2 cells secrete mainly IL-4, IL-5, IL-6, IL-10 and IL-13 [12] (Figure 2.6). Th1 cells cause CMI and promote inflammation, cytotoxicity and delayed-type hypersensitivity. Th2 cells play a role

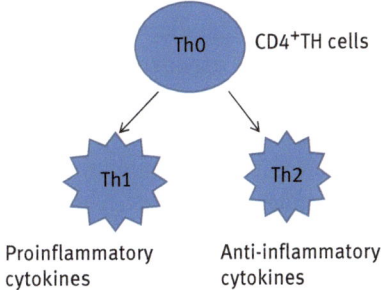

Proinflammatory
cytokines

Anti-inflammatory
cytokines

Figure 2.6: Th1 and Th2 cytokines.

in humoral immunity and downregulate the inflammatory actions of Th1 cells. Cytokines or chemokines can exhibit functional redundancy and one cytokine can influence other groups of cytokines or chemokines, for example, IL-1-like cytokines and IL-6-like cytokines.

Cytokine signaling mechanism by IL-2, IL-4, IL-6, IL-7, IL-10, IL-12, IL-13, IL-15 and IFNs initiates with receptor dimerization on ligand binding following which receptor-associated Janus family tyrosine kinases are activated that phosphorylate the receptor chains and recruit and activate other kinases and transcription factors, including signal transducer and activator of transcription family. This enables rapid translocation of these proteins to the nucleus and stimulation of target gene transcription [12]. IL-1 includes proinflammatory cytokines consisting of IL-1α, IL-1β, IL-1 receptor antagonist and IL-18. IL-1α and IL-1β are produced primarily by mononuclear and epithelial cells upon inflammation, injury and infection. The IL-1 receptor is coded by genes located on human chromosome 2 at a distance from the ligands. IL-18 is a proinflammatory cytokine that is coded by genes in chromosome 11 in humans. IL-2 is expressed by a gene on human chromosome 4 and is secreted by activated T cells. IL-2 and its receptor IL-2 receptor reveal upregulated expression on T cells on antigenic stimulation leading to clonal expansion. IL-4 gene has been traced to be located on human chromosome 5 together with genes coding for granulocyte–macrophage colony-stimulating factor (GM-CSF), IL-3, IL-5, IL-9 and IL-13, of which IL-9 and IL-15 are hematopoietic factors. IL-6 gene is located on human chromosome 7 and is produced by fibroblasts, endothelial cells, macrophages, Th1 cells and B cells. IL-11 gene is located on chromosome 19 in humans and is produced by hematopoietic and nonhematopoietic cell types. Leukemia inhibitory factor (LIF) can bind to CD130 and LIF receptor and plays a role in cell differentiation. Granulocyte colony-stimulating factor is located on chromosome 17 in humans and is synthesized primarily by fibroblasts and monocytes. IL-10, IL-19 and IL-20 genes are located in human chromosome 1 and are known to play a role in suppressing inflammatory responses. IFNs including type I and type II play an important role in pathogen resistance. The pleiotropic cytokines IFN-α and IFN-β are type I IFNs secreted by the virus-infected cells. IFN-y is a type II IFN, secreted by activated Th1 cells and NK cells and the gene is located in chromosome 12 and signals through its own CDw119 receptor and has many biological functions. TNFs act in trimeric shape, which includes TNF-α, TNF-β and lymphotoxin (LT)-β. TNF-α is a proinflammatory cytokine encoded by the gene in human chromosome 6 synthesized primarily by Th1 cells. Tumor necrosis factor beta (TNF-β) or lymphotoxin-α, encoded by gene in human chromosome 6, is a Th1-type cytokine produced by Th1 cells after antigenic or mitogenic stimulation and reveals cytotoxic functions, and it is known to contribute to apoptosis. They bear about 30% protein sequence homology to TNF-α. Both bind to receptors TNF-RI and TNF-RII. TNF-β is known to mediate inflammatory, immunostimulatory and antiviral responses.

2.1.3 Chemokines, their receptors and their genes

Chemokines include low molecular weight proteins (Figure 2.7A) that regulate leukocyte migration through interactions with seven-transmembrane, G protein–coupled receptors (Figure 2.7B). There are four types of chemokines, including C chemokine (Figure 2.7A) [13] examples lymphotactin and SCM-1β, and both of these bind to XCR1. Lymphotactin can attract lymphocytes. The CXC chemokines containing one amino acid between the first two cysteines includes at least 14 ligands (CXCL). They have five receptors (CXCR) and cause chemotaxis of neutrophil. The human CC

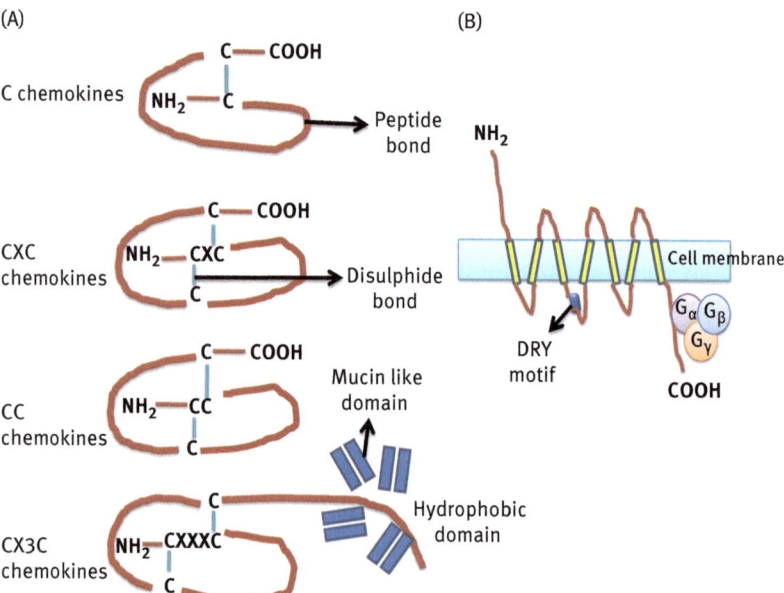

Figure 2.7: Diagrammatic representation of (A) chemokines and (B) chemokine receptors with seven transmembrane domains and a DRY motif in the second intracellular domain. Chemokine receptors are coupled to G proteins with Gα, Gβ and Gγ domains that play a role in signal transduction.

chemokines has more than 27 members (CCL) and can bind to 10 receptors including monocyte chemotactic protein (MCP)-1/CCL2, macrophage inflammatory protein (MIP)-1a/CCL3, MIP-1β/CCL4, RANTES/CCL5 and eotaxin/CCL11. The only chemokine CX3C includes fractalkine is expressed on activated endothelial cells and mediates the attachment of T cells, monocytes and NK cells. Chemokine binds to its receptor leading to its activation by the phosphorylation of carboxyl-terminal serine/threonine residues and dissociation of heterotrimeric G (Gα, Gβ and Gγ) proteins. This in turn leads to mediate multiple signaling pathways through inositol trisphosphate, release of intracellular calcium (Ca^{2+}) from cells, activation of protein kinase C and Ras and Rho proteins. Chemokines mediate the healing of wound and recruit lymphocytes [13].

2.1.4 Oxidative stress

Oxidative stress (Figure 2.8 and Table 2.4) in a system is caused by an agent that disturbs the normal redox state of the cell, leading to changes that can damage the cell, proteins, lipids and DNA and tissue injury, and is manifested by the production of ROS including peroxides and free radicals and affects cellular signaling processes.

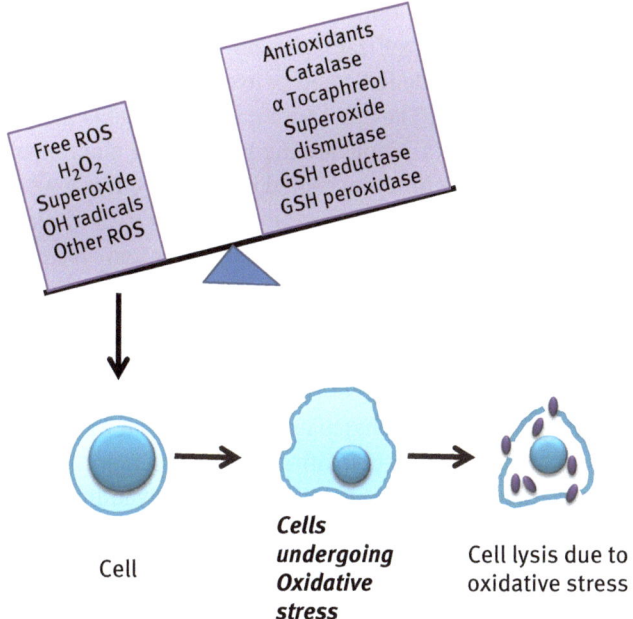

Figure 2.8: Oxidative stress: loss of balance of free radicals and antioxidants. Free radical mediated damage to the cells by oxidative stress.

Table 2.4: Free radicals and nonradical but reactive oxygen species, collectively called reactive oxygen species or ROS.

RADICAL ROS		Non-RADICAL ROS/RNS	
Hydroxyl radical	OH−	Peroxynitrite	ONOO$^-$
Superoxide	O$_2^-$	Hypochlorous acid	HOCl
Nitric Oxide	NO	Hydrogen Peroxide	H$_2$O$_2$
Thyl radical	RS$^-$	Singlet oxygen	1O$_2$
Peroxyl radical	ROO$^-$	Ozone	O$_3$
Lipid peroxyl	LOO	Lipid peroxide	LOOH
Alkyloxyl radical	RO−	Hypobromous acid	HOBr
Sulphonyl radical	ROS$^-$	Organic hydroperoxide	ROOH
Thyl peroxyl radical	RSOO$^-$		
Antioxidant system			
Enzymatic		Nonenzymatic	
Superoxide dismutase	SOD	Sulfiredoxin	Srx
Glutathione	GPx	Glutaredoxin	Grx
Glutathione reductase	GR	Peroxiredoxin	Prx
Glutathione S transferase	GST	Glutathione	GSH
Catalase	CAT	Ceruplasmin	
Thioredoxin peroxidase	TrxPx	Thioredoxin	Trx

DNA damage of the bases and breaks in the double-stranded structure. The ROS includes O_2^- (superoxide radical), OH (hydroxyl radical) and H_2O_2 (hydrogen peroxide). Other free radicals include nitric oxide (NO) and peroxynitrite (ONOO$^-$) and transition metals including iron and copper. The most potent among them is the OH radical with short half-life, reactive and can propagate the free radical chain reactions. Superoxide is produced when an electron is accepted by oxygen but is a weak oxidizing agent. Peroxynitrites are formed in a rapid reaction between O_2^- and nitric oxide (NO). Free radicals are chemical species with one or more unpaired electrons that generate instability and they react with other biological molecules leading to cell death [14].

Free radicals are produced by different biochemical processes on exposure to electromagnetic radiation such as gamma ray. Superoxide dismutase, catalase and glutathione peroxidase are antioxidant enzymes that maintain the cellular homeostasis. It is a natural process by which the host body detoxifies the reactive intermediates or enables repair of the cellular damage. They can remove pathogens by killing them and therefore contribute to cellular defense system [12, 14].

In humans, oxidative stress has been reported to be associated with a host of diseases including cancer, Asperger syndrome, Parkinson's disease, diabetes mellitus, Alzheimer's disease, atherosclerosis, cardiac failure, myocardial infarction, fragile X syndrome, sickle cell disease, vitiligo, autism, infection, chronic fatigue syndrome and depression [12, 14].

2.1.5 Allergy or hypersensitivity

Immune responses against pathogens or foreign particles can lead to tissue injury and disease. Hypersensitivity reactions are classified based on the principal immunologic mechanism leading to tissue injury and disease. Type I or immediate hypersensitivity is a pathologic reaction caused by the release of mediators from mast cells triggered by IgE in response to environmental antigen and the subsequent binding of IgE to high-affinity Fcε receptor (FcεRI) [12] (Figure 2.9A) on mast cells and trigger downstream signaling. Immediate hypersensitivity leads to leakage of mast cells, release of proinflammatory molecules and cause of inflammation. IgE mediates the immediate type of hypersensitivity, which is also known as allergy. Allergy was initially defined by Clemens Von Pirquet as "an altered capacity of the body to react to a foreign substance,"

Type II hypersensitivity (Figure 2.9B) is mediated by IgM and IgG antibodies binding to cell surface antigens followed by activation of complement and Fc receptor-mediated recruitment and activation of leukocytes including neutrophils and macrophages, opsonization and phagocytosis, abnormalities and alteration of cellular functions and cytotoxic reactions.

Type III hypersensitivity (Figure 2.9C) is caused by deposition of immune complex (antigen, IgG, IgM antibody) on membranes of blood vessels, joints, kidneys and skin, observed in serum sickness or, inhaled antigens, extrinsic allergic alveolitis causing complement and Fc-mediated infiltration of leukocytes and ADCC.

Type IV hypersensitivity (Figure 2.9D) involves antigen presentation to T cells by APC including B cells, macrophages and dendritic cells leading to activation of Th and T cytotoxic cells (Tc). This leads to release of cytokines and cell lysis by Tc causing inflammation and cell-mediated lysis. Type IV reactions cause cellular immune response mediated by sensitized lymphocytes.

2.1.6 Apoptosis

Apoptosis (Figure 2.10) is also termed as programmed cell death that occurs as a normal process in multicellular organisms and is a part of the process of normal development of an individual. The development of fingers and toes plays a major role in apoptosis of adjoining tissues. It is a process with distinct morphology identified by blebbing, cell shrinkage, chromatin condensation, fragmentation of DNA and decay of mRNA, and by the development of apoptotic bodies. Unlike necrosis, which is the sudden death of cells by trauma due to sudden cell injury, apoptosis follows an ordered sequence of events.

Apoptosis involves two pathways: Extrinsic and intrinsic pathways. In the *intrinsic pathway*, the cell senses intracellular cell stress and kills itself through activation of caspases. In the *extrinsic pathway* the death ligand binds to cell receptor, recruiting

(A)

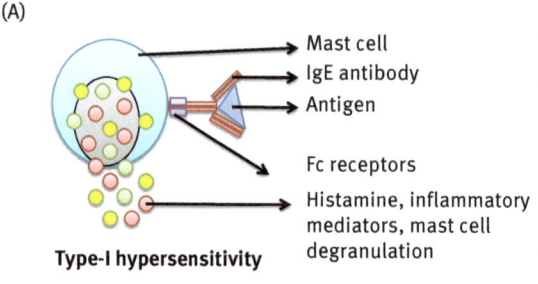

Mast cell
IgE antibody
Antigen

Fc receptors

Histamine, inflammatory mediators, mast cell degranulation

Type-I hypersensitivity

- Mast cell releases vasoactive amines, lipid mediators and cytokines.
- Cytokines mediate inflammation
- Eosinophils and neutrophils play a major role

(B)

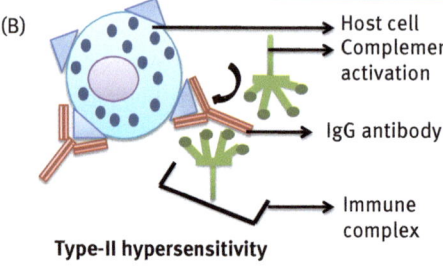

Host cell
Complement activation

IgG antibody

Immune complex

Type-II hypersensitivity

- Mediated by IgM, IgG antibodies against cell surface or extracellular invading foreign antigens
- Antibody-dependent cell-mediated cytotoxicity (ADCC).
- Activation of complement and Fc receptor mediate recruitment and activation of leukocytes including neutrophils and macrophages.
- Opsonization and phagocytosis
- Alteration in cellular functions

(C)

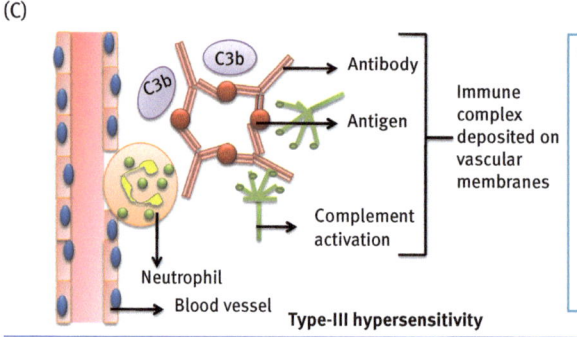

Antibody

Immune complex deposited on vascular membranes

Antigen

Complement activation

Neutrophil
Blood vessel

Type-III hypersensitivity

- Mediated by immune complexes of IgM, IgG antibodies, antigens deposited on vascular membranes that cause complement activation and inflammatory responses.
- Complement and Fc mediated activation and infiltration of neutrophils.

(D)

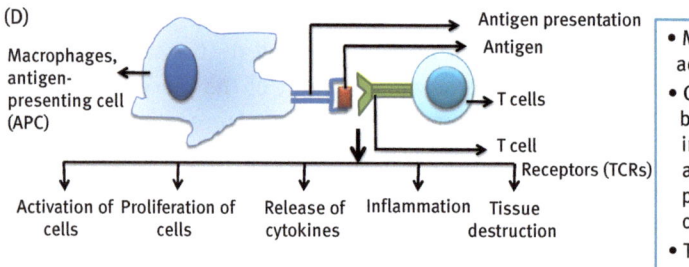

Antigen presentation
Antigen

Macrophages, antigen-presenting cell (APC)

T cells

T cell Receptors (TCRs)

Activation of cells | Proliferation of cells | Release of cytokines | Inflammation | Tissue destruction

Type-IV hypersensitivity

- Macrophage activation
- Cytokine released by Th1 cells, inflammation, activation and proliferation of cells
- T-cell-mediated cell death and tissue destruction

Figure 2.9: Hypersensitivity: (A) Type-I hypersensitivity; (B) Type-II hypersensitivity; (C) Type-III hypersensitivity; and (D) Type-IV hypersensitivity.

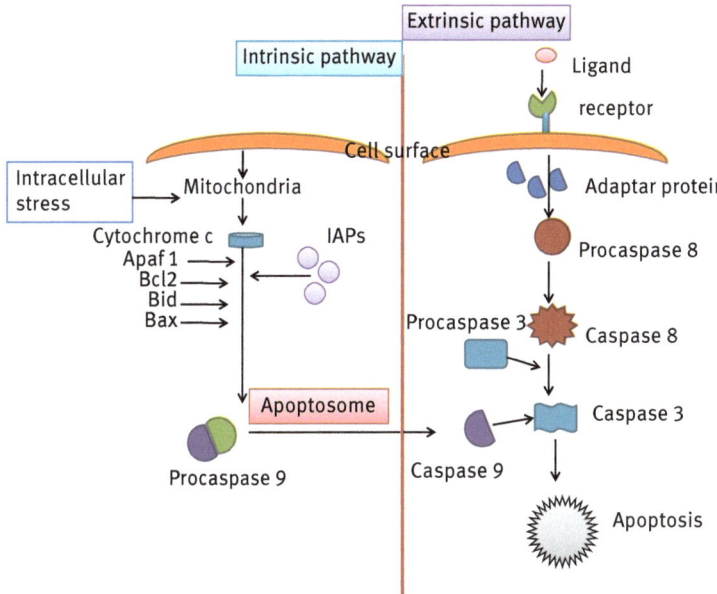

Figure 2.10: Intrinsic and extrinsic apoptosis pathways.

the adapter protein, thereby leading to downstream activation of caspase enzymes leading to cell death. Caspases are proteases that degrade proteins.

While there are group of proteins that activate caspases including Fas receptors and promote apoptosis, other proteins like members of Bcl-2 family inhibit apoptosis.

2.1.7 Intrinsic and extrinsic pathways

2.1.7.1 Intrinsic pathway

Mitochondria are the initiation points for intrinsic pathways of apoptosis. The apoptotic proteins target mitochondria leading to swelling of mitochondria, formation of mitochondrial membrane pores, increasing mitochondrial membrane permeability. Researchers have shown the involvement of nitric oxide to induce apoptosis by altering mitochondrial membrane potential and enhancing its permeability leading to the release of (i) SMACs (second mitochondria-derived activator of caspases) in the cytosol that binds to proteins that inhibit apoptosis (IAPs), thereby inactivating them and enabling the process of apoptosis; and (ii) cytochrome c that regulates the morphological change during apoptosis, which binds to apoptotic protease activating factor-1 and ATP, enabling its binding to procaspase-9 to form the apoptosome which in turn allows the cleavage and activation of procaspase-9 to caspase-9. Caspase-9 then activates effector caspase-3, leading to apoptosis [12].

The mitochondrial outer membrane permeabilization pore (MOMP) is regulated by various proteins, including Bcl-2 family, which promote or inhibit apoptosis. Bax and/or Bak forms the mitochondrial membrane pore, while Bcl-2, Bcl-xL or Mcl-1 inhibit it [12].

2.1.7.2 Extrinsic pathway

Extrinsic pathway is triggered by binding of ligands to receptors, thereby leading to downstream activation of signaling cascades. TNF receptor (TNFR) interaction and Fas or Fas Receptor (Fas R)-Fas ligand interaction may initiate the extrinsic pathways. TNF-α, a cytokine produced by activated macrophages, can bind to two receptors "TNFR1" and "TNFR2" that lead to downstream activation of caspases through the intermediate membrane proteins "TNFR-associated death domain" and "Fas-associated death domain protein" (FADD). cIAP1/2 is an inhibitor to TNF-α signaling by binding to TRAF2, thereby inhibiting apoptosis. FLIP inhibits the activation of caspase-8. The other important pathway include the ligation of Fas or Fas R with FasL leading to activation of a caspase cascade that initiates apoptosis. The fas receptor or *Apo-1* or *CD95* is a transmembrane protein of the TNF family binding to Fas ligand, which leads to the formation of the death-inducing signaling complex, containing FADD, caspase-8 and caspase-10.

Following TNFR1 and Fas activation, a balance between proapoptotic (BAX, BID, BAK or BAD) and anti-apoptotic (Bcl-Xl and Bcl-2) proteins of the Bcl-2 family plays a major role in regulating apoptosis. These proapoptotic homodimers enable the permeability of mitochondrial membrane towards the release of caspase activators including cytochrome c and SMAC (supramolecular activation complex).

Caspases are highly conserved, cysteine-dependent aspartate-specific proteases including the initiator caspases, caspase-2, -8, -9, -10, -11 and -12, and effector caspases including caspase-3, -6 and -7. The initiator caspases are activated by binding to specific oligomeric activator protein which then activate effector caspases through proteolytic cleavage. The active effector caspases then cause proteolytic digestion of intracellular proteins leading to cell death.

2.1.8 Autophagy

Autophagy and lysosomal dysfunction are finding importance in nanomaterial-mediated toxicity. *Autophagy* (or *autophagocytosis*) or "self-devoring" is a natural, regulated, destructive mechanism of the cell that disassembles dysfunctional cellular components, enabling their orderly degradation and recycling. They are of three types: macroautophagy, microautophagy and chaperone-mediated autophagy (CMA). In macroautophagy, dysfunctional cytoplasmic components are isolated in a double-membraned vesicle called as autophagosome that fuses with lysosomes and the contents are degraded and recycled. Microautophagy involving lysosomes in mammals and vacuoles in plants and fungi directly engulf the cytoplasmic components by membrane invagination. In CMA, chaperone-dependent selection of soluble cytosolic proteins are targeted to lysosomes that are translocated across lysosome membrane for degradation.

Autophagy (Figure 2.11) is a process that maintains the health of a cell, since protein degradation is required for normal cell survival. Any disruption to the normal autophagy process leads to misbalance in the cellular homeostasis leading to disease. In disease autophagy is seen as an adaptive stress response seen to be an adaptive response to stress.

By autophagy (macroautophagy), intracellular pathogens and NPs may be degraded. Nanomaterials have been reported to negatively affect these pathways leading to dysfunctions of autophagy and lysosomal pathways, thus leading to immune toxicity [15]. Different formulation and disease affect the pharmacology, pharmacokinetics (PK) and pharmacodynamics (PD). Therefore, their interaction with immune system is important to optimize the drug dose in treatment and reduce the undesirable toxicities of NPs and finds importance from the PK and PD point of view. These studies are carried out on animal models to understand the effect of NPs in a human disease. Preclinical tumor models enable the study of interaction of NP with the immune system, molecules and pathways including the mononuclear phagocyte system, chemokines, hormones and other immune modulators and their role in cancer cure [16].

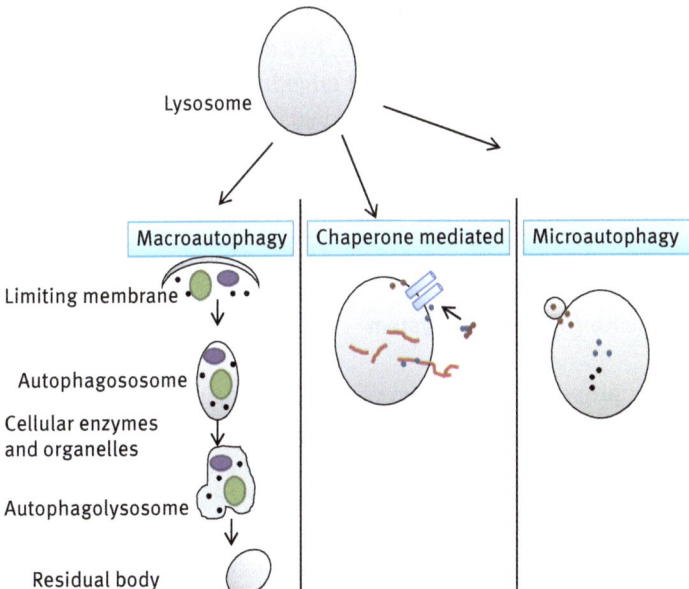

Figure 2.11: Autophagy.

2.2 Immune system and NP interaction

The immune system is a highly complex system that works in coordination of bio-molecules, cells and signaling pathways in order to protect our body against any foreign agents. Any disruption or deviation in the normal functions of the immune system leads to disease. NP–immune system interaction is a complex process involving several pathways and molecules, although some have been understood and considerable knowledge has been gained in the last one and a half decade. Most information remains yet to be studied and calls in for future research. Particularly more studies encompassing the toxic effects of engineered nanodevices, medicines and materials are the need of the hour [3]. Research on NPs has shown their internalization and interaction with the immune system and exhibition of immunomodulatory properties that can either suppress or stimulate the immune system [4–5], leading to enhancement or reduction of intensity of effects. They are known to be engulfed by phagocytic mononuclear cells. The interaction between NPs and the immune cells has been shown to lead to immune suppression or stimulation and understanding of these steps finds importance in inflammatory diseases [6]. Nanomaterials, including polymeric NPs, liposomes, virus-like particles, peptide amphiphiles, micelles, peptide nanofibers and microneedle arrays are being tested for their ability and suitability in design of vaccines, with effective tissue targeting and desirable biocompatibility [17]. NPs find application in cancer immunotherapy for their controlled release, specific targeting and biocompatibility and the properties of composition, morphology, size distribution, charge and immune responses finds importance in their application as cancer drugs [18]. Allergen-specific immunotherapy is the target for treating allergic disorders. NPs due to their improved stability, bioavailability, favorable biodistribution and targeting specific cells but their immunological effects and side effects need to be studied to fulfil their role in allergy vaccines [19]. Nanomaterials and their interaction with albumin and fibrinogen are not desired, which activate the immune system, leading to clearance of the exogenous components [20].

2.2.1 Interaction of NPs with complement system

NPs with diverse surface properties have been observed to regulate the activation of the complement system through different pathways [7]. The effect of different NPs on complement has been found to vary largely. Carbon nanotubes (CNTs) were observed to be opsonized by the complement which in turn enhanced its uptake in U937 cell line that is derived from the human lung lymphoblast without any inflammatory response [10]. IgM Abs were reported to activate complement on dextran-coated NPs with around 250 nm diameter [11]. Peptidoglycan (100 nm size) is a cell wall component of Gram-positive bacteria that could promote complement

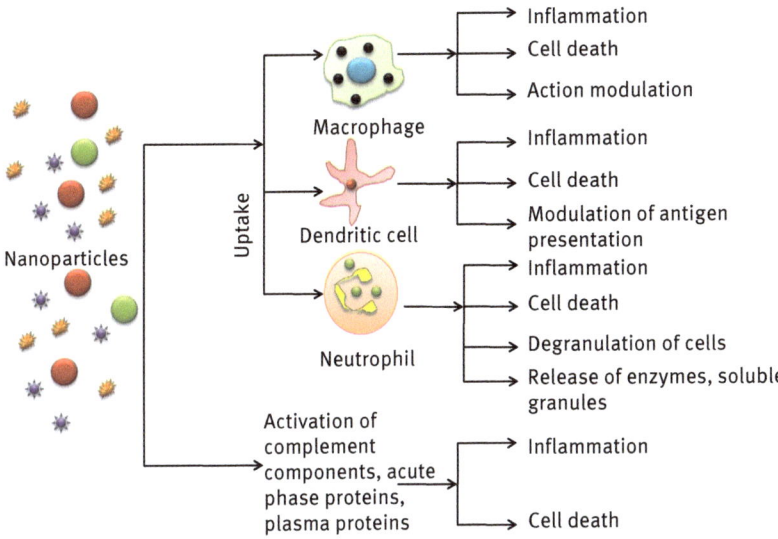

Figure 2.12: Effect of nanoparticles on innate immune system.

activation through the classical pathway [11]. Gold (Au) NPs were reported to produce protein corona, and polyethylene glycol (PEG) coating has been reported to affect the total amount of proteins in the corona of AuNPs but the protein corona composition has not been reported to bear any correlation with NP hematocompatibility [21]. Complement activation-related pseudo allergy (CARPA) is a developing life-threatening condition arising out of exposure to drugs and NP-based drugs, mediated by complement activation but not IgE activation revealing signs of anaphylaxis. Detecting the CAPRA effect of new drugs on serum or plasma derived of healthy individuals and detection of complement activation assays therefore enable screening of new drugs (Figure 2.12) [22].

Iron oxide NPs (IONPs) of 10 nm have been reported to activate complement components C3a, C5a and sC5b-9, induce pro-inflammatory cytokine production in whole blood but did not cause ROS generation or cell death [23].

Spherical AuNPs, gold nanostars and gold nanorods with most biomedical applications could activate the whole complement system, leading to the production of MAC C5b–9 through activation of all three pathways and mediated uptake of all gold nanomaterials (AuNMs) by human U937 promonocytic cells, expressing complement receptors [24]. Complement receptors have been reported to play a role in activation and uptake of AuNM in cells [24].

Even the declared safe intravenous iron nanomedicines designed to treat patients suffering from iron deficiencies suffers from the increased risk of hypersensitivity reactions by affecting the innate immune responses. NPs of size 10 and 30 nm including ferric gluconate, iron sucrose, ferric carboxymaltose and iron isomaltoside 1000 actively taken up by HEK293T cell lines and peripheral blood mononuclear cells in a cholesterol-dependent manner have been reported to activate intracellular TLR-3, TLR-7 and TLR-9 and activated the complement system release of proinflammatory cytokine IL-1β, but not IL-6 [25].

Pristine and derivatized CNT can activate the complement through the classical pathway leading to enhanced uptake by phagocytic cells, and downregulate proinflammatory cytokines TNF-α and IL-1β. Complement deposition led to uptake of CNTs by immune cells overexpressing complement receptors [26]. This study also hints that complement system can recognize patterns in NPs and modulate immune responses [26].

About 20 kDa dextran-coated superparamagnetic iron oxide (SPIO) nanoworms (NWs) with diameter of 111.4 ± 28 nm exposed to both normal and tumor-bearing mice reveal complement-dependent uptake by mouse lymphocytes, neutrophils and monocytes but the role of complement is reported to be highly species dependent [27]. Superparamagnetic IONPs (SPIONs, 10–100 nm) find their applications as contrast agents for biomedical imaging by magnetic resonance imaging. They are composed of a core of y-Fe$_2$O$_3$ (maghemite), Fe$_3$O$_4$ (magnetite) or α-Fe$_2$O$_3$ (hermatite) and a hydrophilic surface coat of polymers, including carboxydextran, chitosan, dextran, phospholipids, PEG and starch [28]. Polymer-coated SPIONs are known to cause complement activation, and CARPA leading to anaphylaxis and occasional death remains a major threat to human applications [28]. About 20 nm poly(lactic-co-glycolic) acid NPs were reported to activate the complement and coagulation cascade at high concentration [29]. Approximately 20 kDa dextran-coated SPIO NWs are revealed to activate C4d and Bb and MAC. In mouse sera, the lectin pathway and alternative pathway of the complement cause the amplification loop and the classical pathway is not reported. In humans, additionally the classical pathway may be involved [30]. Perfluorocarbon (PFOB) NPs are characterized by lipid encapsulations of NPs that find applications in biomedical imaging and as therapeutic agents in preclinical studies. Gadolinium-functionalized PFOB NPs have been reported to lead to complement activation strongly in both humans and mice [31].

2.2.2 Interaction with adaptive immune system

2.2.2.1 Immune stimulation

Adaptive immune system responses are specific with the retention of immunological memory by which they can combat second exposure to the same pathogen. Both the branches of adaptive immune system comprising humoral and cellular immune responses function in a normal individual to combat entry and exposure to foreign pathogen. Recently, exposure to NPs has been reported to stimulate humoral and immune responses.

2.2.2.2 NPs affect production of cytokines

Cytokines are small-sized protein molecules ranging from 5 to 20 kDa molecular weight released by immune cells, macrophages, B and T lymphocyte cells, mast cells, fibroblasts, stromal cells that play an important role in cell signaling by autocrine, paracrine and endocrine and act in immune modulation. They mediate signal transduction by binding to their receptors, and activating downstream signaling events. The Th cells express CD4 on their cell surface and help other immune cells, thereby regulating the immune response by releasing cytokines. They play important roles in B-cell activation and antibody class switching, activation of cytotoxic T cells and enhancing the bactericidal activity of phagocytes and macrophages.

Effector T cells differentiate into Th1 and Th2 cells. Th1 cells affect intracellular bacteria and protozoa, triggered by IL-12 and their release of effector proinflammatory cytokines including IFN-y and IL-2. Th2 cells, on the other hand, are most effective against extracellular parasites including helminths that are triggered by IL-4 and IL-2, thereby leading to their activation and release of effector anti-inflammatory cytokines including IL-4, IL-5, IL-9, IL-10, IL-13 and IL-25. NPs including 250 nm sized polyhydroxylated fullerenes [$C_{60}(OH)_{20}$] with slow cytotoxic effects have been known to stimulate the increase in $CD4^+/CD8^+$ lymphocyte ratio together with increased Th1 cytokines, TNF-α and suppression of Th2 cytokine synthesis [32]. Tumor development can cause shifting of Th1 responses to Th2 responses, and $C_{60}(OH)_{20}$ NPs promote T-cell differentiation to Th1, and lower Th2 cytokines can modulate the Th1/Th2 balance back to Th1 type in mice model of lung cancer and can mediate tumor therapy. Fullerene derivative, polyhydroxylated metallofullerenol $Gd@C_{82}(OH)_{22}$ NPs, has been reported to stimulate macrophages and T cells to release cytokines including IL-2, IL-4, IL-5, TNF-α and IFN-y [32].

2.2.2.3 Adjuvant role

Although NP-induced antigenicity is a relatively new domain and not much information exists on the underlying mechanism, few studies have indicated their role in generating antigenic responses while some studies have negated their antigenic role. Exposure with few C_{60} fullerene conjugated to bovine serum albumin enabled the generation of antibodies. Mice immunized with C_{60} fullerene derivative conjugated to bovine thyroglobulin have been reported to produce fullerene-specific IgG antibodies [33, 34].

NPs have been reported to generate specific antibodies [33] and act as carriers and can stimulate immune response [33–40] by antigen such as polymethylmethacrylate NPs that act as adjuvants for HIV-2 virus vaccine in mice [41]. Aluminum hydroxide NPs acted as adjuvant and activate dendritic cells, production of IgG and IgE [42]. $Al(OH)_3$ NPs induced a strong humoral response [43]). Ultrasmall graphene oxide-supported AuNPs [44] are reported to act as an adjuvant stimulating humoral and cellular immune responses. The adjuvant property of NPs is largely controlled by

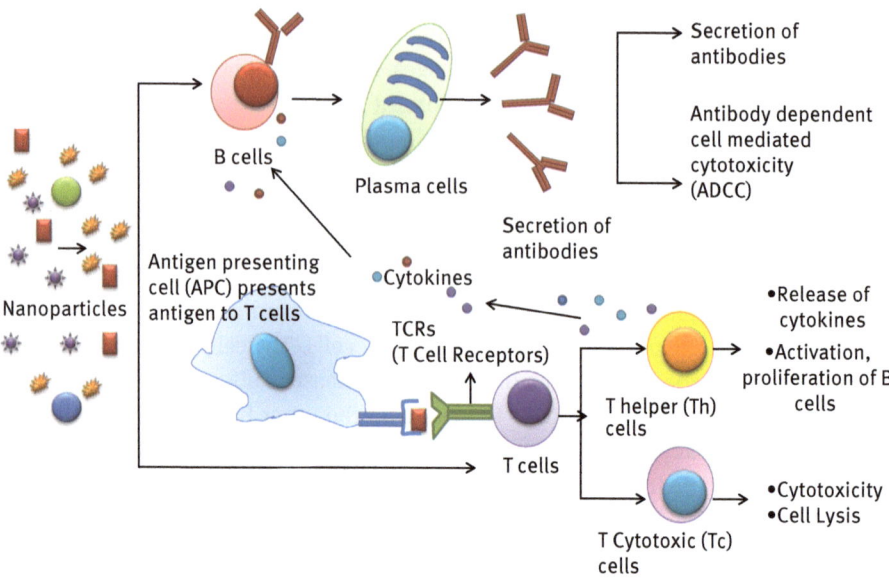

Figure 2.13: Nanoparticles and the adaptive immune system.

their size and surface charge and finds importance in improvisation of vaccines by their slow and controlled properties in releasing antigens [33]. Adjuvants are agents that can trigger the APCs and elicit a strong immune response and find major applications in vaccines. Mesoporous silica NPs (MSNs) [45] with their adjuvant property finds application as vehicles of vaccine delivery but can elicit both humoral and cellular immune responses (Figure 2.13).

2.2.2.4 Allergic responses

NPs that can affect mast cells play a major role in allergic responses [7]. TiO$_2$ NPs have been reported to stimulate mast cells directly to release histamine, contributing to inflammation and toxicity [7]. Research regarding the effects of NPs on the immune system have elucidated its role in causing allergic responses like contact dermatitis. But at the same time, its small size makes its role as a hapten to induce IgE production unlikely. However, its role as an adjuvant in inducing inflammatory cytokine production, antibody synthesis, acting together with bacterial endotoxin and/or surfactants favoring allergic sensitization is reported.

Several NPs have been reported as promoters to allergic reactions [46] including CNTs which when inhaled or administered subcutaneously have been reported to enhance the allergen potential of egg albumin promoting the response has been linked to cytokine response [47]. Epidermal dermatitis with necrolysis was reported following occupational exposure to dendrimer synthesis by producing ROS [48].

IONPs have been reported to increase IgE blood concentrations by intratracheal installation [46]. CNTs [47], dendrimers, magnetite IONPs, titanium dioxide (TiO_2), AuNPs, silver (Ag) NPs and polystyrene NPs have been reported to promote allergy responses [46].

TiO_2 and AuNPs can induce airways hyperreactivity following inhalation, along with bronchoalveolar lavage cells, histology and total IgE alterations [49], and TiO_2 has been reported to aggravate dermatitis [50]. Oral administration of AgNPs in mice leads to overexpression of IL-1, IL-6, IL-4, IL-10, IL-12, TGF-β cytokines, B-cell distribution and increased IgE production [51].

NPs have been reported to play a role in inducing antiallergic effects. Elevated NF-κB p65 in lung tissues of nuclear protein post Ovalbumin (OVA) inhalation has been reported to be reduced after administration of AgNPs [52]. Betamethasone disodium phosphate encapsulated in biocompatible NPs caused antiallergic effects [53]. Chitosan NPs mixed with Imiquimod cream were reported to be effective against allergic asthma [54]. Topical application of cyclosporin A-loaded solid NPs relieved symptoms of atopic dermatitis in an in vivo murine model with alterations of Th2, IL-4 and IL-5 and finds potential applications in treatment of allergy-related skin disorders [55]. Nanogel containing surface-modified NPs improved the skin permeation of ketoprofen and spantide II [56].

2.2.3 Immune suppression

Although few studies have confirmed the immunosuppressive role of nanoparticles, the underlying mechanism is yet to be known. However, this role enables their application in preventing transplant rejection, treatment of inflammatory and autoimmune diseases, and administration and maintenance of immunosuppressive condition [7]. Fe_3O_4 NPs suppressed the humoral response and cytokines in ovalbumin-injected mice, and multiwalled CNTs (MWCNTs) suppressed the overall humoral immunity in mice [57]. Inhaled MWCNTs have been reported to suppress immune function in mice model [58]. CeO_2 NPs [59] exposed to murine macrophages revealed the reduced generation of ROS and inflammatory cytokines IL-6 and TNF-α. Polyvalent dendrimer glucosamine conjugates have been reported to reduce inflammatory cytokines and chemokines in endotoxin-challenged human macrophages and dendritic cells [60]. Fullerene has been reported to inhibit hypersensitivity reaction to allergens [61].

2.2.4 Toxicity

Although oxidative stress, cell damage and apoptosis are hallmark features of nanoparticle-mediated toxicity, reports on nonoxidative stress-mediated cellular damage due to direct interaction of NPs with cellular structures [4] also exist. Although the most common route of uptake of nanomaterials by endocytes enter into the lysosome, making a compartment for its sequestration and degradation, induction of autophagy is another important step. Different NPs have been reported to induce autophagy in different cells but the underlying mechanisms are not well understood. Diverse compositions of NPs are being tested both for their efficacy and safety in human applications. Bioreactivity and toxicity of metal NPs (MNPs) rise from the surface characteristics of hydrophilicity, lipophilicity, catalytic property, surface area, complexity, electronic structure, dimension, binding capacity or coating property and solubility, adsorption of proteins on the NP surface and routes of administration. MNPs designed for applications in diagnosis and therapy are adsorbed by inhalation, contact, ingestion and injection, and they translocate to tissues, encounter the innate immunity system, exert cyto- and genotoxicities, affect cellular functions through modification of receptors, transcription and cytokine production and activate the immune system playing a role in treatment of disorders like cancer [18, 20, 39, 40].

Although few reports exist on the nanotoxicity to the human immune system, their interaction with the immune system initiates with their accumulation in regional lymph nodes and is presented and processed by dendrite cells leading to subsequent immune responses. They are also reported to interact with host proteins and alter their antigenicity and elicit altered immune responses and lead to generation of autoimmune and hypersensitive responses. Single-walled (SW) and MWCNTs promote allergic immune responses in mice [62]; intratracheal installation of iron NPs promotes inflammatory response [63] and asthmatic responses in lungs in mice models [64].

NPs [65] and engineered NPs [66] have been reported to affect the innate immunity [66] and modulate immune responses [67, 68].

Metal-on-metal (MOM) compositions have been largely exploited to detect their safety levels. Complex alloys composed of cobalt–chromium alloys on interaction with the cells have been reported to release cobalt and chromium NPs that lead to detrimental consequences on cellular viability together with the DNA damage and chromosomal aberrations. Hypersensitivity immune reactions are also reported to be increased, leading to both local and systemic reactions termed under the umbrella of arthroprosthetic cobaltism [69] and therefore their application as potential candidates in nanomedicine raises safety concern. Newcastle disease (ND) proves to be fatal to animal health with no treatment available and is threatening to poultry industries. NP-encapsulated ND virus vaccines have been designed with high safety, low toxicity and better immunogenicity by eliciting mucosal and humoral immune responses [70]. Immunotoxicity of both airborne and engineered NPs has been studied to lead to pathological pulmonary inflammation by impairment of both innate and adaptive immunological responses [71–74].

Lipid-based NPs (LNPs) act as delivery devices in cancer, inflammation and infections, but they are reported to interact with different leukocytes leading to

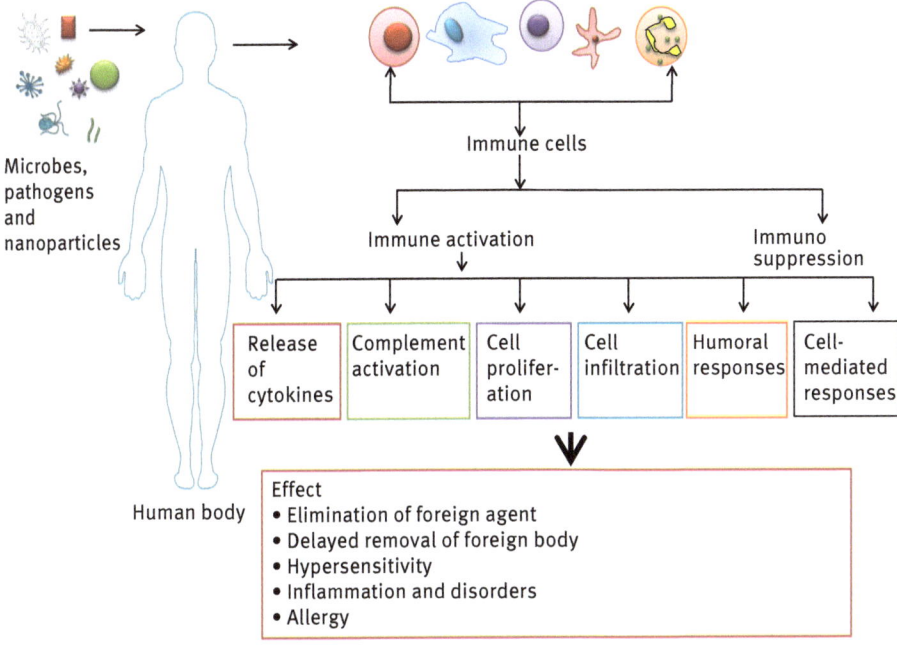

Figure 2.14: Nanoparticle and toxicity of immune system.

suppression or activation of the immune system affecting both the innate and adaptive immune systems [75, 76]. CNTs have been reported to perturb the immune system [77]. Inorganic NPs such as titania, gold and mesoporous silica led to the toxicity [78]. Cellulose nanocrystals and nanofibrillated cellulose exhibit toxicity due to their specific physical characteristics (Figure 2.14) [79].

2.3 Immunomodulatory and proinflammatory responses

MNPs have been reported to cause allergic sensitization manifested by allergic contact dermatitis on exposure to Pd. In vitro studies have revealed that NPs have immunomodulatory effects on cytokine production leading to either Th1 responses mediated by Pl, Pd, Ni, Co-NPs or Th2 responses by Ti, MWCNT and SWCNT. MNPs can act as adjuvants, thereby activating both cellular and humoral immune responses and lead to allergic responses. Lungs in animal models have shown activation of proinflammatory responses with elevated expression on Interleukin-1β (IL-1β), Macrophage inflammatory protein 1 alpha (MIP-1α), Monocyte chemoattractant protein 1 (MCP-1), Macrophage-inflammatory protein 2 (MIP-2), keratinocyte chemoattractant, thymus and activation regulated chemokine (TARC), Granulocyte-macrophage colony-stimulating factor (GM-CSF), (MIP-1α) factors, activated mitogen-activated protein kinases (MAPK) p38 and Jun N-terminal kinases (JNK) and induced oxidative stress. Engineered NPs may lead to allergy responses in lungs, cause proinflammatory disease [53] and modulate the innate and adaptive immune responses. NP-mediated stimulation and suppression can influence their interaction with immune cells to attain desirable immunomodulation and avoid undesirable immunotoxicity [54–56]. Thus, engineered NPs may be designed for medical applications with strategy to bypass immune system recognition or specifically inhibit or enhance the immune responses [56, 57]. While some studies on cytokines have been conducted on the adverse immune response by NPs, some studies are still underway [58–60]. ZnONPs are finding a role in antitumor agents but they have been reported for their cytotoxic, inflammatory, genotoxic and immunotoxic effects and warrant their clinical use [80, 81]. When deposited into the brain, NPs lead to deleterious inflammatory changes that occur in the central nervous system (CNS) with the release of ROS within the brain [82, 83]. IONPs, dendrimers, MSPs, AuNPs and CNTs differ in their distribution and bioavailability based on their physical and chemical properties and after engulfment by immune cells they are known to react with the immune system. Therefore, detecting the biocompatibility is important prior to their selection as drug delivery agents [84, 85].

2.4 Protein corona formation

Protein coronas that are generated on interaction of NPs with the fluids in the body may contribute to immune response [66]. Internalized by immune cells, TiO_2 and AgNPs show accumulation in peripheral lymphoid organs and exert immunomodulatory and immunotoxic effects [67]. Cobalt and nickel NPs have been reported of their inflammatory properties [68]. The size, shape and deformability are reported to influence NP uptake by immune cells and they are found to interact with different cell

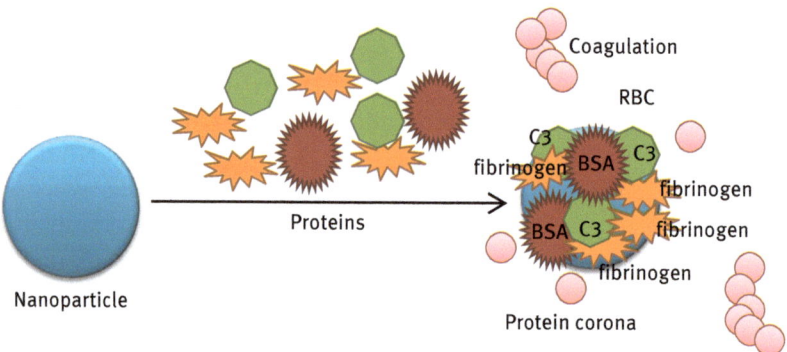

Figure 2.15: Protein corona formation.

types and soluble mediators of the innate immune system thereby leading to immune responses. Complement activation due to corona formation on NPs led to their clearance from the system and thus cause safety. Bacterial lipopolysaccharides have been reported to bind to NPs and lead to immune responses (Figure 2.15) [86]. NPs react with the host system generating the protein "corona" that affect their physicochemical properties and the innate and adaptive immune responses [87, 88], which find applications in designing innovative functional, safe and effective nanotherapies.

2.5 NPs as drugs and imaging devices

Diseases are a major challenge and cause death globally, and their treatment management and diagnosis need utmost care and calls in more research. There are a host of diseases, both infectious and communicable, including infections by virus, bacteria and protozoa, and noninfectious diseases include genetic, brain disorders and disorders of the immune system. Noncommunicable diseases such as cancer, cardiovascular disease, chronic respiratory disease and diabetes mellitus pose a threat to human life globally. Inflammatory diseases including autoimmune diseases and autoinflammatory diseases are known for their hallmark characteristic of imbalance of proinflammatory and anti-inflammatory cytokines that suffer from the lack of medication [89]. Nanomedicines as drugs and imaging devices offer advantage in the treatment of diseases with several disorders [90].

Visceral leishmaniasis (VL) or kala-azar is an infectious disease caused by the protozoan *Leishmania* sp. Symptoms include irregular bouts of fever, loss of weight, swelling of the spleen and liver and anemia, and if left untreated may be fatal and the World Health Organization has suggested the rise in fatality rate in developing countries as 100% within two years [91]. Post-kala-azar dermal leishmaniasis is a post VL complication in endemic areas of the disease with symptoms of hypopigmented macular, maculopapular and nodular rashes after recovering from VL. HIV coinfection is another complicacy associated with VL [92, 93]. Conventional therapy lacks from the disadvantage of side effects and toxicity, and thus NP-mediated therapy holds great promise in treatment of VL [93].

The FDA in the United States has already approved several nanomedicines [94] (Table 2.5), including therapeutic or imaging agents as drugs and some are in the clinical trials. Nanomedicine offers advantage of improved bioavailability, circulation time, stability, increased delivery to tumor site, extended release, controlled delivery, reduced dosages, increased drug loading, faster administration, biodistribution, efficiency and reduced toxicity. The FDA-approved materials [95] range from polymeric, liposomal and nanocrystal formulations, micelles, protein-based NPs and inorganic and metallic particles in clinical trials. Multifunctional materials included under the term theranostics with knowledge in safety and toxicity are required before their human application [96].

Table 2.5: FDA-approved nanomedicines with indicated effects in disorders%.

Indication(s)	Nanomedicine	Composition	Manufacturer	Year of approval
Severe combined immunodeficiency disease	Addgene	PEGADEMASE BOVINE	SIGMA TAU	1990
Crohn's disease, rheumatoid arthritis, psoriatic arthritis, ankylosing spondylitis	CIMZIA	CERTOLIZUMAB PEGOL	UCB INC	2008 2009 2013 2013
Multiple sclerosis	COPAXONE	GLATIRAMER ACETATE	TEVA PHARMS USA	1996
Prostrate cancer	ELIGARD	LEUPROLIDE ACETATE	TOLMAR THERAP	2002
Macular degeneration	Macugen	PEGAPTANIB SODIUM	VALEANT PHARMS LLC	2004
Anemia associated with chronic kidney diseases	MIRCERA	METHOXY POLYETHYLENE GLYCOL-EPOETIN BETA	HOFFMAN-LA ROCHE	2007
Neutropenia	NEULASTA	PEGFILGRASTIM	AMGEN	2002
Chronic hepatitis B, hepatitis C	Pegasys	PEGINTERFERON ALFA-2A	HOFFMAN-LA ROCHE	2002
Chronic hepatitis C in patients with compensated liver disease	PEGINTRON	PEGINTERFERON ALFA-2B	SCHERING	2001
Patients with end-stage renal disease	RENAGEL	SEVELAMER HYDROCHLORIDE	GENZYME	2000
Acromegaly due to excessive secretion of growth hormone	SOMAVERT	PEGVISOMANT	PHARMACIA AND UPJOHN	2003
Acute lymphoblastic leukemia	ONCASPAR	PEGASPARGASE	SIGMA TAU	1994
Gout	KRYSTEXXA	PEGLOTICASE	CREALTA PHARMS LLC	2010
Patients with relapsing forms of multiple sclerosis	PLEGRIDY	PEGINTERFERON BETA-1A	BIOGEN IDEC INC	2014
Control of bleeding episodes and to reduce the frequency of bleeding episodes (prophylaxis) in patients with hemophilia A	ADYNOVATE		BAXALTA US Inc	2015

Kaposi's sarcoma	DAUNOXOME	DAUNORUBICIN CITRATE	GALEN (UK)	1996
Lymphomatous meningitis	DEPOCYT	CYTARABINE	PACIRA PHARMS INC	1999
Adult patients with Philadelphia chromosome-negative (Ph−) acute lymphoblastic leukemia	MARQIBO KIT	VINCRISTINE SULFATE	TALON THERAP	2012
Metastatic adenocarcinoma of the pancreas	ONIVYDE	IRINOTECAN HYDROCHLORIDE	IPSEN INC	2015
Fungal infectious diseases	AMBISOME	AMPHOTERICIN B	ASTELLAS	1997
Post surgery analgesia	DEPODUR	MORPHINE SULFATE	PACIRA PHARMS INC	2004
Macular degeneration	VISUDYNE	VERTEPORFIN	VALEANT LUXEMBOURG	2000
Ovarian cancer, Kaposi's sarcoma, multiple myeloma	DOXIL (LIPOSOMAL)	DOXORUBICIN HYDROCHLORIDE	JANSSEN RES AND DEV	1995
Fungal infections	ABELCET	AMPHOTERICIN B	LEADIANT BIOSCI INC	1995
Respiratory distress syndrome	CUROSURF	PORACTANT ALFA	CHIESI USA INC	1999
Treatment of moderate-to-severe vasomotor symptoms associated with menopause	ESTRASORB	ESTRADIOL HEMIHYDRATE	EXELTIS USA INC	2003
Breast cancer, nonsmall cell lung cancer, pancreatic cancer	ABRAXANE	PACLITAXEL	ABRAXIS BIOSCIENCE	2005
Cutaneous T-cell lymphoma	ONTAK	DENILEUKIN DIFTITOX	EISAI INC	1999
Antiemetic agents	EMEND	APREPITANT	MERCK	2003
Hyperlipidemic disorders	TRICOR	FENOFIBRATE	ABBVIE	2004
Immune suppression	RAPAMUNE	SIROLIMUS	PF PRISM CV	2000
Anti-anorexic	MEGACE	MEGESTROL ACETATE	BRISTOL MYERS SQUIBB	1971
Psychostimulant	AVINZA	MORPHINE SULFATE		2002

(continued)

Table 2.5 (continued)

Indication(s)	Nanomedicine	Composition	Manufacturer	Year of approval
CNS stimulant	FOCALIN XR	DEXMETHYLPHENIDATE HYDROCHLORIDE	NOVARTIS	2005
CNS stimulant	RITALIN	METHYLPHENIDATE HYDROCHLORIDE	NOVARTIS	1995
Spasticity	ZANAFLEX	TIZANIDINE HYDROCHLORIDE	COVIS PHARMA BV	1996
Bone reconstruction	Ostim® Bone Grafting Material	Hydroxyapetite	Heraeus Kulzer, Incorporated	2002
Schizophrenia and schizophrenia affected disorder	INVEGA	PALIPERIDONE	JANSSEN PHARMS	2006
Malignant hyperthermia	RYANODEX	DANTROLENE SODIUM	EAGLE PHARMS	2014
Iron deficiency anemia in adult patients with chronic kidney disease	FERAHEME	FERUMOXYTOL	AMAG PHARMS INC	2009
Hemodialysis-dependent chronic kidney disease, nondialysis-dependent chronic kidney disease, peritoneal dialysis-dependent-chronic kidney disease	VENOFER	IRON SUCROSE	LUITPOLD	2000
Iron deficiency in chronic kidney disease	FERRLECIT	SODIUM FERRIC GLUCONATE COMPLEX	SANOFI AVENTIS US	1999
Iron deficiency in chronic kidney disease	INFED	IRON DEXTRAN	ALLERGAN SALES LLC	1974
Iron-deficient anemia	DEXFERRUM	IRON DEXTRAN	LUITPOLD	1996
Contrast agents in ultrasound	DEFINITY	Perflutren lipid microspheres	LANTHEUS MEDCL	2001
Contrast agents in ultrasound	Optison	Perflutren microspheres stabilized by human serum albumin	GE HEALTHCARE	1998
Magnetic resonance imaging contrast agent	COMBIDEX (ferumoxtran-10)	Iron-dextran colloid	Sinerem (AMAG)	2005

% 87, 88, 89

2.6 Discussion

Currently, there is a dearth of parameters and universal protocols to detect the effect of NPs on the immune system. To add the complicacies are species variability of the immune system in experimental animal models and cell lines, high costs and relatively low throughput of in vivo tests and ethical concerns of animal experiments; however, efforts are being made toward fixed protocols for the testing [74]. Although nanomedicines for VL are being designed for their immunological effects, effects of systemic toxicity need to be investigated further [75]. Likewise, there are host of diseases against which no remedies are available. The main challenge ahead would be the design of NPs in medical use with lowered or removed hypersensitivity and allergy reactions, long-term localized or systemic toxicity and biocompatible units.

References

[1]   Wolfram J, Zhu M, Yang Y, Shen J, Gentile E, Paolino D, Fresta M, Nie G, Chen C, Shen H, Ferrari M, Zhao Y. Safety of nanoparticles in medicine. Curr Drug Targets. 2015;16(14):1671–1681.
[2]   Dobrovolskaia MA. Pre-clinical immunotoxicity studies of nanotechnology-formulated drugs: Challenges, considerations and strategy, J Control Release. 2015 Dec 28;220(Pt B) :571–583.
[3]   Dobrovolskaia MA, Shurin M, Shvedova AA. Current understanding of interactions between nanoparticles and the immune system. Toxicol Appl Pharmacol. 2016 May 15;299:78–89.
[4]   Shvedova AA, Pietroiusti A, Fadeel B, Kagan VE. Mechanisms of carbon nanotube-induced toxicity: focus on oxidative stress. Toxicol Appl Pharmacol. 2012 Jun 1;261(2):121–133.
[5]   Dobrovolskaia MA, McNeil SE. Immunological properties of engineered nanomaterials. Nat Nanotechnol. 2007 Aug;2(8):469–478.
[6]   Song G, Petschauer JS, Madden AJ, Zamboni WC. Nanoparticles and the mononuclear phagocyte system: pharmacokinetics and applications for inflammatory diseases. Curr Rheumatol Rev, 2014;10(1):22–34.
[7]   Peng Q, Li K, Sacks SH, Zhou W. The role of anaphylatoxins C3a and C5a in regulating innate and adaptive immune responses. Inflamm Allergy Drug Targets. 2009 Jul;8(3):236–246.
[8]   Guo RF, Ward PA. Role of C5a in inflammatory responses. Annu Rev Immunol. 2005;23:821–852.
[9]   Veno Kononenko, Mojca Narat, and Damjana Drobne Arh. Nanoparticle interaction with the immune system. Hig Rada Toksikol. 2015;66:97–108.
[10]   Pondman KM, Sobik M, Nayak A, Tsolaki AG, Jäkel A, Flahaut E, Hampel S, Ten Haken B, Sim RB, Kishore U. Complement activation by carbon nanotubes and its influence on the phagocytosis and cytokine response by macrophages. Nanomed. 2014;10:1287–1299.
[11]   Pedersen MB, Zhou X, Larsen EK, Sorensen US, Kjems J, Nygaard JV, Nyengaard JR, Meyer RL, Boesen T, VorupJensen T. Curvature of synthetic and natural surfaces is an important target feature in classical pathway complement activation. J Immunol. 2010;184:1931–1945.
[12]   Janeway CA Jr, Travers P, Walport M, et al. Immunobiology: The Immune System in Health and Disease, 2001, New York: Garland Science. 5th edition.
[13]   Mark J. Cameron and David J. Kelvin. Cytokines, Chemokines and Their Receptors, Madame Curie Bioscience Database [Internet],2000–2013, Austin (TX): Landes Bioscience https://www.ncbi.nlm.nih.gov/books/NBK6294/.

[14] Betteridge DJ. What is oxidative stress?, Metabolism. 2000 Feb. 49(2 Suppl 1):3–8.

[15] Stern ST, Adiseshaiah PP, Crist RM. Autophagy and lysosomal dysfunction as emerging mechanisms of nanomaterial toxicity. Part Fibre Toxicol. 2012 Jun. 14;9:20.

[16] Lucas AT, Madden AJ, Zamboni WC. Challenges in preclinical to clinical translation for anticancer carrier-mediated agents. Wiley Interdiscip Rev Nanomed Nanobiotechnol. 2016 Sep;8(5):642–653.

[17] Yang L, Li W, Kirberger M, Liao W, Ren J. Design of nanomaterial based systems for novel vaccine development. Biomater Sci. 2016 May 26;4(5):785–802.

[18] Zhou X, Liu R, Qin S, Yu R, Fu Y. Current status and future directions of nanoparticulate strategy for cancer immunotherapy. Curr Drug Metab. 2016;17(8):755–762.

[19] Di Felice G, Colombo P. Nanoparticle-allergen complexes for allergen immunotherapy. Int J Nanomed. 2017 Jun. 19;12:4493–4504.

[20] Fornaguera C, Solans C. Methods for the in vitro characterization of nanomedicines-biological component interaction. J Pers Med. 2017 Jan 27;7(1):pii: E2.

[21] Dobrovolskaia MA, Neun BW, Man S, Ye X, Hansen M, Patri AK, Crist RM, McNeil SE. Protein corona composition does not accurately predict hematocompatibility of colloidal gold nanoparticles. Nanomed. 2014;10:1453–1463.

[22] Neun BW, Ilinskaya AN, Dobrovolskaia MA. Analysis of complement activation by nanoparticles. Methods Mol Biol. 2018;1682:149–160.

[23] Wolf-Grosse S, Rokstad AM, Ali S, Lambris JD, Mollnes TE, Nilsen AM, Stenvik J. Iron oxide nanoparticles induce cytokine secretion in a complement-dependent manner in a human whole blood model. Int J Nanomed. 2017 May 23;12:3927–3940.

[24] Quach QH, Kah JC. Non-specific adsorption of complement proteins affects complement activation pathways of gold nanomaterials. Nanotoxicol. 2017 Apr;11(3):382–394.

[25] Verhoef JJF, de Groot AM, van Moorsel M, Ritsema J, Beztsinna N, Maas C, Schellekens H. Iron nanomedicines induce Toll-like receptor activation, cytokine production and complement activation. Biomater. 2017 Mar; 119:68–77.

[26] Pondman KM, Tsolaki AG, Paudyal B, Shamji MH, Switzer A, Pathan AA, Abozaid SM, Ten Haken B, Stenbeck G, Sim RB, Kishore U. Complement deposition on nanoparticles can modulate immune responses by macrophage, B and T cells. J Biomed Nanotechnol. 2016 Jan; 12(1):197–216.

[27] Inturi S, Wang G, Chen F, Banda NK, Holers VM, Wu L, Moghimi SM, Simberg D. Modulatory role of surface coating of superparamagnetic iron oxide nanoworms in complement opsonization and leukocyte uptake. ACS Nano. 2015 Nov. 24;9(11):10758–10768.

[28] Fülöp T, Nemes R, Mészáros T, Urbanics R, Kok RJ, Jackman JA, Cho NJ, Storm G, Szebeni J. Complement activation in vitro and reactogenicity of low-molecular weight dextran-coated SPIONs in the pig CARPA model: Correlation with physicochemical features and clinical information. J Control Release. 2017 Dec. 2;270:268–274.

[29] Fornaguera C, Calderó G, Mitjans M, Vinardell MP, Solans C, Vauthier C. interactions of PLGA nanoparticles with blood components: protein adsorption, coagulation, activation of the complement system and hemolysis studies,Nanoscale. Nanoscale. 2015 Apr 14;7(14):6045–6058.

[30] Banda NK, Mehta G, Chao Y, Wang G, Inturi S, Fossati-Jimack L, Botto M, Wu L, Moghimi SM, Simberg D. Mechanisms of complement activation by dextran-coated superparamagnetic iron oxide (SPIO) nanoworms in mouse versus human serum. Part Fibre Toxicol. 2014 Nov 26;11:64.

[31] Pham CT, Thomas DG, Beiser J, Mitchell LM, Huang JL, Senpan A, Hu G, Gordon M, Baker NA, Pan D, Lanza GM, Hourcade DE. Application of a hemolysis assay for analysis of complement activation by perfluorocarbon nanoparticles. Nanomed. 2014 Apr;10(3):651–660.

[32] Liu Y, Jiao F, Qiu Y, Li W, Qu Y, Tian C, Li Y, Bai R, Lao F, Zhao Y, Chai Z, Chen C. Immunostim-ulatory properties and enhanced TNF-alpha mediated cellular immunity for tumor therapy by $C_{60}(OH)_2 0$ nanoparticles. Nanotechnol. 2009;20:415102.

[33] Zolnik BS, González-Fernández A, Sadrieh N, Dobrovolskaia MA. Nanoparticles and the immune system, Endocrinol. 2010 Feb;151(2):458–465.

[34] Chen BX, Wilson SR, Das N, Coughlin DJ, Erlanger BF. Antigenicity of fullerenes: Antibodies specific for fullerenes and their characteristics. Proc Natl Acad Sci USA. 1998;95:10809–10813.

[35] Dykman LA, Sumaroka MV, Staroverov SA, Zaitseva IS, Bogatyrev VA. Immunogenic properties of the colloidal gold. Biol Bull. 2004;31:75–79.

[36] Andreev SM, Babakhin AA, Petrukhina AO, Romanova VS, Parnes ZN, Petrov RV. Immunogenic and allergenic properties of fullerene conjugates with amino acids and proteins. Dokl Biochem. 2000;370:4–7.

[37] Castignolles N, Morgeaux S, Gontier-Jallet C, Samain D, Betbeder D, Perrin P. A new family of carriers (biovectors) enhances the immunogenicity of rabies antigens. Vaccine. 1996;14:1353–1360.

[38] de Haar C, Hassing I, Bol M, Bleumink R, Pieters R. Ultrafine but not fine particulate matter causes airway inflammation and allergic airway sensitization to co-administered antigen in mice. Clin Exp Allergy. 2006;36:1469–1479.

[39] Niikura K, Matsunaga T, Suzuki T, Kobayashi S, Yamaguchi H, Orba Y, Kawaguchi A, Haegawa H, Kajino K, Ninomiya T, Ijiro K, Sawa H. Gold nanoparticles as a vaccine platform: influence of size and shape on immunological responses in vitro and in vivo. ACS Nano. 2013;7:3926–3938.

[40] Rajananthanan P, Attard GS, Sheikh NA, Morrow WJ. Evaluation of novel aggregate structures as adjuvants: composition, toxicity studies and humoral responses. Vaccine. 1999;17:715–730.

[41] Stieneker F, Kreuter J, Löwer J. High antibody titres in mice with polymethylmethacrylate nanoparticles as adjuvant for HIV vaccines. AIDS. 1991;5:431–435.

[42] Li X, Aldayel AM, Cui Z. Aluminum hydroxide nanoparticles show a stronger vaccine adjuvant activity than traditional aluminum hydroxide microparticles. J control Release. 2014;173:148–157.

[43] Sun B, Ji Z, Liao YP, Wang M, Wang X, Dong J, Chang CH, Li R, Zhang H, Nel AE, Xia T. Engineering an effective immune adjuvant by designed control of shape and crystallinity of aluminum oxyhydroxide nanoparticles. ACS Nano. 2013;7:10834–10849.

[44] Cao, Y, Ma, Y, Zhang, M, Wang, H., Tu, X., Shen, H., Dai, J., Guo, H. and Zhang, Z. Ultrasmall graphene oxide supported gold nanoparticles as adjuvants improve humoral and cellular immunity in mice. Adv Funct Mater. 2014;24:6963–6971.

[45] Mody KT, Popat A, Mahony D, Cavallaro AS, Yu C, Mitter N. Mesoporous silica nanoparticles as antigen carriers and adjuvants for vaccine delivery. Nanoscale. 2013 Jun 21;5(12):5167–5179.

[46] Mocant T, Matea C, Iancu C, Agoston-coldea L, Mocan L, Orasan R. Hypersensitivity and nanoparticles: update and research trends. Clujul Med. 2016;89(2):216–219.

[47] Mitchell LA, Gao J, Wal RV, Gigliotti A, Burchiel SW, McDonald JD. Pulmonary and systemic immune response to inhaled multiwalled carbon nanotubes. Toxicol Sci. 2007;100:203–214.

[48] Toyama T, Matsuda H, Ishida I, Tani M, Kitaba S, Sano S, et al. A case of toxic epidermal necrolysis-like dermatitis evolving from contact dermatitis of the hands associated with exposure to dendrimers. Contact Dermatitis. 2008;59(2):122–123.

[49] Hussain S, Vanoirbeek JA, Luyts K, De Vooght V, Verbeken E, Thomassen LC, et al. Lung exposure to nanoparticles modulates an asthmatic response in a mouse model. Eur Respir J. 2011;37(2):299–309.

[50] Yanagisawa R, Takano H, Inoue K, Koike E, Kamachi T, Sadakane K, et al. Titanium dioxide nanoparticles aggravate atopic dermatitis-like skin lesions in NC/Nga mice. Exp Biol Med (Maywood). 2009;234(3):314–322.

[51] Park E, Bae E, Yi J, Kim Y, Choi K, Lee SH, et al. Repeated-dose toxicity and inflammatory responses in mice by oral administration of silver nanoparticles. Environ Toxicol Pharmacol. 2010;30(2):162–168.

[52] Park HS, Kim KH, Jang S, Park JW, Cha HR, Lee JE, et al. Attenuation of allergic airway inflammation and hyperresponsiveness in a murine model of asthma by silver nanoparticles. Int J Nanomed. 2010;5:505–515.

[53] Matsuo Y, Ishihara T, Ishizaki J, Miyamoto K, Higaki M, Yamashita N. Effect of betamethasone phosphate loaded polymeric nanoparticles on a murine asthma model. Cell Immunol. 2009;260(1):33–38.

[54] Wang X, Xu W, Mohapatra S, Kong X, Li X, Lockey RF, et al. Prevention of airway inflammation with topical cream containing imiquimod and small interfering RNA for natriuretic peptide receptor. Genet Vaccines Ther. 2008 Feb 15;6:7.

[55] Kim ST, Jang DJ, Kim JH, Park JY, Lim JS, Lee SY, et al. Topical administration of cyclosporin A in a solid lipid nanoparticle formulation. Pharmazie. 2009;64(8):510–514.

[56] Shah PP, Desai PR, Patel AR, Singh MS. Skin permeating nanogel for the cutaneous co-delivery of two anti-inflammatory drugs. Biomater. 2012;33(5):1607–1617.

[57] Shen CC, Wang CC, Liao MH, Jan TR. A single exposure to iron oxide nanoparticles attenuates antigen-specific antibody production and T-cell reactivity in ovalbumin-sensitized BALB/c mice. Int J Nanomed. 2011;6:1229–1235.

[58] Mitchell LA, Lauer FT, Burchiel SW, McDonald JD. Mechanisms for how inhaled multiwalled carbon nanotubes suppress systemic immune function in mice. Nat Nanotechnol. 2009;4:451–456.

[59] Hirst SM, Peairs AD, Gogal R, Seal S, Reilly CM. Cerium oxide nanoparticles decrease inflammation in J774 cells. FASEB J. 2008;22(Meeting Abstracts):758.2.

[60] Shaunak S, Thomas S, Gianasi E, Godwin A, Jones E, Teo I, Mireskandari K, Luthert P, Duncan R, Patterson S, Khaw P, Brocchini S. Polyvalent dendrimer glucosamine conjugates prevent scar tissue formation. Nat Biotechnol. 2004;22:977–984.

[61] Ryan JJ, Bateman HR, Stover A, Gomez G, Norton SK, Zhao W, Schwartz LB, Lenk R, Kepley CL. Fullerene nanomaterials inhibit the allergic response. J Immunol. 2007;179:665–672.

[62] Nygaard UC, Hansen JS, Samuelsen M, Alberg T, Marioara CD, Lovik M. Single-walled and multi-walled carbon nanotubes promote allergic immune responses in mice. Toxicol Sci. 2009;109(1):113–123.

[63] Park E, Kim H, Kim Y, Yi J, Choi K, Park K. Inflammatory responses may be induced by a single intratracheal instillation of iron nanoparticles in mice. Toxicol. 2010;275(1):65–71.

[64] Hussain S, Vanoirbeek JA, Luyts K, De Vooght V, Verbeken E, Thomassen LC, et al. Lung exposure to nanoparticles modulates an asthmatic response in a mouse model. Eur Respir J. 2011;37(2):299–309.

[65] Di Gioacchino M, Petrarca C, Lazzarin F, Di Giampaolo L, Sabbioni E, Boscolo P, Mariani-Costantini R, Bernardini G. Immunotoxicity of nanoparticles. 2011 Jan-Mar;24(1 Suppl):65S–71S.

[66] Petrarca C, Clemente E, Amato V, Pedata P, Sabbioni E, Bernardini G, Iavicoli I, Cortese S, Niu Q, Otsuki T, Paganelli R, Di Gioacchino M. Engineered metal based nanoparticles and innate immunity. Clin Mol Allergy. 2015 Jul 15;13(1):13.

[67] Pantic I. Nanoparticles and modulation of immune responses. Sci Prog. 2011;94(Pt 1):97–107.

[68] Zolnik BS, González-Fernández A, Sadrieh N, Dobrovolskaia MA. Nanoparticles and the immune system. Endocrinol. 2010 Feb;151(2):458–465.

[69] Gill HS, Grammatopoulos G, Adshead S, Tsialogiannis E, Tsiridis E. Molecular and immune toxicity of CoCr nanoparticles in MoM hip arthroplasty. Trends Mol Med. 2012 Mar;18(3):145–155.

[70] Sun Y, Zhang Y, Shi C, Li W, Chen G, Wang X, Zhao K. Newcastle disease virus vaccine encapsulated in biodegradable nanoparticles for mucosal delivery of a human vaccine. Hum Vaccin Immunother. 2014;10(8):2503–2506.

[71] Najafi-Hajivar S, Zakeri-Milani P, Mohammadi H, Niazi M, Soleymani-Goloujeh M, Baradaran B, Valizadeh H. Overview on experimental models of interactions between nanoparticles and the immune system. Biomed Pharmacother. 2016 Oct;83:1365–1378.

[72] RomagnaniS. T-cell subsets (Th1 versus Th2). Ann Allergy Asthma Immunol. 2000 Jul;85(1):9–18; quiz 18, 21.

[73] Elsabahy M, Wooley KL. Cytokines as biomarkers of nanoparticle immunotoxicity. Chem Soc Rev. 2013 Jun 21.42(12):5552–5576.

[74] Sruthi S, Mohanan PV. Engineered zinc oxide nanoparticles; biological interactions at the organ level. Curr Med Chem. 2016;23(35):4057–4068.

[75] Peer D. Immunotoxicity derived from manipulating leukocytes with lipid-based nanoparticles. Adv Drug Deliv Rev. 2012 Dec;64(15):1738–48.

[76] Landesman-Milo D, Peer D. Altering the immune response with lipid-based nanoparticles. J Control Release. 2012 Jul 20;161(2):600–608.

[77] Dumortier H. When carbon nanotubes encounter the immune system: desirable and undesirable effects. Adv Drug Deliv Rev. 2013 Dec;65(15):2120–2126.

[78] Fadeel B, Garcia-Bennett AE. Better safe than sorry: Understanding the toxicological properties of inorganic nanoparticles manufactured for biomedical applications. Adv Drug Deliv Rev. 2010 Mar 8;62(3):362–374.

[79] Endes C, Camarero-Espinosa S, Mueller S, Foster EJ, Petri-Fink A, Rothen-Rutishauser B, Weder C, Clift MJ. A critical review of the current knowledge regarding the biological impact of nanocellulose. J Nanobiotechnol. 2016 Dec 1;14(1):78.

[80] Roy R, Das M, Dwivedi PD. Toxicological mode of action of ZnO nanoparticles: Impact on immune cells. Mol Immunol. 2015 Feb;63(2):184–192.

[81] Saptarshi SR, Duschl A, Lopata AL Biological reactivity of zinc oxide nanoparticles with mammalian test systems: an overview. Nanomed (Lond). 2015;10(13):2075–2092.

[82] Bondy SC. Nanoparticles and colloids as contributing factors in neurodegenerative disease. Int J Environ Res Public Health. 2011 Jun;8(6):2200–2211.

[83] Kunzmann A, Andersson B, Thurnherr T, Krug H, Scheynius A, Fadeel B. Toxicology of engineered nanomaterials: focus on biocompatibility, biodistribution and biodegradation. Biochim Biophys Acta. 2011 Mar;1810(3):361–373.

[84] Ali A, Suhail M, Mathew S, Shah MA, Harakeh SM, Ahmad S, Kazmi Z, Alhamdan MA, Chaudhary A, Damanhouri GA, Qadri I. Nanomaterial induced immune responses and cytotoxicity. J Nanosci Nanotechnol. 2016 Jan;16(1):40–57.

[85] Boraschi D, Italiani P, Palomba R, Decuzzi P, Duschl A, Fadeel B, Moghimi SM. Nanoparticles and innate immunity: new perspectives on host defence. Semin Immunol. 2017 Dec;34:33–51.

[86] Inoue K, Takano H. Aggravating impact of nanoparticles on immune-mediated pulmonary inflammation. J Sci World. 2011 Feb 14;11:382–390.

[87] Lee YK, Choi EJ, Webster TJ, Kim SH, Khang D. Effect of the protein corona on nanoparticles for modulating cytotoxicity and immunotoxicity. Int J Nanomed. 2014 Dec 18;10:97–113.

[88] Neagu M, Piperigkou Z, Karamanou K, Engin AB, Docea AO, Constantin C, Negrei C, Nikitovic D, Tsatsakis A. Protein bio-corona: critical issue in immune nanotoxicology. Arch Toxicol. 2017 Mar;91(3):1031–1048.

[89] Lappas CM. The immunomodulatory effects of titanium dioxide and silver nanoparticles. Food Chem Toxicol. 2015 Nov;85:78–83.

[90] Tran TH, Amiji MM. Targeted delivery systems for biological therapies of inflammatory diseases. Expert Opin Drug Deliv. 2015 Mar;12(3):393–414.

[91] http://www.who.int/leishmaniasis/visceral_leishmaniasis/en/.
[92] Dobrovolskaia MA, McNeil SE. Understanding the correlation between in vitro and in vivo immunotoxicity tests for nanomedicines, J Control Release, 2013 Dec 10. 172(2):456–466.
[93]   Want MY, Yadav P, Afrin F. Nanomedicines for therapy of Visceral ieishmaniasis, J Nanosci Nanotechnol, 2016 Mar.16(3):2143–2151.
[94] https://www.fda.gov/Drugs/DrugSafety/default.htm.
[95] Bobo D, Robinson KJ, Islam J, Thurecht KJ, Corrie SR. Nanoparticle-based medicines: A review of FDA-approved materials and clinical trials to date, Pharm Res, 2016 Oct.33(10):2373–2387.
[96] Anselmo AC, Mitragotri S. Nanoparticles in the clinic. Bioeng Transl Med. 2016 Jun 3;1(1):10–29.

3 Nanoparticles and stem cells

Abstract: Stem cells find application in regenerative medicine due to their property of self-renewal and differentiation and are being tested for the treatment of spinal cord, brain injury, kidney, cardiac and muscular system disorders, and also in tissue engineering and modeling. Nanotechnology and its biomedical applications in stem cell thus offer promises towards human welfare. Genetically engineered stem cells are finding applications in delivery of DNA, small interfering RNA, transcription factors and multiple genes replacing the viral vectors. Some of the applications include visualization of stem cells after transplantation by using superparamagnetic iron oxide nanoparticles, labeling by quantum dots, enabling their visualization, trafficking and delivery of stem cells and engineered stem cells. However, their toxic potential affecting stem cell differentiation and cell cycle limits their large-scale applications, and the effect on differentiation of stem cells is also reported from some nanoparticles (NPs). While some research reveals that they are safe to their applicability in stem cells, others do not. At present this field of biology appears tremendously complex. We discuss in this chapter about stem cells, cell cycle, effect of NPs on cell cycle, effect of NPs on cancer stem cells, applications of NPs in visualization and trafficking of stem cells. More research is required to understand the effect of shape and surface characteristics of NPs on stem cells.

Keywords: Mesenchymal stem cells, embryonic stem cells, blastocyst, pluripotent, adult stem cells, multipotent, notch pathway, Wnt signaling, TGF-β cytokines, bone marrow, regenerative medicine, P-gp (CD243), ABCB5, ABCG2 (CDw338), ALDH, alkaline phosphatase, AA4, alpha 6-integrin, monoclonal antibody, antithrombin III, asialo GM1, Bcl-2, beta-galactosidase (β-gal) of ROSA26 mice, beta1-integrin c-kit (CD117), c-Met C1qR(p), END (CD105), PROM1 (CD133), ALCAM (CD166), ITGB1 (CD29), TNFRSF8 (CD30), PECAM-1 (CD31), Siglec-3 (CD33), CD34, CD44, NCAM (CD56), CD73, CD9, CD90, CDCP1, CK19, cyclic CMP, ECMA-7, EDR1, EEC, FGF-4, Flk-2, Flt3/Flk2 Flk1(+), FMS (CD115), FORSE-1, G alpha16, GDF3, Gli2, Gli3, glial fibrillary acidic protein, glycoprotein IB, HAS2 gene, Her5, hMYADM, HSA, hsp25Id2, IL-3Ralpha, integrins, KDR, keratin 15 (aka CK15, cytokeratin 15), keratin 19 (aka CK19, cytokeratin 19, K19), Kit L-selectin (CD62L), Lamin A/C, Lewis X antigen (Le(X)), Lgr5, Lrp4, monosomy 7, mouse orthologue of ARX, MRP4, Msi-1, Musashi-1, mutant, BCRP, nestin, neurofilament, microtubule-associated protein 2, Notch 1, nrp-1, nucleostemin, OC.3, Oct-4, OST-PTP, P-gp/MDR1, p21, p63, p75, PCLP, PCNA, PECAM, phosphorylating-p38, podocalyxin, procalcitonin (PCT), PTPRC, purified LRC, rat liver fatty acid-binding protein/human growth hormone transgenes (Fabpl/hGH), RC1 antigen, Rex-1, Sca-1, SCF, sialyl-lactotetra, SOX10, SOX2, SOX9, SP phenotype, SSEA-1, SSEA-3, SSEA-4, MCM2, MCSP/neuron-glial antigen 2 (NG2) MTCRII Stat, 3, Stat5, Stella, Stra8, Stro-1, tartrate-resistant acid phosphatase (TRAcP), TdT, telomerase reverse transcriptase, thrombomucin, Thy-1/CD 90, Tra-1-60 and Tra-1-81, TWIST1, VEGFR-2, vimentin, X-smoothened, XKrk1, Zac1, induced pluripotent stem cells, cell cycle, CDK, mouse embryonic stem cells (mESCs), silver nanoparticles

https://doi.org/10.1515/9783110579093-003

(AgNPs), reactive oxygen species, Fe_3O_4 nanoparticles, human mesenchymal stem cells (hMSCs), titanium dioxide (TiO_2), Wistar rat, scanning electron microscopy and dynamic light scattering, *qPCR*, cancer stem cells (CSC), graphene, graphene oxide, carbon nanotubes (CNT), activated charcoal, fluorinated graphenes, pluripotent stem cells, activated charcoal (AC), neural progenitor cells, quantum dots (QDs), nanowires, fullerene derivatives (buckyballs), neural stem cells (NSCs), immune-modulatory.

3.1 Introduction

Stem cells are clonogenic cells that are active during early development with the property of self-renewal in maintaining stemness, which enables the maintenance of pluripotency on the one hand and the property of differentiation gives rise to different lineages of cells with specialized function in the body on the other hand (Figure 3.1). Embryonic stem cells located in the inner cell mass of a blastocyst are pluripotent in nature, which can give rise to different types of somatic cell types in the embryo [1, 2]. Stem cell markers are genes and their protein products expressed by the stem cells (Table 3.1) find application in their identification and isolation. The interesting hallmark property lies in its indefinite self-renewal ability and plasticity but the signaling processes involved are not well understood.

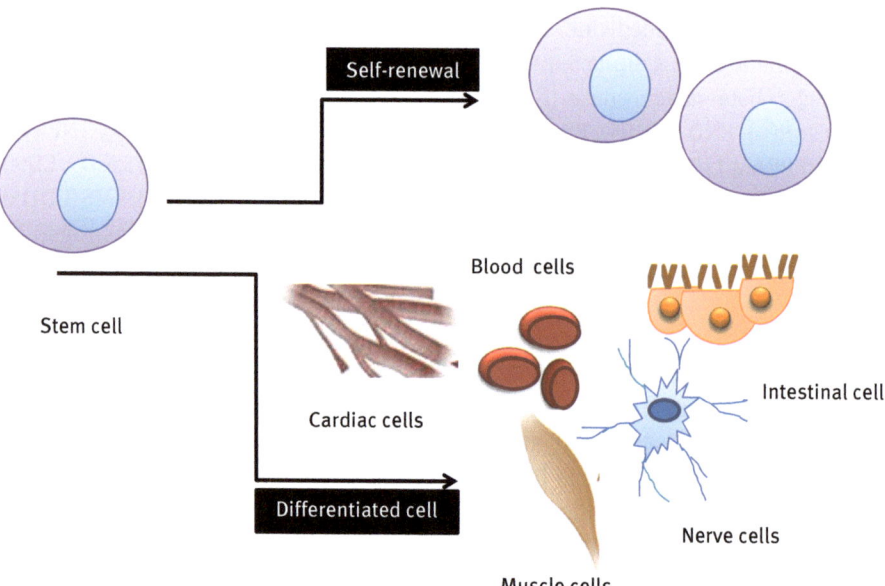

Figure 3.1: Stem cells are endowed with the property to maintain stemness and thus self-renewal and they are also capable of differentiating into different lineages.

Table 3.1: Markers of stem cells.

Name	Function	Location	References
AA4	It is a type I transmembrane protein that plays a role in cell adhesion and cell–cell interactions during hematopoietic and vascular development	Expressed in three major cell types: vascular endothelial cells, aorta-associated hematopoietic clusters and primitive fetal liver hematopoietic progenitors in 3–4 days of development	[7]
P-gp (CD243)	P-glycoprotein 1 is also known as multidrug resistance protein 1 or ATP-binding cassette subfamily B member 1 or CD243 that pumps foreign substances out of the cells	Stem cells	[8]
ABCB5	Transmembrane protein expressed in physiological skin and human malignant melanomas	Stem cells, marker for adult corneal limbal stem cells	[9]
ABCG2 (CDw338)	ATP-binding cassette subfamily G member 2 also called as CDw338 which pumps foreign compounds out of cells	Universal marker of stem cells, found in CSCs	[10]
ALDH	Aldehyde dehydrogenases are a group of enzymes that catalyze the oxidation of aldehydes	Embryonal tissue, adult stem cells isolated from bone marrow, brain, breast and other tissues, CSC	[11]
Alkaline phosphatase	It is a transmembrane-bound enzyme	Stem cell	[12]
Alpha 6-integrin	Function in cell surface adhesion and signaling	Stem cells, mouse spermatogonial stem cells	[13]
Anti-WNT2B monoclonal antibody	Anti-Wingless-type MMTV (mouse mammary tumor virus) integration site family, member 2B antibodies (WNT2B)	Stem cells	[14]
Antithrombin III	Inactivates several enzymes of the coagulation system	Stem cells	[15]
Asialo GM1	Asialo GM1 ganglioside	Stem cells	[16, 17]
Bcl-2	B-cell lymphoma 2 regulates apoptosis	Stem cells	[18, 19]
Beta-galactosidase (β-gal) of ROSA26 mice	Beta-galactosidase	Embryonic stem cells in ROSA26 mice	[20]

(continued)

Table 3.1 (continued)

Name	Function	Location	References
Beta 1-integrin	Cell adhesion and signaling	Mouse spermatogonial stem cells	[13]
c-kit (CD117)	Mast/stem cell growth factor receptor, also known as proto-oncogene c-Kit or tyrosine protein kinase kit or CD117, is a receptor tyrosine kinase protein	Primordial germ cells, spermatogenetic stem cells, HSCs, embryonic stem cells (ESCs), pancreatic progenitor cells	[21–23]
c-Met	Also called tyrosine-protein kinase Met or hepatocyte growth factor receptor (HGFR), coded by *MET* gene and shows tyrosine kinase activity	Pancreatic CSCs of mice, hepatic stem cell	[24–25]
C1qR(p)	C1q receptor that plays a role in complement activation through classical pathway	Stem cells with hepatic and hematopoietic potential	[26]
END (CD105)	Endoglin (ENG) is a type I membrane glycoprotein and is part of the TGF-β receptor complex. It plays a crucial role in angiogenesis.	Mesenchymal stem cell marker	[27]
PROM1 (CD133)	Prominin-1 or CD133 antigen is a member of pentaspan transmembrane glycoproteins	Stem cells, embryonic stem cell-derived progenitors, CSCs	[28–31]
ALCAM (CD166)	Activated leukocyte cell adhesion molecule is a member of the immunoglobulin superfamily, expressed by various cells including colorectal cancer	Mesenchymal stem cells, CSC marker	[32–33]
ITGB1 (CD29)	Integrin beta-1 or CD29 is an integrin unit associated with very late antigen receptors	Porcine spermatogonial stem cell, neural progenitor cells, liver stem cells, mesenchymal stem cells	[34–36]
TNFRSF8 (CD30)	CD30 or TNFRSF8 is a cell membrane protein of the tumor necrosis factor receptor family and tumor marker	Embryonic stem cells	[37]
PECAM-1 (CD31)	Platelet endothelial cell adhesion molecule or CD31 is a protein that plays a key role in removing aged neutrophils from the body	Embryonic stem cells, cardiac progenitor cells, HSCs	[38–39]

Marker	Description	Cell type	Ref.
Siglec-3 (CD33)	CD33 or Siglec-3 (sialic acid binding Ig-like lectin 3) is a transmembrane receptor expressed on cells of myeloid lineages and binds sialic acids	HSC	[40]
CD34	Cell surface glycoprotein functions as a cell–cell adhesion factor and mediates the attachment of stem cells to bone marrow extracellular matrix or directly to stromal cells	HSC	[41]
CD44	Cell surface glycoprotein involved in cell–cell interactions, cell adhesion and migration	Stem cells	[32]
NCAM (CD56)	Neural cell adhesion molecule or CD56 is a homophilic binding glycoprotein expressed on the surface of neurons, glia and skeletal muscle. It plays a role in cell–cell adhesion, neurite outgrowth, synaptic plasticity and learning and memory.	Hepatic progenitor cells, neural progenitor cells, HSC	[42]
CD73	5′-Nucleotidase or ecto-5′-nucleotidase or CD73 is an enzyme that converts adenosine monophosphate(AMP) to adenosine	MSC	[43]
CD9	The protein encoded by this gene is a member of the transmembrane 4 superfamily or tetraspanin family that mediates signal transduction in the regulation of cell development, activation, growth and motility	Spermatogonial stem cell, stem cells	[44]
CD90	Thy-1 or CD90 is a heavily N-glycosylated, GPI-anchored conserved cell surface protein with a single V-like immunoglobulin domain	Marker for different stem cells and for the axonal processes of mature neurons	[32]
CDCP1	CUB domain-containing protein 1	HSC, stem cells	[45]
CK19	Cytokeratin 19	MSC, stem cells	[46]
Cyclic CMP	Cyclic CMP	stem cell, malignancy	[47]
ECMA-7		Stem cell marker	[48]
EDR1	EDR1 (PHC1) Polyhomeotic-like protein 1	Stem cell marker	[49]

(continued)

Table 3.1 (continued)

Name	Function	Location	References
EEC	Endogenous erythroid colonies	Stem cell marker	[50]
FGF-4	Fibroblast growth factors regulate multiple biological functions	Regulation of proliferation and differentiation in embryonic stem cells and tissue stem cells	[51]
Flk-2, Flt3/Flk2	Flk-2/Flt3 receptor tyrosine kinase plays a critical role in maintenance of hematopoietic homeostasis	HSC, stem cells	[52]
Flk1(+)	Flk1	Embryonic stem cells, progenitor vascular endothelial cells	[53]
FMS (CD115)	Colony-stimulating factor 1 receptor or macrophage colony-stimulating factor receptor or CD115 is a cell-surface protein receptor for a cytokine called colony-stimulating factor 1	Stem cells	[54]
FORSE-1	Forebrain surface embryonic-1	Stem cells, neural stem cell progenitors	[55–56]
G alpha16	G proteins play a role in signal transduction from cytokine receptors to intracellular effectors through different pathways	Myeloid and lymphoid progenitors	[57]
GDF3	Growth differentiation factor-3 or Vg-related gene 2 belongs to the TGF-β superfamily	Pluripotent stem cells	[49]
Gli2	Zinc finger protein GLI2 is a transcription factor	Neural progenitor cells, stem cells	[58]
Gli3	Zinc finger protein GLI3 is a member of C2H2-type zinc finger protein subclass of the Gli family	Neural progenitor cells, stem cells	[59]
Glial fibrillary acidic protein	It is an IF protein that is expressed by numerous cell types of the central nervous system, including astrocytes and ependymal cells, which plays a role in functioning of the blood–brain barrier	Stem cells	[60]
Glycoprotein IB	Glycoprotein IB or CD42 is a component of the GPIb-V-IX complex on platelets	Stem cell	[61]

HAS2 gene	Hyaluronan synthases genes	Umbilical cord-derived stem cells	[62]
Her5		Stem cells, neural stem cells	[63]
hMYADM	Myeloid-associated differentiation marker	Stem cells	[64]
HSA	Signal transducer CD24 or heat-stable antigen (HSA) plays a role in cell adhesion	Stem cells	[65]
hsp25	heat shock protein 25	Embryonic stem cells	[66]
Id2	*Id2* is a helix-loop-helix (HLH) transcription factor essential for normal development	Stem cells	[67]
IL-3Rα	Interleukin-3 receptor	Marker for leukemic stem cells, stem cells	[68]
Integrins	Transmembrane receptors that facilitate cell-extracellular matrix adhesion	Stem cells	[69]
KDR	Kinase insert domain receptor is a type III receptor tyrosine kinase or VEGFR-2 is a VEGFR or KDR or Flk1	Stem cells	[70]
Keratin 15 (aka CK15, cytokeratin 15)	Keratin 15 is a type I cytokeratin found in some progenitor basal cells within complex epithelia	Stem cells	[71]
Keratin 19 (aka CK19, cytokeratin 19, K19)	Keratin, type I cytoskeletal protein or CK19 or K19	Stem cells	[71]
Kit	Mast/stem cell growth factor receptor or proto-oncogene c-Kit or tyrosine protein kinase kit or CD117 is a receptor tyrosine kinase protein	Stem cells	[72]
L-selectin (CD62L)	L-selectin or CD62L is a cell adhesion molecule found on lymphocytes and the preimplantation embryo. It belongs to the selectin family that recognizes sialylated carbohydrate groups	Stem cells	[73]

(continued)

Table 3.1 (continued)

Name	Function	Location	References
Lamin A/C	Lamin A/C also known as LMNA belongs to the lamin family of proteins	Embryonic stem cells, stem cells	[74]
Lewis X antigen (Le(X))	Sialyl Lewis^X or sialyl Le^X or SLe^X is a tetrasaccharide carbohydrate with the sequence Neu5Acα2-3Galβ1-4[Fucα1-3]GlcNAcβ, which is usually attached to O-glycans on the surface of cells. It plays a role in cell-to-cell recognition.	Stem cells	[75]
Lgr5	Leucine-rich repeat containing G-protein-coupled receptor 5 or Gpr49	Adult stem cells and cancers	[76]
Lrp4	Low-density lipoprotein receptor-related protein 4	Mouse germ cells, stem cells	[77]
MCM2	DNA replication licensing factor MCM2	Stem cells	[78]
MCSP/neuron-glial antigen 2 (NG2)	Chondroitin sulfate proteoglycan 4 or melanoma-associated chondroitin sulfate proteoglycan or neuron-glial antigen 2 is a chondroitin sulfate proteoglycan	Stem cells	[79]
MTCRII	Metallothionein crypt-restricted immunopositivity indices	Stem cells	[80]
Monosomy 7	Monosomy 7 or partial loss of 7q	Stem cells	[81]
Mouse orthologue of ARX	The ARX gene acts as a transcription factor	Neural progenitor cells	[82]
MRP4	Multidrug resistance-associated proteins (MRPs)	Immature stem cell	[83]
Msi-1	Msi-1, the human homolog of Drosophila Musashi	Stem cell	[84]
Musashi-1		Neural stem cell	[85]
Mutant BCRP	Breast cancer resistance protein	Stem cell	[86]
Nestin	Nestin or neuroectodermal stem cell marker is a type VI IF protein	Neuroepithelial stem cell marker, stem cells	[87]

Neurofilament microtubule-associated protein 2	Plays a role in microtubule assembly, which is an essential step in neuritogenesis	Neural stem cell markers, stem cells	[88]
Notch 1	Notch homolog 1 is a single-pass transmembrane receptor	Stem cells	[89]
nrp-1	Neuropilin-1 is a membrane-bound coreceptor to a tyrosine kinase receptor for both VEGF and semaphorin (SEMA3A) that plays versatile roles in angiogenesis, axon guidance, cell survival, migration and invasion	Neural stem cell marker	[90]
Nucleostemin	Guanine nucleotide-binding protein-like 3 or nucleostemin regulates the cell cycle and affects cell differentiation	Stem cells and cancer cells	[91]
OC.3	Oval cell antigen	Liver stem cell marker, stem cells	[92]
Oct-4	Octamer-binding transcription factor 4 or POU5F1 (POU domain, class 5, transcription factor 1) is a transcription factor that plays a role in self-renewal of undifferentiated embryonic stem cells	Stem cells	[93]
OST-PTP	Osteotesticular protein tyrosine phosphatase	Germ stem cell marker	[94]
P-gp/MDR1	P-gp/multidrug resistance gene	Stem cell marker	[95]
p21	p21Cip1 or cyclin-dependent kinase inhibitor 1 or CDK-interacting protein 1 is a cyclin-dependent kinase inhibitor that is capable of inhibiting all cyclin/CDK complexes	Stem cell marker	[96]
p63	Tumor protein p63 or transformation-related protein 63	Keratinocyte stem cells	[97]
p75	p75 neurotrophin receptor, growth factors that stimulate neuronal cells to survive and differentiate	Stem cells	[98]
PCLP	Podocalyxin-like protein is a sialomucin-type membrane protein structurally related to CD34 and endoglycan	HSC	[99]

(continued)

Table 3.1 (continued)

Name	Function	Location	References
PCNA	Proliferating cell nuclear antigen is a DNA clamp that acts as a processivity factor for DNA polymerase δ in eukaryotic cells and is essential for replication	Stem cells	[100]
PECAM	Platelet endothelial cell adhesion molecule (PECAM-1) or CD31. The *PECAM1* gene plays a key role in removing aged neutrophils from the body	HSC	[101]
Phosphorylating-p38	Phosphorylating-p38 mitogen-activated protein kinase	Intestinal stem cells	[102]
Podocalyxin	Podocalyxin is the sialoglycoprotein	Human embryonic and induced pluripotent stem cells	[103]
Procalcitonin (PCT)	Procalcitonin is a peptide precursor of the hormone calcitonin, the latter being involved with calcium homeostasis	Stem cells	[104]
PTPRC	Protein tyrosine phosphatase receptor type C is an enzyme also known as CD45 antigen	Stem cells	[105]
Purified LRC	Epidermal label-retaining cells	Epidermal stem cells	[106]
Rat liver fatty acid-binding protein/ human growth hormone transgenes (Fabpl/hGH)	Rat liver fatty acid-binding protein/human growth hormone transgenes	Stem cells	[107]
RC1 antigen	RC1 antigen	Neural stem cells	[108]
Rex-1	Rex1 (Zfp-42) is a known marker of pluripotency and is usually found in undifferentiated embryonic stem cells	Neural, embryonic stem cell markers	[109]
Sca-1	Stem cells antigen-1 18-kDa mouse GPI-anchored cell surface protein of the LY6 gene family	Parenchymal stem cell marker	[110]

SCF	Stem cell factor, KIT ligand (KL or steel factor) is a cytokine that binds to the c-KIT receptor (CD117) and plays a role in hematopoiesis, spermatogenesis and melanogenesis	Stem cells	[111]
Sialyl-lactotetra	Cell surface glycoconjugate	Undifferentiated pluripotent stem cells	[112]
SOX10	Transcription factor SOX-10	Stem cells, progenitor cells, mesenchymal stem cells	[113]
SOX2	SRY (sex-determining region Y)-box 2 or SOX2 is a transcription factor that is essential for maintaining self-renewal, or pluripotency, of undifferentiated embryonic stem cells. SOX2 has a critical role in maintenance of embryonic and neural stem cells	Stem cells	[114]
SOX9	Transcription factor SOX-9 recognizes the sequence CCTTGAG along with other members of the HMG-box class DNA-binding proteins, acts during chondrocyte differentiation and with steroidogenic factor 1 regulates transcription of the anti-Müllerian hormone gene	Stem cells	[115]
SP phenotype	Side population (SP)	Stem cells	[116]
SSEA-1	Stage-specific embryonic antigen-1	Stem cells	[117]
SSEA-3	Stage-specific embryonic antigen 3 (SSEA-3) is a glycosphingolipid, specifically an oligosaccharide composed of five carbohydrate units connected to a sphingolipid	Stem cells	[118]
SSEA-4	SSEA-4 is an early embryonic glycolipid antigen	Marker for undifferentiated pluripotent human embryonic stem cells	[119]
Stat3	Signal transducer and activator of transcription 3 (STAT3) is a transcription factor, which is phosphorylated by Janus kinases in response to cytokines and growth factors and translocates to the cell nucleus, where they act as transcription activators	Stem cells and CSCs	[120]

(continued)

Table 3.1 (continued)

Name	Function	Location	References
Stat5	Signal transducer and activator of transcription 5 (STAT5) involved in cytosolic signaling	Stem cells, embryonic stem cells	[121]
Stella	It encodes a protein with a SAP-like domain and a splicing factor motif-like structure, suggesting possible roles in chromosomal organization or RNA processing	Germ stem cells and pluripotent stem cells	[122]
Stra8	Stimulated by retinoic acid gene 8	Stem cells	[123]
Stro-1	STRO-1 is a cell surface antigen expressed by stromal elements in the human bone marrow	MSC marker	[124]
Tartrate-resistant acid phosphatase (TRAcP)	Tartrate-resistant acid phosphatase (TRAP or TRAPase), also called acid phosphatase 5 tartrate resistant (ACP5), is a glycosylated monomeric metalloprotein enzyme expressed in mammals	Stem cells	[125]
TdT	Terminal deoxynucleotidyl transferase (TdT), also known as DNA nucleotidyl terminal transferase, is a specialized DNA polymerase expressed in immature, pre-B, pre-T lymphoid cells, and acute lymphoblastic leukemia/lymphoma cells	Stem cells, HSC, CSCs	[126]
Telomerase reverse transcriptase	It is an enzyme	Stem cells	[127]
Thrombomucin	Cell surface protein	Stem cells	[128]
Thy-1/CD 90	Thy-1 cell surface antigen	Stem cells	[129]
Tra-1-60 and Tra-1-81	Cell surface antigen	Stem cells	[130]

TWIST1	Twist-related protein 1 (TWIST1) also known as class A basic helix-loop-helix protein 38 (bHLHa38) is a basic helix-loop-helix transcription factor	Stem cells	[131]
VEGFR-2	VEGFRs are receptors for VEGF	Stem cells	[132]
Vimentin	*Vimentin* is a type III IF protein that is expressed in the mesenchymal cell	Neural stem cells, stem cells	[133]
X-smoothened	Smoothened, frizzled class receptor	Stem cells	[58]
XKrk1	Xenopus c-kit-related receptor tyrosine kinase	Stem cells	[134]
Zac1	*Zac1* (*Lot1*) paternally expressed imprinted genes	Stem cells	[60]

CSCs, cancer stem cells; HSCs, hematopoietic stem cells; GPI, glycophosphatidylinositol; Flk1, fetal liver kinase 1; TGF-β, transforming growth factor beta; IF, intermediate filament; VEGFR2, vascular endothelial growth factor receptor 2; VEGF, vascular endothelial growth factor; CK19, cytokeratin 19; K19, keratin 19; MSC, mesenchymal stem cell.

3.2 Adult stem cells

Adult stem cells are found in adult tissues that are multipotent and can give rise to related cell types (Table 3.2), replenish dead cells and regenerate damaged tissues. Signaling event controlling the stem cell proliferation has been attributed to the notch pathway in different stem cells, including hematopoietic, neural and mammary stem cells, regulation of stem cell regeneration by Wnt signaling and by transforming growth factor beta (TGF-β) cytokines [3–5].

All the different types of cells of the hematopoietic system develop from the hematopoietic stem cells in the bone marrow (BM) and was discovered by Frieden-stein and others. Studies revealed that the BM stroma has the potential to develop into osteoblasts, adipocytes and chondrocytes in vitro, indicating the presence of nonhematopoietic BM multipotent precursor cells, which in turn led to the under-standing of other types of progenitor cells called as the mesenchymal stem cells (MSC). Later such multipotent precursors were identified from many adult and embry-onic tissues located in the stroma of many adult tissues. Identification of MSCs has been hindered quite some time due to the extremely low tissue frequency and dearth of suitable MSC-specific immunophenotype. Cultured human mesenchymal stromal cells are known to express cell surface markers including CD73, CD90 and CD105 and lack endothelial or hematopoietic cell markers including CD31, CD34 and CD45 [6]. BM MSCs differentiate into osteoblasts, adipocytes and reticular cells that support the development of hematopoietic cells and immune cells including dendritic cells, T cells and B cells. Mesenchymal stromal cells have been known to regulate the func-tion of B and T lymphocytes, dendritic cells and natural killer cells [6].

Table 3.2: Types of stem cells.

Types of cells	Characteristics
Embryonic stem cells	Derived from blastocyst
	Pluripotent that differentiates into almost all different types of cells
	Indefinite self-renewal property
Induced pluripotent stem cells	Indefinite self-renewal property
	Generated from reprogrammed somatic cells
Adult stem cells	Located in adult tissues
	Includes mesenchymal stem cells, neural stem cells, adipose-derived stem cells, intestinal stem cells, mammary stem cells, HSCs, endothelial stem cells, olfactory stem cells, neural crest stem cells, testicular stem cells
	Also called somatic stem cells
	Multipotent and can give rise to related cells

It is this property of self-renewal and differentiation into different lineages of cells by the stem cells offers us the advantage in understanding the biology of development and they also find application in regenerative medicine and gene therapy.

Nanotechnology is being used to exploit the stem cells in different ways. Recent studies have enabled the understanding of their effect on stem cells that affect the cell cycle and toxicity, effect on cancer stem cells (CSC), the role of stem cells in delivery of nanoparticles (NPs) and delivery of MSCs with regenerative potential maintaining their niche through NPs. The following sections discuss the latest area of research in the biomedical field.

Stem cells (Table 3.2) identified by different markers are enlisted in Table 3.1.

3.3 Cell cycle

The cell cycle involves signaling events leading to cell division controlled by multiple checkpoints regulating the cycle stages controlled by cyclin-dependent kinases (CDKs) such as serine and threonine kinases that phosphorylate their substrates, thereby the cell progresses through stages G1, S, G2 and M. They regulate cell cycle by binding to the activating subunit "cyclin." The fluctuating intracellular cyclin level and stable CDK levels throughout the cell cycle enable activation of CDKs. Different CDK/cyclin pairs enable progression through cell cycle phase and these phase-specific cyclin expression enables CDK activation in cell cycle events. Regulatory phosphorylation and dephosphorylation of CDK enable proper cell cycle events.

A normal, classical model of cell cycle events includes CDK–cyclin complexes including CDKs (CDK2, CDK4 and CDK6), a mitotic CDK1 (orCDC2) and 10 cyclins that belong to four different classes (A-, B-, D- and E-type cyclins) [135, 136]. A cell cycle is initiated by CDK4 and CDK6 binding leading to kinase activation in G1 [135, 136]. Cyclin degradation and CDK inactivation controls the progress of the cell cycle. The end of G1 phase and the onset of S phase are marked by the increase in cyclin E-CDK2. The end of S phase and initiation of G2 involve cyclin A-CDK2 and cyclin A-CDK1. Increase in cyclin B indicates the initiation of mitotic events and decrease in cyclin B indicates the end of M phase. CDK1 inactivation due to decreased cyclin B marks the completion of cell cycle. This cyclin–CDK complex plays a role in controlling the cell cycle events and finds application in clinical therapeutics.

Inhibitor proteins including INK4 proteins comprising INK4A, INK4B, INK4C, INK4D and Cip/Kip family proteins comprising p21, p27 and p57 regulate the cell cycle by inhibiting continuous progression of the cell cycle, thus enabling maintenance of cell cycle homeostasis [6] (Figure 3.2). Inhibitory activity of cyclin-dependent kinase inhibitor, a tumor repressor, suppresses the growth through pRb. p27 mutation renders it unable to bind to CDK–cyclin complexes leading to excessive stem and progenitor cell proliferation observed in tumors. Loss of function of p16^{Ink4a} gene by mutation, deletion or hypermethylation [137] leads to uncontrolled progress of G1. Loss of p27 in G2 has been associated with several tumors of bladder, breast and lungs.

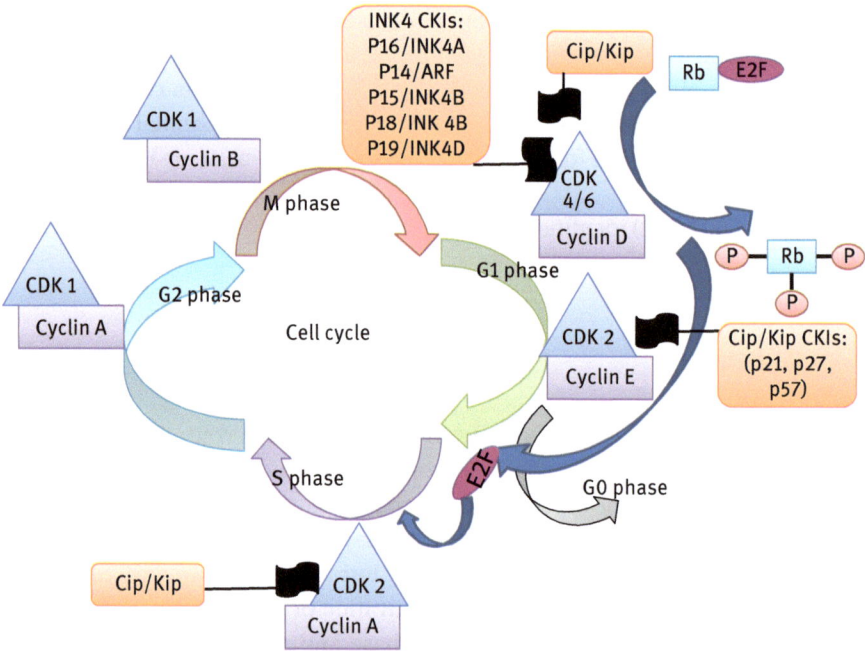

Figure 3.2: Cell cycle. Cyclin-dependent kinases (CDKs) and cyclins play an important role in signaling events of the cell cycle. Cyclin-dependent kinase inhibitors (CKIs), INK4, Cip/Kip inhibitors are the chief regulators. E2F is a transcription factor.

3.3.1 NP and its effect on stem cell cycle

The effect of NPs on stem cell self-renewal and proliferation is not well understood. Mouse embryonic stem cells (mESCs) exposed to silver NPs (AgNPs) have been reported to affect the mESCs morphology and caused cell cycle arrest at G1 and S phases through inhibition of the hyperphosphorylation of retinoblastoma (Rb) protein, reduced Oct4A isoform expression for the pluripotency of mESCs and induction of OCT4B-265, OCT4B-190, OCT4B-164 isoforms indicative of stress in stem cells. Reactive oxygen species generation is associated with the overall toxicity by AgNPs, which was reduced by polysaccharide-coated AgNPs [138]. Fe_3O_4-engineered NPs exerted toxic effects on human MSCs (hMSCs) by reducing cell viability, altered nuclear morphology, loss of mitochondrial membrane potential, oxidative stress, cell death and decreased G0–G1 phased cells. Nanoparticles have been known to affect the hMSC mitochondrial membrane potential, upregulated expression of CYP1A, TNF3, TNFSF10, E2F1 and CCNC genes causing metabolic stress, altered cell cycle, oxidative stress in hMSCs [139].

Titanium dioxide (TiO_2) NPs affect the cell membrane integrity, adhesion, migration, proliferation and osteogenic differentiation of MSC by affecting the alkaline phosphatase, osteocalcin (OC) and osteopontin (OPN) expression in Wistar rat animal model [140].

3.3.2 NP safety and toxicity on stem cells

The silicon dioxide (SiO_2) NPs find application in the food industry. Based on their size, shape and physical properties, when ingested they exhibit metabolic toxicity and related chronic disorders. When exposed to hMSCs, they affect the normal cell cycle, and change the cell morphology such as cytoplasmic organization and nuclear morphology. This is revealed by scanning electron microscopy, dynamic light scattering, 3-(4,5-dimethylthiazol-2-yl)-2,5-diphenyltetrazolium bromide cell viability assay, fluorescent microscopy and flow cytometric methods. Expression of cell cycle regulating genes including proliferating cell nuclear antigen, early growth response protein 1 (EGR1), E2F transcription factor (E2F1), cyclin D1, cyclin C, and cyclin D3, cell cycle proteins and cell cycle-dependent metabolic stress through EGR1, CCND (cyclin D1), and E2F1 (transcription factor) genes studied for expression by quantitative polymerase chain reaction (qPCR) were indicative of their toxicity that affected the cell cycle of stem cells by causing metabolic stress [6]. Alginate-enclosed, chitosan-conjugated, calcium phosphate, iron-saturated bovine lactoferrin (Fe-bLf) nanocarriers/nanocapsules (NCs) have shown promises of improved sustained drug release and induction of apoptosis by downregulating survivin which is a apoptosis inhibitor protein inhibiting caspases and preventing cell deaths, in CSCs, thus revealing promising application in cancer therapy [141]. Activation of both extrinsic and intrinsic apoptotic pathways by targeting survivin in cancer cells and CSCs was reported to be a hallmark feature of this nanoformulation [141].

Graphene is a two-dimensional nanomaterial known for its properties of electroconductivity together with its derivatives including graphene oxide, carbon nanotubes (CNT), activated charcoal and fluorinated graphenes. Three-dimensional graphene foams have promising applications in human life particularly from the point of view of biomedicine as biosensors or theranostics. It has been reported to lead to adhesion of stem cell, their growth, expansion and differentiation but not affecting the cell viability in vitro, thus providing promising applications in engineering of tissues, regenerative, translational and personalized medicine. But there are few studies on the effects and nanotoxicity of graphene and its derivatives on stem cells and nothing is known about their behavior in clinical conditions [142]. In vitro three-dimensional models using pluripotent stem cells have enabled detection of toxic effects of different NPs on stem cells [143].

Transplantation of scaffolds of biomaterials encasing human embryonic stem cells (hESCs) has been proved to be promising in restoration of damaged nerves and neurological disorders. Carbon-based biomaterials including CNTs and graphene have enabled the stem cell attachment and differentiation, thus enabling transplantation of stem cells, but their application is limited by their nanotoxic effects. A natural coal-based activated charcoal composite has been reported to support and promote differentiation and growth of matured neuronal cells with comparable structural characteristics such as axonal length and density, and functional characteristics indicating active synapses from hESCs and could also enable concentration of growth factors and cell adhesion proteins, thus allowing attachment and hESC differentiation [144]. hESCs have been applied in detection of nanotoxicology by gold NPs (AuNPs) and observed that AuNPs could cause loss of viability and loss of cohesiveness and detachment, and affect pluripotency and neuronal differentiation. DNA methylation of hESCs is proved to be toxic [145].

AgNPs in hESC-derived neural stem cells (NSCs) or neural progenitor cells known to differentiate into neurons caused oxidative stress, dysfunctional neurogenesis, cell cycle arrest (Figure 3.3) and apoptosis by miRNA transcription studies [146].

Gametogenesis and development of progeny can be affected by the effect of toxic agents on germ cells or somatic nursing cells. CNTs, nanowires, fullerene derivatives (buckyballs) and quantum dots (QDs) despite their advantages show their toxic effects when exposed to mouse spermatogonial stem cell line in a concentration-dependent manner while their soluble salts had no significant effect. AuNPs revealed most toxicity while molybdenum trioxide (MoO_3) NPs when tested on same cells revealed less toxicity [147]. NSCs could however be visualized by iron oxide NPs (IONPs) after transplantation through MRI. While nanotoxicity of PMA-coated IONPs has been reported from murine cancer cell line, dimercaptosuccinic acid-coated IONPs provide a safer alternative [148].

Although AgNPs of 30 nm diameter is reported of cytotoxic effects on human MSCs, they do not affect the osteogenic differentiation, thus promising their application in bone tissue graft designing (Figure 3.3) [149].

Nanoparticles	Effect on stem cells	References
Silver nanoparticles	Cell cycle arrest at G1 and S phases	134
Fe_3O_4 nanoparticles	Cell death, decreased G0–G1 phased cells	135
Titanium nanoparticles	Reduced proliferation, osteogenic differentiation of MSC	136
Silica nanoparticles	Toxicity, stress, affected stem cells and cell cycle	130
Nanoparticles (carbon-based, metal-based; dendrimers and composite nanomaterials (178)	Toxic effects on pluripotent stem cells (PSCs)	139
Activated charcoal composite biomaterial	Promote hESCs differentiation	140
Gold nanopartciles	Affect pluripotency, neuronal differentiation	141
Silver nanoparticles	Cell cycle arrest, neural progenitor cells (NPCs) differentiate into neurons	142
Silver nanoparticles	Cytotoxic effects	145

Figure 3.3: Toxic effects of nanoparticles on stem cells.

3.3.3 NPs and CSCs

Tumors sometimes contain drug-resistant cells termed as CSCs, in nondividing cells arrested in the G0 states that show properties of self-renewal and differentiation into different tumor cells and cause relapse and metastasis. They are known to overexpress drug transporters. NP delivery of small molecules, small interfering RNA (siRNA) and antibodies can affect the embryonic signaling pathways in CSCs, inhibit drug efflux transporters that target metabolism and affect the viability in CSCs with promising application in targeting CSCs [150].

Surface markers including CD44, epithelial cell adhesion molecule, aldehyde dehydrogenase and nucleolin [151], CD90 and CD133 [152, 153] signaling pathways including Notch, Hedgehog and Wnt [151, 153], PI3K/Akt intracellular signaling pathway that regulates cell cycle, Janus tyrosine kinase- signal transducer and activator of transcription signaling pathway involved in immunity, cell division, cell death and tumor formation processes and Ras/ERK signaling pathways [151] and TGF-β (152) have been reported for their aberrant expression in CSCs affecting their proliferation and survival. Synergistic drug combination targeting CSCs through lipid-based NPs is a promising target to CSCs [151] and plays a probable role in inhibiting reprogramming of cancer cell [152].

Aptamers or chemical antibodies include small-sized, low-toxic, biocompatible and low immunogenicity single-stranded nucleic acid either DNA or RNA of antibodies and show great potential in targeting CSCs and cancer therapy [154]. Polymeric NPs are showing great potential in delivery of specific antibodies or ligands to target CSCs and thus offer promises to their application in cancer therapy [155]. Prodrugs, micelles, liposomes and NPs of biodegradable polymers due to their biocompatibility, targeted delivery, prolonged exposure, escape of autophagy provide promises in targeting CSCs [156]. CSCs in glioblastoma multiforme and blood–brain barrier add to the complexity of therapy in this major brain tumor which can be circumvented by the application of NPs [157].

Tumor necrosis factor-related apoptosis-inducing ligand (TRAIL) gene delivery alone or in combination with chemotherapeutic agents are recent technologies with application in targeting chemotherapy-resistant CSCs. NPs are finding applications in TRAIL gene delivery, thereby inducing apoptosis in CSCs [158].

3.3.4 MSCs and delivery of NPs

MSCs have been known for their role in tissue regeneration due to their proliferative properties, capabilities of differentiation, inducing chemotactic and immune-modulatory activity. NP devices find applications in designing of the stem cell niche that enables maintenance of MSCs and keeping their stemness intact. Targeted delivery of bioactive molecules with the property to maintain MSC enhancing their regenerative potential and designing imaging devices enabling monitoring of distribution, toxicity and therapy of transplanted stem cell are being designed [159].

MSCs have been exploited as targeted carriers of anticancer drug-loaded NPs in delivery of drugs to tumors and metastatic diseases [160].

3.4 Stem cells in delivery of nucleotides, siRNA

Stem cells with less toxicity and immunogenicity are considered to be biosafe agents finding applications in delivery of nucleotides and siRNA circumventing the problems associated with delivery by viral vectors [161]. Genetic manipulations of MSCs have further added to their therapeutic efficiency [162]. Engineered and synthetic transcription factors, genetically engineered zinc finger proteins, effectors and deficient cas9 proteins regulating cell fate are finding applications as better tools in therapy [163].

3.5 Visualization of stem cells by NPs

Stem cell-based therapies have been proven to be promising in cancer, cardiovascular diseases, brain disorders, spinal cord injury and others. In the recent years, considerable progress has been made by scientists across the globe to show that stem cells coated or labeled with NPs are being able to be visualized by applying magnetic resonance imaging (MRI) approaches and noninvasive methods. Superparamagnetic IONPs (SPIONs) when targeted to culture of human neural progenitor cells (hNPCs) revealed the property of attraction to external magnetic field and enhanced brain retention in a rat model of traumatic brain injury with blood–brain barrier disruption (BBBD). Focused ultrasound (FUS)-BBBD has been reported to enable the entry of stem cells to the brain from the bloodstream with low retention but SPION-loaded hNPCs after FUS-mediated BBBD have been reported to enhance their retention in the brain by MRI approaches. A powerful Halbach magnetic array revealed their retention in deeper brain areas, emphasizing the importance of this noninvasive strategy to deliver SPION-labeled stem cells in the brain and its deeper regions [164]. Ferumoxytol NPs labeled on porcine NSCs when transplanted into the spinal cord ventral horn with magnetic resonance guidance have been observed to enable visualization by MRI [165]. ZAIS (ZnS-AgInS$_2$) coated with ZnS or ZnS-coated ZAIS (ZnS-ZAIS) carboxylated NPs or ZZC is known for its fluorescence. ZZC and octa-arginine (R8) peptides or R8-ZZC complex has been proved to be efficient in labeling of adipose tissue-derived stem cells with low cytotoxicity with no adverse effect of inflammatory cytokine release or potential of differentiation of stem cells and could be detected by in vivo imaging system [166]. Fluorophore-doped siloxane core nanoprobes with dual imaging functions of (1)H MRI and/or fluorescence have enabled the visualization of neural progenitor/stem cells [167]. Gadolinium-based

Trimetasphere® (endohedral metallofullerenes such as trimetallic nitride endohedral metallofullerenes) has enabled the tracking of human amniotic fluid-derived stem cells within the lung tissue in vivo [168]. Real-time visualization of glioblastoma stem cells could be possible by the use of NPs observed using in situ transmission electron microscopy [169]. SPION coated with aminopolyvinyl alcohol has been shown to enable the tracking of human mesenchymal stromal cells in vivo by MRI-based approaches [170]. Poly-(l-lactide) NPs loaded with iron have been reported to enable visualization of mesenchymal stromal cells by MRI-based methods [171]. Ferumoxytol-labeled human NSCs have enabled their visualization under MRI [172]. SPIONs have enabled the visualization of human dental pulp stem cells under MRI [173].

Self-assembled peptide amphiphile SPIONs that are used to label rat MSCs have enabled better contrasting agents in MRI for stem cell labeling and tracking due to their biocompatibility with potential application in tissue engineering and drug delivery applications [174].

3.6 Labeling by QDs

The properties of MSCs with low immunogenicity, regulating immune modulations find application in transplantation, regenerative medicine and treatment of neural, kidney, cardiac, endocrinal, muscle and skeletal tissues and spinal cord injury [175] and have been found suitable for the short-term trafficking of kidney stem cells and mouse embryo cells [176]. QDs with narrow emission spectra and resistant to photobleaching effects are finding applications in labeling, thereby monitoring and homing MSCs as in transplantation of pancreatic islets [177]. Silica-coated cadmium selenide QDs enabled the visualization of stem cells and cancer cells observed under coupled plasma-optical emission spectroscopy and confocal laser scanning microscopy [177] and have not been reported to affect mESCs and kidney stem cells [178].

3.7 Discussion

According to the US Environmental Protection Agency, NPs are comprised of four categories such as carbon based, metal based, dendrimers and nanomaterial composites [179] and their various applications have enriched their use in stem cells with the potential of self-renewal and differentiation. Genetically engineered NPS with their applications in regenerative medicine and is a step toward advancement of therapy to a host of disorders.

Visualization of these stem cells at the site of delivery or activity by tagging them with NPs and then visualization using MRI or fluorescence offers immense advantages in understanding and monitoring of stem cells.

Yet toxicity associated with some NPs are a cause of concern and therefore future research is focused toward generation of nontoxic, compatible NPs with applications enabling maintenance of stem cell properties and its beneficial properties in transplantation.

References

[1]   Nombela-Arrieta C, Ritz J, Leslie E, Silberstein LE. The elusive nature and function of mesenchymal stem cells. Nat Rev Mol Cell Biol. 2011;12:126–131.

[2]   Vazin T, Freed WJ. Human embryonic stem cells. derivation, culture, and differentiation: a review. Restor Neurol Neurosci. 2010 Jan 1;28(4):589–603.

[3]   Dontu G, Jackson KW, McNicholas E, Kawamura MJ, Abdallah WM, Wicha MS. Role of notch signaling in cell-fate determination of human mammary stem/progenitor cells. Breast Cancer Res. 2004;6(6):R605–R615.

[4]   Beachy PA, Karhadkar SS, Berman DM. Tissue repair and stem cell renewal in carcinogenesis. Nature. 2004;432(7015):324–331.

[5]   Sakaki-Yumoto M, Katsuno Y, Derynck R. TGF-β family signaling in stem cells. Biochim Biophys Acta. 2013;1830(2):2280–2296.

[6]   Periasamy VS, Athinarayanan J, Akbarsha MA, Alshatwi AA. Silica nanoparticles induced metabolic stress through EGR1, CCND, and E2F1 genes in human mesenchymal stem cells. Appl Biochem Biotechnol. 2015 Jan;175(2):1181–1192.

[7]   Petrenko O, Beavis A, Klaine M, Kittappa R, Godin I, Lemischka IR. The molecular characterization of the fetal stem cell marker AA4. Immunity. 1999 Jun;10(6):691–700.

[8]   Islam MO, Kanemura Y, Tajria J, Mori H, Kobayashi S, Shofuda T, Miyake J, Hara M, Yamasaki M, Okano H. Characterization of ABC transporter ABCB1 expressed in human neural stem/progenitor cells. FEBS Lett. 2005 Jul 4;579(17):3473–3480.

[9]   Ksander BR, Kolovou PE, Wilson BJ, Saab KR, Guo Q, Ma J, McGuire SP, Gregory MS, Vincent WJ, Perez VL, Cruz-Guilloty F, Kao WW, Call MK, Tucker BA, Zhan Q, Murphy GF, Lathrop KL, Alt C, Mortensen LJ, Lin CP, Zieske JD, Frank MH, Frank NY. ABCB5 is a limbal stem cell gene required for corneal development and repair. Nature.2014;511(7509):353–357.

[10]   Ding XW, Wu JH, Jiang CP. ABCG2: A potential marker of stem cells and novel target in stem cell and cancer therapy. Life Sci. 2010 Apr 24;86(17–18):631–637.

[11]   Moreb JS. Aldehyde dehydrogenase as a marker for stem cells. Curr Stem Cell Res Ther. 2008 Dec;3(4):237–246.

[12]   Kateřina Štefková, Jiřina Procházková, Jiří Pacherník. Alkaline phosphatase in stem cells. Stem Cells Int Vol. 2015;Article ID 628368. doi: 10.1155/2015/628368

[13]   Takashi Shinohara, Mary R Avarbock, Ralph L Brinster. β1- and α6-integrin are surface markers on mouse spermatogonial stem cells. Proc Natl Acad Sci U S A. 1999 May 11;96(10):5504–5509.

[14]   Katoh M. WNT2B: comparative integromics and clinical applications (review). Int J Mol Med. 2005;16(6):1103–1108.

[15]   Gordon B, Haire W, Ruby E, et al. Factors predicting morbidity following hematopoietic stem cell transplantation. Bone Marrow Transplant. 1997;19(5):497–501.

[16]   Harris MT, Schwarting GA, Stout RD. Selective expression of asialo GM1 on maturational subsets of lymphocytes in normal and athymic mice. Thymus. 1981;3(3):153–167. PMID 6171918.

[17] Yu RK, Suzuki Y, Yanagisawa M. Membrane glycolipids in stem cells. FEBS Lett. 2010;584(9):1694–1699.

[18] Polakowska RR, Piacentini M, Bartlett R, Goldsmith LA, Haake AR. Apoptosis in human skin development: morphogenesis, periderm, and stem cells. Develop Dyn. 1994;199(3):176–188.

[19] Nam HoonCho, Young KyuPark Young TaeKim, Hyunwon Yang, Sei Kwang Kim. Lifetime expression of stem cell markers in the uterine endometrium. Fertil Steril. 2004;81(2): 403–407.

[20] Aoyama N, Molin DGM, Mentink MMT, Koerten HK, de Ruiter MC, Gittenberger-de Groot AC, Poelmann RE. Changing intracellular compartmentalization of β-galactosidase in the ROSA26 reporter mouse during embryonic development: a light- and electron-microscopic study. Anat Rec. 2004;279A:740–748.

[21] Lei Zhang, Jiangjing Tang, Christopher J Haines, Huai Feng, Liangxue Lai, Xiaoming Teng, Yibing Han. C-kit expression profile and regulatory factors during spermatogonial stem cell differentiation. BMC Dev Biol, 2013; 13: 38.

[22] Wenxiu Zhao, Xiang Ji Fangfang Zhang, Liang Li and Lan Ma Embryonic stem cell markers, Molecules, 2012; 17: 6196–6236.

[23] Fengxia Ma, Fang Chen, Ying Chi, Shaoguang Yang, Shihong Lu, and Zhongchao Han. Isolation of pancreatic progenitor cells with the surface marker of hematopoietic stem cells. Int J Endocrinol. 2012; 8: Article ID 948683.

[24] Chenwei Li, Jing Jiang Wu, Mark Hynes, Joseph Dosch, Bedabrata Sarkar, Theodore H. Welling, Marina Pasca di Magliano, Diane M. Simeone. c-Met is a marker of pancreatic cancer stem cells and therapeutic target gastroenterology, Gastroenterology. 2011; .e5 141 (6): 2218–2227.

[25] Suzuki A, Zheng YW, Fukao K, Nakauchi H, Taniguchi H. Liver repopulation by c-met-positive stem/progenitor cells isolated from the developing rat liver, Hepatogastroenterology, 2004 Mar-Apr;51(56):423–6.

[26] Danet GH, Luongo JL, Butler G, et al. C1qRp defines a new human stem cell population with hematopoietic and hepatic potential. Proc Natl Acad Sci U S A. 2002;99(16):10441 5.

[27] Nagatomo K, Komaki M, Sekiya I, Sakaguchi Y, Noguchi K, Oda S, Muneta T, Ishikawa I. Stem cell properties of human periodontal ligament cells. J Periodontal Res. 2006 Aug; 41(4):303–10.

[28] Peichev M, Naiyer AJ, Pereira D, et al. Expression of VEGFR-2 and AC133 by circulating human CD34(+) cells identifies a population of functional endothelial precursors. Blood. 2000; 95(3): 952–8.

[29] Zhou L, Wei X, Cheng L, Tian J, Jiang JJ. *CD133*, one of the markers of cancer stem cells in hep-2 cell line. The Laryngoscope. 2007; **117**(3): 455–60.

[30] Kania G, Corbeil D, Fuchs J, et al. Somatic stem cell marker prominin-1/CD133 is expressed in embryonic stem cell-derived progenitors. Stem Cells. 2005; **23**(6): 791–804.

[31] Holmberg Olausson K, Maire CL, Haidar S, Ling J, Learner E, Nistér M, Ligon KL. Prominin-1 (CD133) defines both stem and non-stem cell populations in CNS development and gliomas. PLoS One. 2014 Sep 3;9(9):e106694. eCollection 2014.

[32] Ghaneialvar H, Soltani L, Rahmani HR, Lotfi AS, Soleimani M. Characterization and classification of mesenchymal stem cells in several species using surface markers for cell therapy purposes. Indian J Clin Biochem. 2018 Jan;33(1):46–52.

[33] Dana H, Marmari V, Mahmoodi G, Mahmoodzadeh H, Ebrahimi M, et al. CD166 as a stem cell marker? A potential target for therapy colorectal cancer?. J Stem Cell Res Ther. 2016; 1(6): 00041.

[34] Park MH, Kim MS, Yun JI, Choi JH, Lee E, Lee ST. Integrin heterodimers expressed on the surface of porcine spermatogonial stem cells. DNA Cell Biol. 2018. Jan 25.

[35] Pruszak J, Ludwig W, Blak A, Alavian K, Isacson O.CD15, CD24, and CD29 define a surface biomarker code for neural lineage differentiation of stem cells, Stem Cells. 2009 Dec. 27(12):2928–40.

[36]   Pruszak J, Ludwig W, Blak A, Alavian K, Isacson O.CD15, CD24, and CD29 define a surface biomarker code for neural lineage differentiation of stem cells, Stem Cells. 2009 Dec.27(12):2928–40.

[37]   Mateizel I, Spits C, Verloes A, Mertzanidou A, Liebaers I, Sermon K. Characterization of CD30 expression in human embryonic stem cell lines cultured in serum-free media and passaged mechanically. Hum Reprod. 2009 Oct;24(10):2477–89.

[38]   Heike T, Yoshimoto M, Baba S, Doi H, Nakahata T Identification of cardiac stem cells with FLK1, CD31, and VE-cadherin expression during embryonic stem cell differentiation, Iida M. FASEB J. 2005 Mar;.19(3):371–8.

[39]   Christina I. Baumann, Alexis S. Bailey, Weiming Li, Michael J. Ferkowicz, Mervin C. Yoder, and William H. Fleming,PECAM-1 is expressed on hematopoietic stem cells throughout ontogeny and identifies a population of erythroid progenitors. Blood. 2004; 104:1010–1016.

[40]   Blakolmer K, Jaskiewicz K, Dunsford HA, Robson SC. Hematopoietic stem cell markers are expressed by ductal plate and bile duct cells in developing human liver. Hepatology. 1995 Jun;21(6):1510–6.

[41]   Ning H, Lin G, Lue TF, Lin CS. Neuron-like differentiation of adipose tissue-derived stromal cells and vascular smooth muscle cells. Differentiation. 2006 Dec; 74 (9–10): 510–518.

[42]   Van Den Heuvel MC, Slooff MJ, Visser L, Muller M, De Jong KP, Poppema S, Gouw AS. Expression of anti-OV6 antibody and anti-N-CAM antibody along the biliary line of normal and diseased human livers. Hepatology. 2001 Jun; 33(6):1387–1393.

[43]   Mazor M, Cesaro A, Ali M, Best TM, Lespessaille E, Toumi H. Progenitor cells from cartilage: Grade specific differences in stem cell marker expression. Int J Mol Sci. 2017 Aug 12;18(8):.

[44]   Abdul Wahab AY, Md Isa ML, Ramli R. Spermatogonial stem cells protein identification in in vitro culture from non-obstructive azoospermia patient. Malays J Med Sci. 2016 May; 23(3):40–8.

[45]   Morii E, Ikeda JI, Ezoe S, Xu JX, Nakamichi N, Tomita Y, Shibayama H, Kanakura Y, Aozasa K. Role of DNA methylation for expression of novel stem cell marker CDCP1 in hematopoietic cells. Kimura H. Leukemia. 2006 Sep; 20(9):1551–1556.

[46]   Khodabandeh Z, Vojdani Z, Talaei-Khozani T, Bahmanpour S. Hepatogenic differentiation capacity of human Wharton's jelly mesenchymal stem cell in a co-culturing system with endothelial cells in matrigel/collagen scaffold in the presence of fetal liver extract. Int J Stem Cells. 2017 Nov 30;10(2):218–226.

[47]   Scavennec J, Carcassonne Y, Gastaut JA, Blanc A, Cailla HL. Relationship between the levels of cyclic cytidine 3′:5′-monophosphate, cyclic guanosine 3′:5′-monophosphate, and cyclic adenosine 3′:5′-monophosphate in urines and leukocytes and the type of human leukemias. Cancer Res. 1981 Aug;41(8):3222–3227.

[48]   Boulter CA, Wagner EF. The effects of v-src expression on the differentiation of embryonal carcinoma cells. Oncogene. 1988 Mar; 2(3):207–14.

[49]   Giuliano CJ, Kerley-Hamilton JS, Bee T, Freemantle SJ, Manickaratnam R, Dmitrovsky E, Spinella MJ. Retinoic acid represses a cassette of candidate pluripotency chromosome 12p genes during induced loss of human embryonal carcinoma tumorigenicity. Biochim Biophys Acta. 2005 Oct 15 ;1731(1):48–56.

[50]   Reid CD. The significance of endogenous erythroid colonies (EEC) in haematological disorders, Blood Rev. 1987 Jun; 1(2):133–140.

[51]   Kosaka N, Sakamoto H, Terada M, and Ochiya T. Pleiotropic function of FGF-4: Its role in development and stem cells. Developmental Dynamics. 2009; 238:265–276.

[52]   Christensen JL, Weissman IL.Flk-2 is a marker in hematopoietic stem cell differentiation: A simple method to isolate long-term stem cells. Proc Natl Acad Sci U S A. 2001 Dec 4; 98(25):14541–6.

[53] Yamashita J, Itoh H, Hirashima M, Ogawa M, Nishikawa S, Yurugi T, Naito M, Nakao K, Nishikawa S. Flk1-positive cells derived from embryonic stem cells serve as vascular progenitors. Nature. 2000 Nov 2; 408(6808):92–96.

[54] Rappold I, Ziegler BL, Köhler I, Marchetto S, Rosnet O, Birnbaum D, Simmons PJ, Zannettino AC, Hill B, Neu S, Knapp W, Alitalo R, Alitalo K, Ullrich A, Kanz L, Bühring HJ. Functional and phenotypic characterization of cord blood and bone marrow subsets expressing FLT3 (CD135) receptor tyrosine kinase. Blood. 1997 Jul 1; 90(1):111–25.

[55] Belzile JP, Stark TJ, Yeo GW, Spector DH. Human cytomegalovirus infection of human embryonic stem cell-derived primitive neural stem cells is restricted at several steps but leads to the persistence of viral DNA. J Virol. 2014 Apr;88(8):4021–4039.

[56] Pan J, Yeger H, Cutz E. Neuronal developmental marker FORSE-1 identifies a putative progenitor of the pulmonary neuroendocrine cell lineage during lung development. J Histochem Cytochem. 2002 Dec.;50(12):1567–1578.

[57] Pfeilstöcker M, Karlic H, Paukovits J, Anzenberger G, Louda N, Salamon J, Mühlberger H, Strobl H, Pittermann E, Heinz R. In vivo and in vitro effects of cytokines and the hemoregulatory peptide dimer (pEEDCK)2(pyroGlu-Glu-Asp-Cys-Lys)2 on G alpha16-positive hematopoiesis. Leukemia. 1999 Apr;13(4):590–594.

[58] Perron M, Boy S, Amato MA, Viczian A, Koebernick K, Pieler T, Harris WA. A novel function for Hedgehog signalling in retinal pigment epithelium differentiation. Development, 2003 Apr; 130(8):1565–1577.

[59] Perron M, Boy S, Amato MA, Viczian A, Koebernick K, Pieler T, Harris WA. A novel function for Hedgehog signalling in retinal pigment epithelium differentiation. Development. 2003 Apr;130(8):1565–1577.

[60] Valente T, Junyent F, Auladell C. Zac1 is expressed in progenitor/stem cells of the neuroectoderm and mesoderm during embryogenesis: differential phenotype of the Zac1-expressing cells during development. Dev Dyn. 2005 Jun; 233(2):667–679.

[61] Tschöpe D, Langer E, Schauseil S, Rösen P, Kaufmann L, Gries FA. Klin Wochenschr. Increased platelet volume–sign of impaired thrombopoiesis in diabetes mellitus. 1989 Feb 15.67(4):253–259.

[62] Grskovic B, Pollaschek C, Mueller MM, Stuhlmeier KM. Expression of hyaluronan synthase genes in umbilical cord blood stem/progenitor cells. Biochim Biophys Acta. 2006 Jun; 1760(6):890–895.

[63] Chapouton P, Adolf B, Leucht C, Tannhäuser B, Ryu S, Driever W, Bally-Cuif L. Her5 expression reveals a pool of neural stem cells in the adult zebrafish midbrain. Development. 2006 Nov; 133(21):4293–303.

[64] Wang Q, Li N, Wang X, Shen J, Hong X, Yu H, Zhang Y, Wan T, Zhang L, Wang J, Cao X. Membrane protein hMYADM preferentially expressed in myeloid cells is up-regulated during differentiation of stem cells and myeloid leukemia cells. Life Sci. 2007 Jan 9; 80(5):420–429.

[65] Anzai H, Nagayoshi M, Obata M, Ikawa Y, Atsumi T. Self-renewal and differentiation of a basic fibroblast growth factor-dependent multipotent hematopoietic cell line derived from embryonic stem cells. Dev Growth Differ. 1999 Feb;41(1):51–158.

[66] Stahl J, Wobus AM, Ihrig S, Lutsch G, Bielka H. The small heat shock protein hsp25 is accumulated in P19 embryonal carcinoma cells and embryonic stem cells of line BLC6 during differentiation. Differentiation. 1992 Sep;51(1):33–37.

[67] Gultice AD, Selesniemi KL, Brown TL. Hypoxia inhibits differentiation of lineage-specific Rcho-1 trophoblast giant cells. Biol Reprod. 2006 Jun;74(6):1041–1050.

[68] Testa U, Riccioni R, Diverio D, Rossini A, Lo Coco F, Peschle C. Interleukin-3 receptor in acute leukemia. Leukemia. 2004 Feb;18(2):219–226.

[69]   Hall PE, Lathia JD, Miller NG, Caldwell MA, ffrench-Constant C. Integrins are markers of human neural stem cells. Stem Cells. 2006 Sep;24(9):2078–2084.

[70]   Ziegler BL, Valtieri M, Porada GA, De Maria R, Müller R, Masella B, Gabbianelli M, Casella I, Pelosi E, Bock T, Zanjani ED, Peschle C. KDR receptor: A key marker defining hematopoietic stem cells. Science. 1999 Sep 3 ;285(5433):1553–1558.

[71]   Regauer S. Extramammary Paget's Disease–a proliferation of adnexal origin? Histopathology. 2006 May; 48(6):723–729.

[72]   Taoudi S, Morrison AM, Inoue H, Gribi R, Ure J, Medvinsky A. Progressive divergence of definitive haematopoietic stem cells from the endothelial compartment does not depend on contact with the foetal liver. Development. 2005 Sep;132(18):4179–4191.

[73]   Perry SS, Wang H, Pierce LJ, Yang AM, Tsai S, Spangrude GJ. L-selectin defines a bone marrow analog to the thymic early T-lineage progenitor. Blood. 2004 Apr 15; 103(8):2990–2996. Epub 2003 Dec 30.

[74]   Constantinescu D, Gray HL, Sammak PJ, Schatten GP, Csoka AB. Lamin A/C expression is a marker of mouse and human embryonic stem cell differentiation. Stem Cells. 2006 Jan;24(1):177–185.

[75]   Muramatsu T, Muramatsu H. Carbohydrate antigens expressed on stem cells and early embryonic cells. Glycoconj J. 2004; 21 (1–2): 41–45.

[76]   Barker N, van Es JH, Kuipers J, Kujala P, van den Born M, Cozijnsen M, Haegebarth A, Korving J, Begthel H, Peters PJ, Clevers H. Identification of stem cells in small intestine and colon by marker gene Lgr5. Nature. 2007 Oct 25;449(7165):1003–1007.

[77]   Yamaguchi YL, Tanaka SS, Kasa M, Yasuda K, Tam PP, Matsui Y. Expression of low density lipoprotein receptor-related protein 4 (Lrp4) gene in the mouse germ cells. Gene Expr Patterns. 2006 Aug;6(6):607–612.

[78]   Mohan A, Kandalam M, Ramkumar HL, Gopal L, Krishnakumar S. Stem cell markers: ABCG2 and MCM2 expression in retinoblastoma. Br J Ophthalmol. 2006 Jul;90(7):889–893.

[79]   Legg J, Jensen UB, Broad S, Leigh I, Watt FM. Role of melanoma chondroitin sulphate proteoglycan in patterning stem cells in human interfollicular epidermis. Development. 2003 Dec; 130(24):6049–6063.

[80]   Donnelly ET, Bardwell H, Thomas GA, Williams ED, Hoper M, Crowe P, McCluggage WG, Stevenson M, Phillips DH, Hewer A, Osborne MR, Campbell FC. Metallothionein crypt-restricted immunopositivity indices (MTCRII) correlate with aberrant crypt foci (ACF) in mouse Br J Cancer. 2005 Jun 20;92(12):2160–2165.

[81]   Hutter JJ Jr, Hecht F, Kaiser-McCaw B, Hays T, Baranko P, Cohen J, Durie B. Bone marrow monosomy 7: hematologic and clinical manifestations in childhood and adolescence. Hematol Oncol. 1984 Jan-Mar;2(1):5–12.

[82]   Colombo E, Galli R, Cossu G, Gécz J, Broccoli V. Mouse orthologue of ARX, a gene mutated in several X-linked forms of mental retardation and epilepsy, is a marker of adult neural stem cells and forebrain GABAergic neurons. Dev Dyn. 2004 Nov.231(3):631–639.

[83]   Steinbach D, Wittig S, Cario G, Viehmann S, Mueller A, Gruhn B, Haefer R, Zintl F, Sauerbrey A. The multidrug resistance-associated protein 3 (MRP3) is associated with a poor outcome in childhood ALL and may account for the worse prognosis in male patients and T-cell immunophenotype. Blood. 2003 Dec 15;102(13):4493–4498.

[84]   Clarke RB, Spence K, Anderson E, Howell A, Okano H, Potten CS.A putative human breast stem cell population is enriched for steroid receptor-positive cells. Dev Biol. 2005 Jan 15; 277(2):443–456.

[85]   Wakasaki T, Niiro H, Jabbarzadeh-Tabrizi S, Ohashi M, Kimitsuki T, Nakagawa T, Komune S, Akashi K. Musashi-1 is the candidate of the regulator of hair cell progenitors during inner ear regeneration. BMC Neurosci. 2017 Aug 16;18(1):64.

[86]   Staud F, Pavek P. Breast cancer resistance protein (BCRP/ABCG2). Int J Biochem Cell Biol. 2005 Apr;37(4):720–725.

[87]   Loo DT, Althoen MC, Cotman CW. Differentiation of serum-free mouse embryo cells into astrocytes is accompanied by induction of glutamine synthetase activity. J Neurosci Res. 1995 Oct 1 ;42(2):184–191.

[88]   Okawa H, Okuda O, Arai H, Sakuragawa N, Sato K. Amniotic epithelial cells transform into neuron-like cells in the ischemic brain. Neuroreport. 2001 Dec 21; 12(18):4003–7.

[89]   Venkatesh K, Reddy LVK, Abbas S, Mullick M, Moghal ETB, Balakrishna JP, Sen D. NOTCH signaling is essential for maturation, self-renewal, and tri-differentiation of in vitro derived human neural stem cells. Cell Reprogram. 2017 Dec; 19(6):372–383.

[90]   Imaoka S, Mori T, Kinoshita T, Bisphenol A causes malformation of the head region in embryos of Xenopus laevis and decreases the expression of the ESR-1 gene mediated by notch signaling. Biol Pharm Bull. 2007 Feb;30(2):371–4.

[91]   Kafienah W, Mistry S, Williams C, Hollander AP. Nucleostemin is a marker of proliferating stromal stem cells in adult human bone marrow. Stem Cells. 2006 Apr;24(4):1113–1120.

[92]   Sigal SH, Brill S, Reid LM, Zvibel I, Gupta S, Hixson D, Faris R, Holst PA. Characterization and enrichment of fetal rat hepatoblasts by immunoadsorption ("panning") and fluorescence-activated cell sorting. Hepatology. 1994 Apr;19(4):999–1006.

[93]   Dana Zeineddine, Aya Abou Hammoud, Mohamad Mortada, Hélène Boeuf. The Oct4 protein: more than a magic stemness marker. Am J Stem Cells. 2014; 3(2): 74–82.

[94]   Maduro MR, Davis E, Davis A, Lamb DJ. Osteotesticular protein tyrosine phosphatase expression in rodent testis. J Urol. 2002 May;167(5):2282–2283.

[95]   Tokura Y, Shikami M, Miwa H, Watarai M, Sugamura K, Wakabayashi M, Satoh A, Imamura A, Mihara H, Katoh Y, Kita K, Nitta M. Augmented expression of P-gp/multi-drug resistance gene by all-trans retinoic acid in monocytic leukemic cells. Leuk Res. 2002 Jan;26(1):29–36.

[96]   Ju Z, Choudhury AR, Rudolph KL. A dual role of p21 in stem cell aging. Ann N Y Acad Sci. 2007 Apr;1100:333–44.

[97]   Pellegrini G, Dellambra E, Golisano O, Martinelli E, Fantozzi I, Bondanza S, Ponzin D, McKeon F, De Luca M. p63 identifies keratinocyte stem cells. Proc Natl Acad Sci U S A. 2001 Mar 13;98(6):3156–3161.

[98]   Tomellini E, Lagadec C, Polakowska R, Le Bourhis X. Role of p75 neurotrophin receptor in stem cell biology: more than just a marker. Cell Mol Life Sci. 2014 Jul;71(13):2467–2481.

[99]   Kerosuo L, Juvonen E, Alitalo R, Gylling M, Kerjaschki D, Miettinen A. Podocalyxin in human haematopoietic cells. Br J Haematol. 2004 Mar;124(6):809–818.

[100] Leung AY, Leung JC, Chan LY, Ma ES, Kwan TT, Lai KN, Meng A, Liang R. Proliferating cell nuclear antigen (PCNA) as a proliferative marker during embryonic and adult zebrafish hematopoiesis. Histochem Cell Biol. 2005 Aug;124(2):105–111.

[101] Baumann CI, Bailey AS, Li W, Ferkowicz MJ, Yoder MC, Fleming WH. PECAM-1 is expressed on hematopoietic stem cells throughout ontogeny and identifies a population of erythroid progenitors. Blood. 2004 Aug 15; 104(4):1010–1016.

[102] Fu XB, Xing F, Yang YH, Sun TZ, Guo BC. Activation of phosphorylating-p38 mitogen-activated protein kinase and its relationship with localization of intestinal stem cells in rats after ischemia-reperfusion injury. World J Gastroenterol. 2003 Sep;9(9):2036–2039.

[103] Toyoda H, Nagai Y, Kojima A, Kinoshita-Toyoda A. Podocalyxin as a major pluripotent marker and novel keratan sulfate proteoglycan in human embryonic and induced pluripotent stem cells. Glycoconj J. 2017 Dec;34(6):817–823.

[104] Ortega M, Rovira M, Filella X, Martínez JA, Almela M, Puig J, Carreras E, Mensa J. Prospective evaluation of procalcitonin in adults with non-neutropenic fever after allogeneic hemato-poietic stem cell transplantation. Bone Marrow Transplant. 2006 Mar; 37(5):499–502.

[105] Herrera MB, Bruno S, Buttiglieri S, Tetta C, Gatti S, Deregibus MC, Bussolati B, Camussi G. Isolation and characterization of a stem cell population from adult human liver. Stem Cells. 2006 Dec;24(12):2840–2850.

[106] Braun KM, Watt FM. Epidermal label-retaining cells: Background and recent applications. J Investig Dermatol Symp Proc. 2004 Sep; 9(3):196–201.

[107] Rubin DC, Swietlicki E, Roth KA, Gordon JI. Use of fetal intestinal isografts from normal and transgenic mice to study the programming of positional information along the duodenal-to-colonic axis. J Biol Chem. 1992 Jul 25;267(21):15122–15133.

[108] Nakafuku M, Nakamura S. Establishment and characterization of a multipotential neural cell line that can conditionally generate neurons, astrocytes, and oligodendrocytes in vitro. J Neurosci Res. 1995 Jun 1;41(2):153–168.

[109] Lamoury FM, Croitoru-Lamoury J, Brew BJ. Undifferentiated mouse mesenchymal stem cells spontaneously express neural and stem cell markers Oct-4 and Rex-1. Cytotherapy. 2006;8(3):228–242.

[110] Holmes C, Stanford WL. Concise review: Stem cell antigen-1: Expression, function, and enigma. Stem Cells. 2007 Jun;25(6):1339–1347.

[111] Ceponis A, Konttinen YT, Takagi M, Xu JW, Sorsa T, Matucci-Cerinic M, Santavirta S, Bankl HC, Valent P. Expression of stem cell factor (SCF) and SCF receptor (c-kit) in synovial membrane in arthritis: correlation with synovial mast cell hyperplasia and inflammation. J Rheumatol. 1998 Dec;25(12):2304–2314.

[112] Barone A, Säljö K, Benktander J, Blomqvist M, Månsson JE, Johansson BR, Mölne J, Aspegren A, Björquist P, Breimer ME, Teneberg S. Sialyl-lactotetra a novel cell surface marker of undifferentiated human pluripotent stem cells. J Biol Chem. 2014 Jul 4;289(27): 18846–18859.

[113] Christopher Dravis, Benjamin T. Spike, J. Chuck Harrell, Claire Johns, Christy L. Trejo, E. Michelle Southard-Smith, Charles M. Perou, and Geoffrey M. Wahl. Sox10 regulates stem/ progenitor and mesenchymal cell states in mammary epithelial cells. Cell Rep. 2015 Sep 29; 12(12): 2035–2048.

[114] Ellis P, Fagan BM, Magness ST, Hutton S, Taranova O, Hayashi S, McMahon A, Rao M, Pevny L. SOX2, a persistent marker for multipotential neural stem cells derived from embryonic stem cells, the embryo or the adult, Dev Neurosci. 2004 Mar-Aug; 26 (2–4): 148–65.

[115] Kadaja M, Keyes BE, Lin M, Pasolli HA, Genander M, Polak L, Stokes N, Zheng D, Fuchs E. SOX9: Stem cell transcriptional regulator of secreted niche signaling factors. Genes Dev, 2014 Feb 15; 28(4):328–41.

[116] Smalley MJ, Clarke RB. The mammary gland "side population": A putative stem/progenitor cell marker?. J Mammary Gland Biol Neoplasia. 2005 Jan; 10(1):37–47.

[117] Son MJ, Woolard K, Nam DH, Lee J, Fine HA. SSEA-1 is an enrichment marker for tumor-initiating cells in human glioblastoma. Cell Stem Cell. 2009 May 8; 4(5):440–52.

[118] Wenxiu Zhao, Xiang Ji, Fangfang Zhang, Liang Li and Lan Ma. Embryonic Stem Cell Markers. Molecules. 2012 17: 6196–6236.

[119] Gang EJ, Bosnakovski D, Figueiredo CA, Visser JW, Perlingeiro RC. SSEA-4 identifies mesenchymal stem cells from bone marrow. Blood. 2007 Feb 15; 109(4):1743–51.

[120] Sarani Ghoshal, Bryan C. Fuchs, Kenneth K. Tanabe. STAT3 is a key transcriptional regulator of cancer stem cell marker CD133 in HCC. Hepatobiliary Surg Nutr. 2016 Jun; 5(3): 201–203.

[121] Nemetz C, Hocke GM. Transcription factor stat5 is an early marker of differentiation of murine embryonic stem cells. Differentiation. 1998 Mar; 62(5):213–20.

[122] Wongtrakoongate P, Jones M, Gokhale PJ, Andrews PW. STELLA facilitates differentiation of germ cell and endodermal lineages of human embryonic stem cells. PLoS One. 2013; 8(2):e56893.

[123] Xu X, Pantakani DVK, Lührig S, Tan X, Khromov T, Nolte J, et al. Stage-specific germ-cell marker genes are expressed in all mouse pluripotent cell types and emerge early during induced pluripotency. PLoS ONE. 2011; 6(7): e22413.

[124] Hongxiu Ning, Guiting Lin, Tom F. Lue, Ching-Shwun Lin. Mesenchymal stem cell marker Stro-1 is a 75kd endothelial antigen. Biochem Biophys Res Commun. 2011 Sep 23; 413(2): 353–357.

[125] Snipes RG, Lam KW, Dodd RC, Gray TK, Cohen MS. Acid phosphatase activity in mononuclear phagocytes and the U937 cell line: monocyte-derived macrophages express tartrate-resistant acid phosphatase. Blood. 1986 Mar; 67(3):729–34.

[126] Bollum FJ. Terminal deoxynucleotidyl transferase as a hematopoietic cell marker. Blood. 1979 Dec; 54(6):1203–1215.

[127] Robert K. Montgomery, Diana L. Carlone, Camilla A. Richmond, Loredana Farill, Mariette E, Kranendonk G, Daniel E. Henderson, Nana Yaa Baffour-Awuah, Dana M. Ambruzs, Laura K. Fogli, Selma Algra, David T. Breault. Mouse telomerase reverse transcriptase (mTert) expression marks slowly cycling intestinal stem cells. Proc Natl Acad Sci U S A. 2011 Jan 4; 108(1): 179–184.

[128] Kelly M. McNagny, Inger Pettersson, Fabio Rossi, Ingo Flamme, Andrej Shevchenko, Matthias Mann, and Thomas Graf. Thrombomucin, a novel cell surface protein that defines thrombocytes and multipotent hematopoietic progenitors. J Cell Biol. 1997 Sep 22; 138(6):1395–407.

[129] Shaikh MV, Kala M, Nivsarkar M. CD90 a potential cancer stem cell marker and a therapeutic target. Cancer Biomark. 2016;16(3):301–7.

[130] Schopperle WM, DeWolf WC. The TRA-1-60 and TRA-1-81 human pluripotent stem cell markers are expressed on podocalyxin in embryonal carcinoma. Stem Cells 2007 Mar; 25(3):723–730.

[131] Izadpanah, M.H., Abbaszadegan, M.R., Fahim, Y. et al. Ectopic expression of TWIST1 upregulates the stemness marker OCT4 in the esophageal squamous cell carcinoma cell line KYSE30, Cell Mol Biol Lett, (2017). 22: 33.

[132] Yao X, Ping Y, Liu Y, Chen K, Yoshimura T, Liu M, Gong W, Cheng C, Niu Q, Guo D, Zhang X, Wang JM, Bian X, Ribatti D. Vascular endothelial growth factor receptor 2 (VEGFR-2) plays a Key role in vasculogenic mimicry formation, neovascularization and tumor initiation by glioma stem-like cells. PLoS One. 2013; 8(3): e57188.

[133] Juan Zhang, Jianwei Jiao. Molecular biomarkers for embryonic and adult neural stem cell and neurogenesis. Biomed Res Int. 2015. 2015: 727542.

[134] Baker CV, Sharpe CR, Torpey NP, Heasman J, Wylie CC.A Xenopus c-kit-related receptor tyrosine kinase expressed in migrating stem cells of the lateral line system. Mech Dev. 1995 Apr; 50 (2–3): 217–28.

[135] Schwartz, G.K. Dickson, M.A. Cell cycle, CDKs and cancer: A changing paradigm. Nat Rev Cancer. 2009; 9(3): 153–66.

[136] Bloom, J. Cross, F.R. Multiple levels of cyclin specificity in cell-cycle control. *Nature Reviews* Mol Cell Biol. 2007; 8: 149–160.

[137] Junan Li, Ming Jye Poi, Ming-Daw Tsai. The regulatory mechanisms of tumor suppressor P16^{INK4A} and relevance to cancer biochemistry. 2011 Jun 28; 50(25): 5566–5582.

[138] Rajanahalli P, Stucke CJ, Hong Y. The effects of silver nanoparticles on mouse embryonic stem cell self-renewal and proliferation. Toxicol Rep. 2015 May 16;2:758–764.

[139] Periasamy VS, Athinarayanan J, Alhazmi M, Alatiah KA, Alshatwi AA. Fe$_3$O$_4$ nanoparticle redox system modulation via cell-cycle progression and gene expression in human mesenchymal stem cells. Environ Toxicol. 2016 Aug; 31(8):901–12.

[140] Hou Y, Cai K, Li J, Chen X, Lai M, Hu Y, Luo Z, Ding X, Xu D. Effects of titanium nanoparticles on adhesion, migration, proliferation, and differentiation of mesenchymal stem cells. Int J Nanomed. 2013;8:3619–30.

[141] Kanwar JR, Mahidhara G, Roy K, Sasidharan S, Krishnakumar S, Prasad N, Sehgal R, Kanwar RK. Fe-bLf nanoformulation targets survivin to kill colon cancer stem cells and maintains absorption of iron, calcium and zinc. Nanomedicine (Lond). 2015 Jan; 10(1):35–55.

[142] Menaa F, Abdelghani A, Menaa B. Graphene nanomaterials as biocompatible and conductive scaffolds for stem cells: Impact for tissue engineering and regenerative medicine, J Tissue Eng Regen Med, 2015 Dec.9(12):1321–38.

[143] Handral HK, Tong HJ, Islam I, Sriram G, Rosa V, Cao T. Pluripotent stem cells: An in vitro model for nanotoxicity assessments, J Appl Toxicol, 2016 Oct.36(10):1250–8.

[144] Chen EY, Wang YC, Mintz A, Richards A, Chen CS, Lu D, Nguyen T, Chin WC. Activated charcoal composite biomaterial promotes human embryonic stem cell differentiation toward neuronal lineage. J Biomed Mater Res A. 2012 Aug; 100(8):2006–17.

[145] Senut MC, Zhang Y, Liu F, Sen A, Ruden DM, Mao G. Size-dependent toxicity of gold nanoparticles on human embryonic stem cells and their neural derivatives. Small. 2016 Feb 3; 12(5):631–46.

[146] Oh JH, Son MY, Choi MS, Kim S, Choi AY, Lee HA, Kim KS, Kim J, Song CW, Yoon S. Integrative analysis of genes and miRNA alterations in human embryonic stem cells-derived neural cells after exposure to silver nanoparticles. Toxicol Appl Pharmacol. 2016 May 15; 299:8–23.

[147] Braydich-Stolle L, Hussain S, Schlager JJ, Hofmann MC. In vitro cytotoxicity of nanoparticles in mammalian germline stem cells. Toxicol Sci. 2005 Dec; 88(2):412–9.

[148] Joris F, Valdepérez D, Pelaz B, Wang T, Doak SH, Manshian BB, Soenen SJ, Parak WJ, De Smedt SC, Raemdonck K. Choose your cell model wisely: The in vitro nanoneurotoxicity of differentially coated iron oxide nanoparticles for neural cell labeling. Acta Biomater. 2017 Jun; 55:204–213.

[149] Liu X, He W, Fang Z, Kienzle A, Feng Q. Influence of silver nanoparticles on osteogenic differentiation of human mesenchymal stem cells. J Biomed Nanotechnol. 2014 Jul; 10(7):1277–85.

[150] Zhao Y, Alakhova DY, Kabanov AV. Can nanomedicines kill cancer stem cells?. Adv Drug Deliv Rev. 2013 Nov; 65 (13–14): 1763–83.

[151] Cruz AF, Fonseca NA, Moura V, Simoes S, Moreira JN. Targeting cancer stem cells and non-stem cancer cells: The potential of lipid-based nanoparticles. Curr Pharm Des. 2017 Nov 14.

[152] Lu B, Huang X, Mo J, Zhao W. Drug delivery using nanoparticles for cancer stem-like cell targeting. Front Pharmacol .2016 Apr 12;7:84.

[153] Hong IS, Jang GB, Lee HY, Nam JS. Targeting cancer stem cells by using the nanoparticles. Int J Nanomed. 2015 Sep 10; 10(Spec): 251–60.

[154] Zhou G, Latchoumanin O, Bagdesar M, Hebbard L, Duan W, Liddle C, George J, Qiao L. Aptamer-based therapeutic approaches to target cancer stem cells. Theranostics. 2017 Sep 13; 7(16):3948–3961.

[155] Li B, Li Q, Mo J, Dai H. Drug-loaded polymeric nanoparticles for cancer stem cell targeting. Front Pharmacol. 2017 Feb 14. 8:51.

[156] Gao J, Li W, Guo Y, Feng SS. Nanomedicine strategies for sustained, controlled and targeted treatment of cancer stem cells. Nanomedicine (Lond). 2016 Dec; 11(24):3261–3282.

[157] Kim SS, Harford JB, Pirollo KF, Chang EH. Effective treatment of glioblastoma requires crossing the blood-brain barrier and targeting tumors including cancer stem cells: The promise of nanomedicine. Biochem Biophys Res Commun. 2015 Dec 18;468(3):485–9.

[158] Naoum GE, Tawadros F, Farooqi AA, Qureshi MZ, Tabassum S, Buchsbaum DJ, Arafat W. Role of nanotechnology and gene delivery systems in TRAIL-based therapies. Ecancermedicalscience. 2016 Aug 1; 10:660.

[159] Nanotechnology for mesenchymal stem cell therapies.Corradetti B, Ferrari M. J Control Release. 2016 Oct; 28;240:242–250.

[160] Gao Z, Zhang L, Hu J, Sun Y. Mesenchymal stem cells: a potential targeted-delivery vehicle for anti-cancer drug, loaded nanoparticles. Nanomedicine. 2013 Feb;9(2):174–84.

[161] Vaseghi G, Rafiee L, Javanmard SH. Non-viral delivery systems for breast cancer gene therapy. Curr Gene Ther. 2017; 17(2):147–153.

[162] Wang W, Xu X, Li Z, Lendlein A, Ma N. Genetic engineering of mesenchymal stem cells by non-viral gene delivery. Clin Hemorheol Microcirc. 2014; 58(1):19–48.

[163] Rathnam C, Chueng SD, Yang L, Lee KB. Advanced gene manipulation methods for stem cell theranostics. Theranostics. 2017; Jul 8; 7(11):2775–2793.

[164] Shen WB, Anastasiadis P, Nguyen B, Yarnell D, Yarowsky PJ, Frenkel V, Fishman PS. Magnetic enhancement of stem cell-targeted delivery into the brain following MR-guided focused ultrasound for opening the blood-brain barrier. Cell Transplant. 2017 Jul; 26(7):1235–1246.

[165] Lamanna JJ, Urquia LN, Hurtig CV, Gutierrez J, Anderson C, Piferi P, Federici T, Oshinski JN, Boulis NM. Magnetic resonance imaging-guided transplantation of neural stem cells into the porcine spinal cord. Stereotact Funct Neurosurg. 2017;95(1):60–68.

[166] Ogihara Y, Yukawa H, Kameyama T, Nishi H, Onoshima D, Ishikawa T, Torimoto T, Baba Y. Labeling and in vivo visualization of transplanted adipose tissue-derived stem cells with safe cadmium-free aqueous ZnS coating of ZnS-AginS2 nanoparticles. Sci Rep. 2017 Jan 6; 7:40047.

[167] Addington CP, Cusick A, Shankar RV, Agarwal S, Stabenfeldt SE, Kodibagkar VD. Siloxane nanoprobes for labeling and dual modality functional imaging of neural stem cells. Ann Biomed Eng. 2016 Mar; 44(3):816–27.

[168] Murphy SV, Hale A, Reid T, Olson J, Kidiyoor A, Tan J, Zhou Z, Jackson J, Atala A. Use of trimetasphere metallofullerene MRI contrast agent for the non-invasive longitudinal tracking of stem cells in the lung. Methods. 2016 Apr 15; 99:99–111.

[169] Pohlmann ES, Patel K, Guo S, Dukes MJ, Sheng Z, Kelly DF. Real-time visualization of nanoparticles interacting with glioblastoma stem cells. Nano Lett. 2015 Apr 8; 15(4):2329–35.

[170] Schulze F, Dienelt A, Geissler S, Zaslansky P, Schoon J, Henzler K, Guttmann P, Gramoun A, Crowe LA, Maurizi L, Vallée JP, Hofmann H, Duda GN, Ode A. Amino-polyvinyl alcohol coated superparamagnetic iron oxide nanoparticles are suitable for monitoring of human mesenchymal stromal cells in vivo. Small. 2014 Nov 12; 10(21):4340–4351.

[171] Vernikouskaya I, Fekete N, Bannwarth M, Erle A, Rojewski M, Landfester K, Schmidtke-Schrezenmeier G, Schrezenmeier H, Rasche V. Iron-loaded PLLA nanoparticles as highly efficient intracellular markers for visualization of mesenchymal stromal cells by MRI. Contrast Media Mol Imaging. 2014 Mar-Apr; 9(2):109–21.

[172] Gutova M, Frank JA, D'Apuzzo M, Khankaldyyan V, Gilchrist MM, Annala AJ, Metz MZ, Abramyants Y, Herrmann KA, Ghoda LY, Najbauer J, Brown CE, Blanchard MS, Lesniak MS, Kim SU, Barish ME, Aboody KS, Moats RA. Magnetic resonance imaging tracking of ferumoxytol-labeled human neural stem cells: Studies leading to clinical use. Stem Cells Transl Med. 2013 Oct; 2(10):766–75.

[173] Struys T, Ketkar-Atre A, Gervois P, Leten C, Hilkens P, Martens W, Bronckaers A, Dresselaers T, Politis C, Lambrichts I, Himmelreich U. Magnetic resonance imaging of human dental pulp stem cells in vitro and in vivo. Cell Transplant. 2013; 22(10):1813–29.

[174] Gu L, Li X, Jiang J, Guo G, Wu H, Wu M, Zhu H. Stem cell tracking using effective self-assembled peptide-modified superparamagnetic nanoparticles. Nanoscale. 2018 Jun 19; (10):15967–15979.

[175] Mannucci S, Calderan L, Quaranta P, Antonini S, Mosca F, Longoni B, Marzola P, Boschi F. Quantum dots labelling allows detection of the homing of mesenchymal stem cells administered as immunomodulatory therapy in an experimental model of pancreatic islets transplantation. J Anat. 2017 Mar; 230(3):381–388.

[176] Hossain MA, Chowdhury T, Bagul A. Imaging modalities for the in vivo surveillance of mesenchymal stromal cells. J Tissue Eng Regen Med. 2015 Nov;9(11):1217–24.

[177] Vibin M, Vinayakan R, John A, Fernandez FB, Abraham A. Effective cellular internalization of silica-coated CdSe quantum dots for high contrast cancer imaging and labelling applications. Cancer Nanotechnol. 2014; 5(1):1.

[178] Rak-Raszewska A, Marcello M, Kenny S, Edgar D, Sée V, Murray P. Quantum dots do not affect the behaviour of mouse embryonic stem cells and kidney stem cells and are suitable for short-term tracking. PLoS One. 2012 ;7(3):e32650.

[179] EPA US. Nanotechnology White Paper. U.S. Environmental Protection Agency. EPA 100/B-07/001. Washington, DC. 2007.

4 Nanoparticles and brain

Abstract: Brain disorders affect millions of people across the globe. In 2017, World Health Organization has included stroke as a neurological disorder. Treatment of brain disorders has challenges in targeting the blood–brain barrier (BBB) due to its anatomy and physiology. Other brain barriers also pose a challenge to conventional modes of therapy. However, nanoparticles (NPs) with their small size can cross the BBB and prove to be a major promise in targeting the brain disorders. However, different NPs have revealed their diverse toxic effects. Therefore, their use needs more research prior to their application in humans. In this chapter, we discuss (i) how NPs can harm the health of the brain and (ii) different NPs in targeting brain disorders and their toxic effects.

Keywords: Blood–brain barrier, brain, cytokine, human blood, nanoparticle, PBCA, toxicity, Al_2O_3, titanium dioxide nanoparticles, ZnO, metal oxide nanoparticles, morphological changes, nanotoxicity, oxidative stress, brain homogenate, plasma, protein corona, silica, myelin-like bodies, PSD-95, Synapsin I, synaptophysin, PEGylation, central nervous system, graphene-based nanomaterials, BV2, P25, environmental nanotoxicity, neurotoxicity, calcium homeostasis, excitotoxicity, MK-801, reactive oxygen species, cell death, free radicals, NMDA receptor, zinc homeostasis, iron oxide, magnetic nanoparticles, theranostic, behavioral tests, CNP, m*yelin*-associated glycoprotein, m*yelin* oligodendrocyte glycoprotein, DNA methylation, autophagy, endothelial cells, mesoporous silica nanoparticles, neurons

4.1 Introduction

Disorders of the brain and their treatment remain a major domain of research in the current world. Brain diseases are of different forms. Infections, trauma, stroke, seizures and tumors are some of the major categories of brain diseases. Table 4.1 shows an overview of various diseases of the brain.

4.2 Blood–brain barrier

Targeting a brain disorder suffers from the challenges of crossing the blood–brain barrier (BBB). It is formed by the endothelium of brain capillary and controls exchange of molecules between blood and cerebrospinal fluid, thus maintaining cerebral homeostasis. Located at choroid plexus epithelium at capillary level between blood and cerebral tissue, the BBB comprises (i) tight junctions and (ii) polarized expression of many transport systems. The brain endothelial cells lack fenestrations, consist of

https://doi.org/10.1515/9783110579093-004

Table 4.1: Diseases of the brain.

Infectious diseases	
Meningitis	Viral infection affecting the lining of brain or spinal cord, neck stiffness, headache, fever
Encephalitis	Viral infection affecting brain tissues; meningitis and encephalitis occurring together is called as meningoencephalitis
Brain abscess	Bacterial infection in the brain
Seizures including epilepsy	Abnormal and excessive electrical activities in the brain; head injuries, brain infections and strokes may lead to epilepsy
Trauma	
Concussion	A brain injury with temporary disturbance in brain function, unconsciousness, confusion and memory problems
Traumatic brain injury	Leads to permanent brain damage
Intracerebral hemorrhage	Bleeding inside the brain or stroke due to high blood pressure
Tumors and increased pressure	
Brain tumor	Any abnormal tissue growth inside the brain, malignant (cancerous) or benign
Glioblastoma	Aggressive, cancerous brain tumor
Pseudotumor cerebri (false brain tumor)	Increased skull pressure, vision changes, headaches, dizziness and nausea
Hydrocephalus	An abnormally increased amount of cerebrospinal fluid in skull
Normal pressure hydrocephalus	A form of hydrocephalus with problems in walking, with dementia and urinary incontinence
Vascular (blood vessels) conditions	Brain diseases connected with blood
Stroke	Interruptions in blood flow and oxygen supply to brain tissues
Ischemic stroke	A blood clot in an artery blocking blood flow and causing a stroke
Hemorrhagic stroke	Bleeding in the brain creates congestion and pressure on brain tissues disrupting blood flow
Transient ischemic attack	A temporary interruption of blood flow and oxygen, but get cured completely without damage to brain tissue
Brain aneurysm	An artery in the brain that swells to form a balloon and rupture causing a stroke due to bleeding
Subdural hematoma	Bleeding on the brain surface; exerts pressure on the brain causing neurological disorders

Table 4.1 (continued)

Brain diseases: autoimmune conditions	
Vasculitis	An inflammation of the blood vessels of the brain; confusion, seizures, headaches and unconsciousness can occur
Multiple sclerosis	The immune system damages nerves causing muscle spasm, fatigue and weakness
Neurodegenerative diseases	
Parkinson's disease	Nerves in the brain degenerate leading to problems in movement and coordination, a tremor of hands, stiffness of limbs and trunk, slowness of movement and unstable posture
Huntington's disease	An inherited nerve disorder, degeneration of brain cells, dementia and difficulty controlling movements
Pick's disease (frontotemporal dementia)	Brain nerve destruction, personality changes, inappropriate behavior, difficulty with speech and loss of memory and intellectual ability are symptoms
Amyotrophic lateral sclerosis	Nerves controlling muscle function are steadily and rapidly destroyed, paralysis and inability to breathe
Dementia:	A decline in cognitive function due to death or malfunction of nerve cells in the brain
Alzheimer's disease	Degeneration of nerves in brain causing memory loss

tight junctions, express membrane receptors that enable active transport of nutrients to the brain and remove xenobiotics from the cerebral and vascular compartments. Brain endothelium has controlled permeability toward molecules from plasma and expressed high transendothelial electrical resistance. Prohibiting the entry of xenobiotics protects the brain and thus forms a major limitation to conventional therapeutics. Nanodrugs owing to their size hold a great promise in targeting brain disorders, but it is important to understand the physiology of BBB in development of nanodrugs

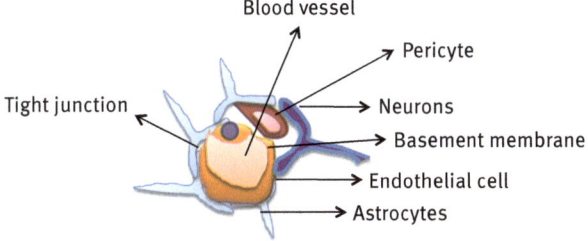

Figure 4.1: Blood–brain barrier.

to target brain disorders. Loss of function or malfunction of BBB leads to neurological disorders as then the brain vasculature is open to the brain draining in all items from the blood (Figure 4.1).

4.3 Nanotechnology and application in research of brain disorders

The major problems associated with brain disorders encompass inflammation, injury, tumors, neurodegeneration, specialized barriers and delivery. Therapy to brain disorders suffers due to the presence of BBB in the brain that prevents the entry of harmful substances in the brain. Delivery of medicines in treatment of brain disorders is thus a difficult domain, and thus research in brain disorders focus around the development of agents that can cross the BBB toward the development of new and innovative treatments of stroke, Alzheimer's disease (AD) and multiple sclerosis (MS). Stoke has been categorized as a neurological disorder by the World Health Organization in 2017. Stroke or brain injury affects the BBB at the site of injury, which can affect the neural tissue.

Imaging is a way to study the BBB that is affected due to brain diseases. Conventional imaging suffers limitations in not visualizing the details in the BBB accurately, revealing tumor size, location, type and response to therapy. Research on improved imaging techniques are also needed to study the barrier at sites of injury and tumors. Study of BBB in neurodegenerative diseases, including Parkinson's disease (PD), playing a role in neurodegeneration or aging process needs further research. In addition to BBB, other specialized brain barriers function individually or/and in unison, but their role in disease and control of drug delivery is limited due to the lack of experimental models of these specialized barriers, and thus little is understood about their function.

4.4 NPs and brain health

Nanoparticles (NPs) offer promises to the drawbacks in conventional modes of therapy of brain disorders. Drug-loaded poly(n-butylcyanoacrylate) (PBCA) NPs prepared by nanoemulsion method coated with polysorbate 80 (PS80) by adsorption has been reported to enhance the concentration of drugs in the cerebrum in rat model. When the drug was tested on human whole blood of normal individual, ex vivo experiments revealed the release of inflammatory cytokine interleukin-8. PBCA-NPs maintained cell viability and integrity with no inflammation or organ damage in rats in vivo, thus proving its efficacy as a drug delivery system to overcome the BBB [1]. TiO$_2$NPs (titanium dioxide nanoparticles), ZnONPs (zinc oxide nanoparticles) and Al_2O_3NPs (aluminum oxide nanoparticles) when tested over erythrocytes, brain and liver of male mouse led to their translocation in the cytoplasm and nucleus as observed by transmission electron microscopic studies and production of oxidative stress (OS) with release of reactive oxygen species (ROS) and altered antioxidant enzymes activities. Release of dopamine and norepinephrine in brain cerebral cortex and increased brain OS were indicative of their neurotoxic nature. Toxicity is found in ZnONPs followed by Al_2O_3NPs and TiO_2NPs [2]. ZnONP toxicity was reported by protein corona, from blood and brain, and proteins were absorbed onto these NPs affecting the biological processes (Figure 4.2) [3].

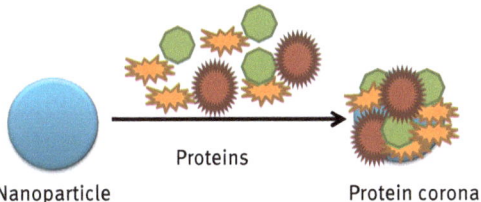

Nanoparticle   Proteins   Protein corona

Figure 4.2: Protein corona formation.

SiO_2NPs could adsorb the proteins on its surface termed as the protein corona when in contact with blood and brain in rat animal model, and size, charge and surface chemistry of NPs have been reported to play a role in the number and type of proteins adsorbed to NPs. Using liquid chromatography-tandem mass spectrometry plasma proteins were reported to be adsorbed to both negatively charged 20 and 100 nm SiO_2NPs and arginine-coated SiO_2NPs. The proteins in the corona that indicate probable biological pathways are affected [4]. It has been emphasized on a holistic study including aspects of biochemical, cellular and immunological processes to understand the mechanisms of nanotoxicity [5]. TiO_2NPs due to their photocatalytic properties and the related surface affect BBB [6].

Silver nanoparticles (AgNPs) have been reported to be toxic from in vitro and in vivo studies. Although the underlying mechanism of AgNP toxicity is not well understood, their migration and accumulation into the brain have been reported to cause neurotoxicity. Rats exposed to 10 nm sized AgNPs have been reported to reveal ultrastructural changes manifested by blurred synapses and increased number of synaptic vesicles in the center of the presynaptic cleft and their disruption and degeneration leading to release of neutrophil and myelin-like structures from fragmented membranes and organelles, decrease in presynaptic protein synapsin I and synaptophysin, PSD-95 protein leading to impairment of nerve function and affected cognitive processes [7]. Polyethylene glycol (PEG)-coated graphene oxide (rGO) when exposed to rat model revealed the concentration-dependent toxicity on astrocytes and rat brain endothelial cells leading to cell death with downregulation of astrocytes (GFAP, connexin-43), the endothelial tight (occludin) and adherens (β-catenin) junctions and basal lamina (laminin), increase in ROS leading to oxidative damage of the BBB [8].

TiO$_2$ photo-reactivity leads to brain damage by generation of ROS causing OS. Brain cultures of immortalized mouse microglia (BV2), rat dopaminergic (DA) neurons (N$_2$7), and primary cultures of embryonic rat striatum, when exposed to Degussa P$_2$5, a commercially available TiO$_2$ nanomaterial, revealed release of ROS, up-regulation of inflammatory, apoptotic, and cell cycling pathways and down-regulation of energy metabolism, increase of intracellular ATP and caspase 3/7 activity and neuronal apoptosis [9]. AgNPs lead to neuronal cell death when exposed to rat cerebellar granule cells leading to increase in intracellular calcium, release of ROS, decrease in mitochondrial potential and through the activation of NMDA (N-methyl-d-aspartate) receptors that play a role in memory function maintaining synaptic plasticity [10]. In primary neuronal cultures AgNPs exert their toxicity by destabilization of the calcium homeostasis through glutamatergic NMDA receptors. Zinc nanoparticles (ZnNP) supplementation could positively influence toxic effects by AgNPs by an inhibitory effect on NMDA-sensitive calcium channels [11]. Iron nanoparticles (FeNPs) are reported to cross the BBB and find applications in neuroimaging devices, but their role in neurotoxicity is not known. Surface-coated different types of FeNPs including DMSA-Fe$_2$O$_3$, DMSA-Fe$_3$O$_4$, PEG-Fe3O$_4$ and PEG-Au-Fe$_3$O$_4$ lead to increased OS, inflammation and proapoptotic agents in neuroglia, macrophages, lymphocytes and endothelial cells [12]. Carbon nanotubes (CNTs) with their property of tissue penetrance, small dimensions, and biopersistence can rapidly translocate in different parts of the body through the blood stream showing accumulation in the lungs and brain and later reaching the liver and kidneys when exposed in CD1 mice. Single-walled carbon nanotubes could induce systemic cell proliferation, indicating a dynamic response of cells of both bone marrow and the immune system, transient accumulation in the lungs, spleen, and kidneys with increase in aspartate aminotransferase, alanine aminotransferase and bilirubinemia, with liver damage and lead to M1 macrophage-driven inflammation [13]. Oral administration of AgNP (10 nm) could cause cytotoxicity, when exposed to adult rats and induces changes in cerebral myelin, depression and hyperalgesia, increase body weight and body temperature, morphological alterations in myelin sheaths and alter expression of myelin-specific proteins CNP, (2',3'-cyclic nucleotide 3'-phosphodiesterase), myelin-associated glycoprotein and myelin oligodendrocyte glycoprotein [14]. Systemic administration of TiO$_2$NPs could lead to their accumulation in the brain and induce brain dysfunction, generation of OS, apoptosis, inflammatory response, genotoxicity, direct impairment of cell components, disrupted signaling pathways, dysregulated neurotransmitters, synaptic plasticity, autophagy and DNA methylation [15]. Quantum dots (QDs) with potential applications in medicine but they are also known to exert their toxicity by release of ROS, altered cytoplasmic calcium Ca^{2+} balance. Cadmium selenide (CdSe) QDs could elevate cytoplasmic calcium levels in primary cultures of hippocampal neurons [16]. Mesoporous silica NPs with biomedical applications and drug delivery to different human body areas.

Exposure to endothelial cells lead to protein corona formation revealed mitochondrial activity and membrane integrity, induced cell autophagy, increase in NO production, increase excitability [17].

4.5 NPs and targeting of brain disorders

Although oxytocin finds importance in treatment of autism spectrum disorder, they suffer from the limitation of not crossing the BBB and requires frequent dosing due to its rapid metabolism in blood. Polymeric NP formulations of 100–278 nm in diameter of poly(lactic-co-glycolic acid) (PLGA) or bovine serum albumin conjugated to either transferrin or rabies virus glycoprotein as targeting ligands were reported to have promising delivery of oxytocin [18].

Gliomas are the most common types of intracranial malignant tumor, which suffers from effective treatment as therapeutic agents show limited penetration through the BBB. Gelatin siloxane NPs (GSNPs) conjugated to human immunodeficiency virus-derived Tat, tumor-targeting aptamer (TTA)1 and PEG to generate Tat-TTA1-PEG-GS NPs have been reported to transfer therapeutic agents efficiently in glioma [19].

Huperzine A (HupA)-loaded, mucoadhesive and targeted PLGA NPs with surface modification by lactoferrin (Lf)-conjugated N-trimethylated chitosan (TMC) forming a composition of HupA Lf-TMC NPs, with size 153.2 ± 13.7 nm, have been reported for its efficient intranasal delivery of HupA to the brain in the treatment of AD [20].

NPs including PGLA, PEG-PGLA and liposomes are finding applications in targeting of neuroinflammation [21].

Solid lipid nanoparticles (SLNs), nanostructured lipid carriers (NLC), liposomes and nanoemulsions have been reported for efficient nose to brain delivery of therapeutics through inhalation routes [22]. Doxorubicin (DOX)-loaded glycolipid-like NP has been reported to provide an effective delivery of DOX in tumor region in glioblastoma [23]. Peptides and proteins as therapeutic agents have been reported to show promises when delivered by polymer- and lipid-based nanocarriers [24] by the nose to brain routes initiated by inhalation [24].

PS80-coated and Atorvastatin-loaded PLGA-block-PEG NPs of size 30–172 nm have been used for the delivery of Atorvastatin to the brain [25]. AgNPs have been used in drug delivery therapy of AD in brain tissues at cellular and subcellular levels of the brain [26]. Engineered RNA NPs as aptamers, siRNA, miRNA, ribozymes derived from the bacteriophage phi29 DNA packaging motor pRNA, imaging as fluorophore and radiolabels are finding applications in targeting intracranial brain tumors in mice [27]. Highly branched dendrimers are recently finding application as potential nonviral vectors for the efficient delivery of drugs and nucleic acids to the brain and cancer cells [28]. Biocompatible natural NPs engineered with natural polymers including polysaccharides and proteins enable the delivery of different agents including anticancer drugs, analgesics, anti-Alzheimer's drugs, protease inhibitors and other macromolecules to the brain [29].

Odorranalectin, a lectin conjugated on PEG-PLGA NPs, find application in targeting of disorders of the CNS [30]. SLNs and NLCs with their inherent ability to cross the BBB have been finding application in targeting of brain diseases including brain cancer, ischemic stroke, AD, PD, MS, glioblastoma multiforme and neurodegenerative diseases [31].

Traditional Chinese medicine Scutellarin (SCU) has been tested to be delivered in the treatment of ischemic cerebrovascular disease in the brain by SCU-loaded HP-β-CD/chitosan NPs through the nasal route [32]. Inorganic NPs like ~80 nm gold nanospheres find application as epidermal growth factor receptors on cancer cell targets [33]. Curcumin encapsulated in PLGANPs, with its anti-amyloidogenic activity, can remove the β-amyloid (Aβ) formation which is a hallmark of AD. Encapsulated curcumin modified with g7 ligand for crossing of BBB has been reported to be effective in targeting AD [34]. Gint4.T aptamer polymeric NPs have been reported to recognize platelet-derived growth factor receptor β and can cross the BBB targeting brain disorders [35]. NPs have found application in delivery of siRNA-based therapeutics in PD [36].

Pathological Lewy bodies containing α-synuclein (α-syn) is predominant in PD. Targeting of α-syn expression in neurons in PD has been possible by the use of magnetic Fe3O4NPs coated with oleic acid molecules, photoimmobilization of N-isopropylacrylamide derivative onto oleic acid molecules and adsorption of short hairpin RNA [37]. Magnetic NPs find importance in the delivery of Neuropilin-1 to target the treatment of gliomas (Table 4.2) [38].

4.6 NPs in imaging of the brain

Conventional expensive and invasive methods of positron emission tomography (PET) for viewing the senile plaques, which are characteristic extracellular deposits of Aβ peptides in AD brain, have been improvised by the application of neuroprotective, nontoxic, dual-modal nanoprobe composed by Aβ-specific fluorescence cyanine sensors with Superparamagnetic iron oxide nanoparticles (SPIONs) viewed by near-infrared magnetic resonance imaging (MRI) contrast agent fluorescence imaging with high sensitivity and selectivity to Aβ species. This finds importance as a tool in efficient diagnosis and detection of AD [40]. NP sensors comprising neurotransmitter acetylcholine (Ach) catalyzing enzymes and pH-sensitive gadolinium contrast agents colocalized onto the surface of polymer NPs have been engineered for detection of Ach in the living brain by MRI. The sensors function by enzymatic hydrolysis of Ach leading to decrease in pH detected by the pH-sensitive gadolinium chelate [41]. A noninvasive in vivo neuroimaging of exosomes by 5 nm glucose-coated

Table 4.2: NP in brain targeting.

Nanoparticles	Disease	Delivered agent	References
Polymeric NP	ASD	Neuropeptide oxytocin	[18]
Gelatin siloxane NPs conjugated to human immunodeficiency virus-derived Tat, tumor-targeting aptamer 1 and PEG	Gliomas	Therapeutic agents/DNA	[19]
HupA Lf-TMC NPs	AD	Huperzine A	[20]
PGLA, PEG-PGLA, liposomes	Neuroinflammation	Therapeutic agents	[21]
SLN and NLC, liposomes, nanoemulsions	Brain disorders	Therapeutic agents	[22]
DOX-loaded glycolipid-like NPs	Glioblastoma	DOX	[23]
Polymer- and lipid- based nanocarriers	AD, PD	Therapeutic peptides and proteins	[24]
Polysorbate 80-coated and Atorvastatin-loaded PLGA-block-PEG	Brain disorder	Atorvastatin	[25]
Silver NPs	AD	Drugs	[26]
RNA NPs	Intracranial brain tumors	Drugs and imaging agents	[27]
Dendrimers	Brain cancers	Drugs, therapeutic agents	[28]
Natural NPs	Brain cancer, AD	Anticancer drugs, analgesics, anti- Alzheimer's drugs, protease inhibitors, and other macromolecules	[29]
SLNs and NLCs	Brain cancer, ischemic stroke, AD, PD, multiple sclerosis, glioblastoma multiforme, neurodegenerative diseases	Drugs and therapeutic agents	[31]

(continued)

Table 4.2 (continued)

Nanoparticles	Disease	Delivered agent	References
Traditional Chinese medicine SCU-loaded HP-β-CD/chitosan NPs	Ischemic cerebrovascular disease	Traditional Chinese medicine SCU	[32]
~80 nm gold nanospheres	Cancer, brain tumor	Drugs and therapeutic agents	[33]
Curcumin encapsulated in PLGA NPs	AD	Curcumin	[34]
Gint4.T aptamer polymeric NPs	Cross BBB in brain disorders	Therapeutics	[35]
NPs	PD	siRNA based therapeutics	[36]
Magnetic Fe3O4 NPs coated with oleic acid molecules and photoimmobilization of N-isopropylacrylamide derivative onto oleic acid molecules and adsorption of shRNA	PD	Therapeutics, shRNA targeting expression of α-syn in PD	[37]
Magnetic NPs	Gliomas	Neuropilin-1	[38]
Lf-conjugated SLN loaded with Dox	Brain disorders	Dox	[39]

Note: ASD, autism spectrum disorder; DOX, doxorubicin; Dox, docetaxel; AD, Alzheimer's disease; PD, Parkinson's disease; BBB, blood–brain barrier; PLGA, poly(lactic-co-glycolic acid); shRNA, short hairpin RNA; SLN, solid lipid nanoparticles; SCU, Scutellarin; Lf, lactoferrin; NLCs, nanostructured lipid carriers; PEG, polyethylene glycol; NPs, nanoparticles.

gold NP labeling finds therapeutic application and imaging by computed tomography (CT) [42]. Delivery of neurotrophic factors, including glial cell-line-derived neurotrophic factor (GDNF), by MRI-guided focused ultrasound and brain penetrating NPs has been reported to enable targeted GDNF transgene expression in the brain treatment in PD [43].

High photoluminescence quantum yield, photostability CdSe@ZnS/ZnS core–multishell quantum dots (CM-QDs) were constructed by self-assembly of CM-QDs and pH-responsive methoxy PEG-poly(β-amino ester/amidoamine)-dodecylamine (mPEG-PAEA-DDA) multiblock copolymers forming CM-QDs-loaded mPEG-PAEA-DDA micelles enabled fluorescence imaging by pH-triggering and has been reported to find application in detection of cerebral ischemia in rat model [44].

Mass spectrometry imaging of tissue implanted with 8 nm AgNP matrix enabled imaging of lipids, including galactoceramides, diacylglycerols, ceramides, phosphatidylcholines, cholesteryl ester and cholesterol in positive mode and phosphatidylethanolamides, sulfatides, phosphatidylinositol and sphingomyelins in negative mode in rodent models of disease and injury [45]. To optimize MRI of antibody-conjugated SPIONs for detecting Aβ plaques and activated microglia in a 3X transgenic mouse model of AD are being designed [46]. NPs are also designed with applications in improved CT scan and MRI methods [47]. NPs find application as brain targeting and imaging of MS, an autoimmune neurodegenerative disease [48]. Brain mapping in epilepsy is finding a new dimension using superparamagnetic NPs thereby improving the tissue contrast of MRI, thus improving the resection of epileptic foci [49].

Core–shell magnetic NPs and screen-printed carbon electrodes have been designed for detection of *Legionella pneumophila* SeroGroup 1 infection [50].

Systemic lupus erythematosus (SLE) is a systemic autoimmune disease affecting many organs including kidneys and central nervous system. Novel MRI and PET molecular imaging using targeted iron oxide NPs and radioactive ligands, respectively, are funding applications in detection of SLE and associated inflammation [51].

AD diagnosis by amyloid imaging has been improved by biocompatible curcumin magnetic NPs of size <100 nm by MRI [52]. Pediatric brain tumors (PBTs), a leading cause of mortality in children, have been tested with improved imaging and diagnosis by manganese-containing Prussian blue NPs detected by MRI and fluorescence-based imaging of PBTs [53].

4.7 Discussion

Brain disorders and their treatment is the major cause of concern globally. Although NPs offer great promises in delivery of materials across BBB and its study, yet the nanotoxicity studies indicate that they are still unsafe for medicinal use. Several studies with different NPs are underway in animal models to check their effect on the host system. Holistic study including aspects of biochemical, cellular, and immunological processes to understand the mechanisms of nanotoxicity [5].The major limitations to the study of other barriers to the brain and the effects of NP on them are the nonavailability of suitable experimental model system that form the scope of future research.

References

[1]   Kolter M, Ott M, Hauer C, Reimold I, Fricker G. Nanotoxicity of poly(n-butylcyano-acrylate) nanoparticles at the blood-brain barrier, in human whole blood and in vivo. J Control Release. 2015 Jan; 10(197):165–179.

[2]   Shrivastava R, Raza S, Yadav A. Effects of sub-acute exposure to TiO2, ZnO and Al2O3 nanoparticles on oxidative stress and histological changes in mouse liver and brain. Drug Chem Toxicol. 2014 Jul;37(3):336–347.

[3]   Shim KH, Hulme J, Maeng EH, Kim MK, An SS. Analysis of zinc oxide nanoparticles binding proteins in rat blood and brain homogenate. Int J Nanomed. 2014 Nov 15;9 Suppl 2:217–224.

[4]   Analysis of SiO$_2$ Nanoparticles binding proteins in rat blood and brainhomogenate. Int J Nanomed. 2014 Dec 15;9 Suppl 2:207–215.

[5]   Kardos J, Jemnitz K, Jablonkai I, Bóta A, Varga Z, Visy J, Héja L. The janus facet of nanomaterials. Biomed Res Int. 2015;317184.

[6]   Rollerova E, Tulinska J, Liskova A, Kuricova M, Kovriznych J, Mlynarcikova A, Kiss A, Scsukova S. Titanium dioxide nanoparticles: Some aspects of toxicity/focus on the development. Endocr Regul. 2015 Apr; 49(2):97–112.

[7]   Skalska J, Frontczak-Baniewicz M, Strużyńska L. Synaptic degeneration in rat brain after prolonged oral exposure to silver nanoparticles. Neurotoxicol. 2015 Jan;46:145–154.

[8]   Mendonça MC, Soares ES, de Jesus MB, Ceragioli HJ, Batista ÂG, Nyúl-Tóth Á, Molnár J, Wilhelm I, Maróstica MR Jr, Krizbai I, da Cruz-Höfling MA. PEGylation of reduced graphene oxide induces toxicity in cells of the blood-brain barrier: An in vitro and in vivo study. Mol Pharm. 2016 Nov 7; 1311:3913–3924.

[9]   Long TC, Tajuba J, Sama P, Saleh N, Swartz C, Parker J, Hester S. Nanosize titanium dioxide stimulates reactive oxygen species in brainmicroglia and damages neurons in vitro. Environ Health Perspect. 2007 Nov;11511:1631–1637.

[10]   Ziemińska E, Stafiej A, Strużyńska L. The role of the glutamatergic NMDA receptor in nanosilver-evoked neurotoxicity in primary cultures of cerebellar granule cells. Toxicol. 2014 Jan 6;315:38–48.

[11]   Ziemińska E, Strużyńska L. Zinc Modulates. Nanosilver-induced toxicity in primary neuronal cultures. Neurotox Res. 2016 Feb;29(2):325–343.

[12]   Kim Y, Kong SD, Chen LH, Pisanic TR 2nd, Jin S, Shubayev VI. In vivo nanoneurotoxicity screening using oxidative stress and neuroinflammation paradigms. Nanomed. 2013 Oct;9(7):1057–1066.

[13]   Principi E, Girardello R, Bruno A, Manni I, Gini E, Pagani A, Grimaldi A, Ivaldi F, Congiu T, De Stefano D, Piaggio G, de Eguileor M, Noonan DM, Albini A. Systemic distribution of single-walled carbon nanotubes in a novel model: Alteration of biochemical parameters, metabolic functions, liver accumulation, and inflammation in vivo. Int J Nanomed. 2016 Sep 1;11:4299–4316.

[14]   Dąbrowska-Bouta B, Zięba M, Orzelska-Górka J, Skalska J, Sulkowski G, Frontczak-Baniewicz M, Talarek S, Listos J, Strużyńska L. Influence of a low dose of silver nanoparticles on cerebral myelin and behavior of adult rats. Toxicol. 2016 Jul; 1:363–364:29–36.

[15]   Song B, Zhang Y, Liu J, Feng X, Zhou T, Shao L. Unraveling the neurotoxicity of titanium dioxide nanoparticles: Focusing on molecular mechanisms. Beilstein J Nanotechnol. 2016 Apr 29;7:645–654.

[16]   Tang M, Wang M, Xing T, Zeng J, Wang H, Ruan DY. Mechanisms of unmodified CdSe quantum dot-induced elevation of cytoplasmic calcium levels in primary cultures of rat hippocampal neurons. Biomater. 2008 Nov;.29(33):4383–4391.

[17]   Orlando A, Cazzaniga E, Tringali M, Gullo F, Becchetti A, Minniti S, Taraballi F, Tasciotti E, Re F. Mesoporous silica nanoparticles trigger mitophagy in endothelial cells and perturb neuronal network activity in a size- and time-dependent manner. Int J Nanomed. 2017 May 8;12:3547–3559.

[18]   Zaman RU, Mulla NS, Braz Gomes K, D'Souza C, Murnane KS, D'Souza MJ. Nanoparticle formulations that allow for sustained delivery and brain targeting of the neuropeptide oxytocin. Int J Pharm. 2018. Jul 19;548(1):698–706.

[19]   Lin XN, Tian X, Li W, Sun J, Wei F, Feng W, Huang ZC, Tian XH. Highly efficient glioma targeting of tat peptide-TTA1 aptamer-polyephylene glycol-modified gelatin-siloxane nanoparticles. J Nanosci Nanotechnol. 2018 Apr 1;18(4):2325–2329.

[20]   Meng Q, Wang A, Hua H, Jiang Y, Wang Y, Mu H, Wu Z, Sun K. Intranasal delivery of Huperzine A to the brain using lactoferrin-conjugated N-trimethylated chitosan surface-modified PLGA nanoparticles for treatment of Alzheimer's disease. Int J Nanomed. 2018 Feb 1;13:705–718.

[21] Poupot R, Bergozza D, Fruchon S. Nanoparticle-based strategies to treat neuro-inflammation. Materials (Basel). 2018 Feb 9;11(2):

[22] Cunha S, Almeida H, Amaral MH, Lobo JMS, Silva AC. Intranasal lipid nanoparticles for the treatment of neurodegenerative diseases. Curr Pharm Des. 2017 Nov 27.

[23] Wen L, Tan Y, Dai S, Zhu Y, Meng T, Yang X, Liu Y, Liu X, Yuan H, Hu F. VEGF-mediated tight junctions pathological fenestration enhances doxorubicin-loaded glycolipid-like nanoparticles traversing BBB for glioblastoma-targeting therapy. Drug Deliv. 2017 Nov;24(1):1843–1855.

[24] Samaridou E, Alonso MJ. Nose-to-brain peptide delivery – The potential of nanotechnology. Bioorg Med Chem. 2018 Jun 1;26(10):2888–2905.

[25] Simşek S, Eroğlu H, Kurum B, Ulubayram K. Brain targeting of Atorvastatin loaded amphiphilic PLGA-b-PEG nanoparticles. J Microencapsul. 2013;30(1):10–20.

[26] Aliev G, Daza J, Herrera AS. del Carmen Arias Esparza M, Morales L, Echeverria V, Bachurin SO, Barreto GE. Nanoparticles as alternative strategies for drug delivery to the Alzheimer brain: Electron microscopy ultrastructural analysis. CNS Neurol Disord Drug Targets. 2015; 14(9):1235–1242.

[27] Lee TJ, Haque F, Vieweger M, Yoo JY, Kaur B, Guo P, Croce CM. Functional assays for specific targeting and delivery of RNA nanoparticles to brain tumor. Methods Mol Biol. 2015; 1297:137–152.

[28] Somani S, Dufès C. Applications of dendrimers for brain delivery and cancer therapy. Nanomed (Lond). 2014 Oct;9(15):2403–2414.

[29] Elzoghby AO, Abd-Elwakil MM, Abd-Elsalam K, Elsayed MT, Hashem Y, Mohamed O. Natural polymeric nanoparticles for brain-targeting: Implications on drug and gene delivery. Curr Pharm Des. 2016; 22(22):3305–3323.

[30] Wen Z, Yan Z, He R, Pang Z, Guo L, Qian Y, Jiang X, Fang L. Brain targeting and toxicity study of odorranalectin-conjugated nanoparticles following intranasal administration. Drug Deliv. 2011 Nov; 18(8):555–561.

[31] Tapeinos C, Battaglini M, Ciofani G. Advances in the design of solid lipid nanoparticles and nanostructured lipid carriers for targeting brain diseases. J Control Release. 2017 Oct 28;264:306–332.

[32] Liu S, Ho PC. Intranasal administration of brain-targeted HP-β-CD/chitosan nanoparticles for delivery of scutellarin, a compound with protective effect in cerebral ischaemia. J Pharm Pharmacol. 2017 Nov; 69(11):1495–1501.

[33] Feng Q, Shen Y, Fu Y, Muroski ME, Zhang P, Wang Q, Xu C, Lesniak MS, Li G, Cheng Y. Self-assembly of gold nanoparticles shows microenvironment-mediated dynamic switching and enhanced brain tumor targeting. Theranostics. 2017 Apr 10;7(7):1875–1889.

[34] Barbara R, Belletti D, Pederzoli F, Masoni M, Keller J, Ballestrazzi A, Vandelli MA, Tosi G, Grabrucker AM. Novel curcumin loaded nanoparticles engineered for blood–brain barrier crossing and able to disrupt A-beta aggregates. Int J Pharm. 2017 Jun 30;526 (1–2):413–424.

[35] Monaco I, Camorani S, Colecchia D, Locatelli E, Calandro P, Oudin A, Niclou S, Arra C, Chiariello M, Cerchia L, Comes Franchini M. Aptamer functionalization of nanosystems for glioblastoma targeting through the blood–brain barrier. J Med Chem. 2017 May 25; 60(10):4510–4516.

[36] Cortes H, Alcala-Alcala S, Avalos-Fuentes A, Mendoza-Munoz N, Quintanar-Guerrero D, Leyva-Gomez G, Floran B. Nanotechnology. As potential tool for siRNA delivery in Parkinson's disease, Curr Drug Targets. 2017 Nov 30; 18(16):1866–1879.

[37] Niu S, Zhang LK, Zhang L, Zhuang S, Zhan X, Chen WY, Du S, Yin L, You R, Li CH, Guan YQ[1]. Inhibition by multifunctional magnetic nanoparticles loaded with alpha-synuclein RNAi plasmid in a Parkinson's disease model. Theranostics. 2017 Jan 1; 7(2):344–356.

[38] Chen L, Zhang G, Shi Y, Qiu R, Khan AA. Neuropilin-1 (NRP-1) and magnetic nanoparticles, a potential combination for diagnosis and therapy of gliomas. Curr Pharm Des. 2015; 21(37):5434–5449.

[39] Singh I, Swami R, Pooja D, Jeengar MK, Khan W, Sistla R. Lactoferrin bioconjugated solid lipid nanoparticles: A new drug delivery system for potential brain targeting. J Drug Target. 2016; 24(3):212–223.

[40] Li Y, Xu D, Chan HN, Poon CY, Ho SL, Li HW, Wong MS. Dual-modal NIR-fluorophore conjugated magnetic nanoparticle for imaging amyloid-β species in vivo. Small. 2018 Jul; 14(28):e1800901.

[41] Luo Y, Kim EH, Flask CA, Clark HA. Nanosensors for the chemical imaging of acetylcholine using magnetic resonance imaging. ACS Nano. 2018 Jun 6; doi:10.1021/acsnano.8b01640.

[42] Betzer O, Perets N, Angel A, Motiei M, Sadan T, Yadid G, Offen D, Popovtzer R. In vivo neuroimaging of exosomes using gold nanoparticles. ACS Nano. 2017 Nov 28;11(11): 10883–10893.

[43] Mead BP, Kim N, Miller GW, Hodges D, Mastorakos P, Klibanov AL, Mandell JW, Hirsh J, Suk JS, Hanes J, Price RJ. Novel focused ultrasound gene therapy approach noninvasively restores dopaminergic neuron function in a rat Parkinson's disease model. Nano Lett. 2017 Jun 14; 17(6):3533–3542.

[44] Yang HY, Fu Y, Jang MS, Li Y, Yin WP, Ahn TK, Lee JH, Chae H, Lee DS. CdSe@ZnS/ZnS quantum dots loaded in polymeric micelles as a pH-triggerable targeting fluorescence imaging probe for detecting cerebral ischemic area. Colloids Surf B Biointerfaces.2017 Jul 1;155:497–506.

[45] Muller L, Baldwin K, Barbacci DC, Jackson SN, Roux A, Balaban CD, Brinson BE, McCully MI, Lewis EK, Schultz JA, Woods AS. Laser desorption/ionization mass spectrometric imaging of endogenous lipids from rat brain tissue implanted with silver nanoparticles. J Am Soc Mass Spectrom. 2017 Aug;28(8):1716–1728.

[46] Tafoya MA, Madi S, Sillerud LO. Superparamagnetic nanoparticle-enhanced MRI of Alzheimer's disease plaques and activated microglia in 3X transgenic mouse brains: Contrast optimization. J Magn Reson Imaging. 2017 Aug; 46(2):574–588.

[47] Mehta A, Ghaghada K, Mukundan S Jr. Molecular imaging of brain tumors using liposomal contrast agents and nanoparticles. Magn Reson Imaging Clin N Am. 2016 Nov;24(4):751–763.

[48] Ghalamfarsa G, Hojjat-Farsangi M, Mohammadnia-Afrouzi M, Anvari E, Farhadi S, Yousefi M, Jadidi-Niaragh F. Application of nanomedicine for crossing the blood-brain barrier: Theranostic opportunities in multiple sclerosis. J Immunotoxicol. 2016 Sep;13(5):603–619.

[49] Pedram MZ, Shamloo A, Alasty A, Ghafar-Zadeh E. Toward epileptic brain region detection based on magnetic nanoparticle patterning. Sensors (Basel). 2015 Sep 22;15(9):24409–24427.

[50] Martín M, Salazar P, Jiménez C, Lecuona M, Ramos MJ, Ode J, Alcoba J, Roche R, Villalonga R, Campuzano S, Pingarrón JM, González-Mora JL. Rapid Legionella pneumophila determination based on a disposable core-shell Fe_3O_4@poly(dopamine) magnetic nanoparticles immunoplatform. Anal Chim Acta. 2015 Aug 5; 887:51–58.

[51] Thurman JM, Serkova NJ. Non-invasive imaging to monitor lupus nephritis and neuropsychiatric systemic lupus erythematosus. Version 2. F1000Res. 2015 Jun 16 [revised 2015 Oct 27];4:153.

[52] Cheng KK, Chan PS, Fan S, Kwan SM, Yeung KL, Wáng YX, Chow AH, Wu EX, Baum L. Biomaterials Curcumin-conjugated magnetic nanoparticles for detecting amyloid plaques in Alzheimer's disease mice using magnetic resonance imaging (MRI). 2015 Mar;44:155–172.

[53] Dumont MF, Yadavilli S, Sze RW, Nazarian J, Fernandes R. Manganese-containing Prussian blue nanoparticles for imaging of pediatric brain tumors. Int J Nanomed. 2014 May 23;9:2581–2595.

5 Nanoparticles, neurotoxicity and safety in application to neurological disorders

Abstract: Due to unintentional exposures through air or at work or intentional exposure to nanoparticles (NPs), humans and animals are at a risk of neurodegeneration and suffer from neurotoxicity and associated inflammation stress, and altered biochemical and genetic expression. Due to their property of crossing the blood–brain bareer, they have been proved of their potential neurotoxic effects on the central nervous sytem. In this chapter, we discuss the different effects of NPs inducing neutotoxicty in animal cell lines and human system.

Keywords: oxidative stress (OS), single-walled nanotubes (SWNTs), fullerene (C$_{60}$), cadmium selenide (CdSe), quantum dots (QDs), carbon black (CB), dye-doped silica nanospheres (NSs), blood–brain barrier (BBB), blood–cerebrospinal fluid barrier (BCSFB), central nervous system (CNS), liposomes, micelles, dendrimers, nanogels, nanoemulsions, nanosuspensions, epigenetic, iron oxide nanoparticles (ION), cancer, silver nanoparticles (AgNPs), superparamagnetism, Parkinson's disease (PD), Alzheimer's disease (AD), neurodegenerative disorders, human immunodeficiency virus (HIV), HIV-associated neurocognitive disorders (HAND), magnetic nanoparticles (MNPs), quantum dots (QDs), titanium dioxide nanoparticles (TiO$_2$NPs), selenium nanoparticles, cerium oxide nanoparticles (CeO$_2$NPs), neurotoxicity, polyamidoamine (PAMAM) dendrimers, multi-walled carbon nanotubes (MWCNTs), *Ruditapes philippinarum*, *Chlamys farreri*, superoxide dismutase (SOD), catalase (CAT), malondialdehyde (MDA), silica nanoparticles (SiO$_2$NPs), zebrafish, chitosan nanoparticles (CSNPs), *Danio rerio*, *Paracentrotus lividus*, copper oxide nanoparticles (CuONPs), cerebral astrocytes, N-vinyl-2-pyrrolidone-based nanoparticles (Me-PSs), *Pelophylax ridibundus*, 3-mercaptopropionic acid, gene ontology, Kyoto Encyclopedia of Genes and Genomes (KEGG), *Caenorhabditis elegans* (*C. elegans*), cadmium telluride (CdTe), zinc oxide nanoparticle (ZnONP), *Prochilodus lineatus*, *Pheretima hawayana*, acetylcholinesterase (AChE) activities, interleukin 6 (IL-6), glial fibrillary acidic protein (GFAP), nickel oxide nanoparticles, carbon black nanoparticles (CBNPs), silica nanoparticles (SiO$_2$NPs), perivascular macrophages, Sprague Dawley (SD) rats, vascular endothelial growth factor (VEGF), brain microvessel endothelial cells (BMECs), diesel exhaust NPs, dentate gyrus, hippocampus, amyloid beta (Aβ) production, embryonic neural stem cells (NSCs), cerebellar granule cells (CGCs)

5.1 Introduction

Nanomedicine holds promise to treat, prevent and diagnose neurodegenerative diseases. Two major barriers including blood–brain barrier (BBB) and blood–cerebrospinal fluid barrier (BCSFB) pose a threat to the delivery of drugs and agents.

https://doi.org/10.1515/9783110579093-005

Nanoparticles (NPs), liposomes, micelles, dendrimers, nanogels, nanoemulsions and nanosuspensions are being tested for the delivery of therapeutics in central nervous system (CNS). Many of these nanomedicines have been known to cross the BBB models by endocytosis and/or transcytosis, but further researches to test their neurotoxicity need to be carried out to confirm their applicability [1] (Figure 5.1).

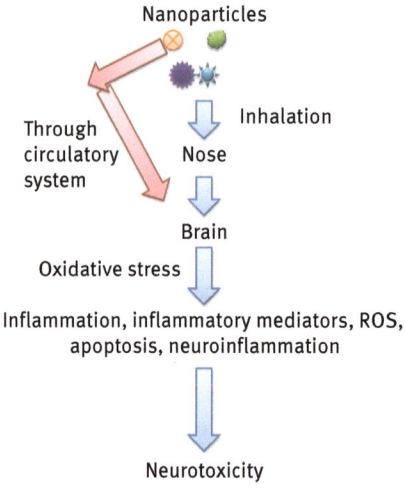

Figure 5.1: NP-mediated neurotoxicity.

An ideal medicine would involve design of non-toxic, biodegradable and biocompatible materials. Therefore, understanding the toxic effects of nanomaterials on the CNS and systemic toxicity is important to design nanomedicine. The major drawback of NPs in application of drug targetting in brain-related diseases is neurotoxicity that leads to memory deficit, behavioral changes, changes in structure and membrane potential of neurons and induces cytotoxicity, genotoxicity and epigenetic changes [2]. NPs have the potential to penetrate human body through various pathways and eventually cross the BBB to potentially cause neurotoxicity, neuroinflammation and neurodegeneration of the CNS [3,4]. Metals are known for their toxicity [5], but for NPs including SWNTs, C_{60}, CdSeNPs and QDs, carbon black (CB) and dye-doped silica nanospheres (NSs) have found application in medicine, engineering in the last one and half decade. Titanium dioxide nanoparticles (TiO_2NPs) have been known to induce neurotoxicity leading to neural damage in vivo through disruption of the BBB, induction of neuroinflammation, increase in OS and protein aggregation. While silica nanoparticles (SiO_2NPs) find application in biomedical products, underlying mechanism of their toxic effects is poorly understood.

Superparamagnetism and biocompatible properties of iron oxide nanoparticles (IONs) and their easy translocation to different organs including crossing the BBB has enabled their interesting applications in biomedical field in treatment of cancer, however, concerns over their cytotoxic effects exist. But their large-scale application suffers from concerns of acute toxicity, genotoxicity, immunotoxicity, neurotoxicity and reproductive toxicity [6]. Impact of silver nanoparticles (AgNPs) on the environment, aquatic organisms and cell lines, their side effects in mammalian organisms and neurotoxic effects and accumulation in mammalian brain are being studied globally prior to their application for drug delivery in humans [7].

Parkinson's disease (PD) and Alzheimer's disease (AD) are known to affect human life globally. The physiology of PD is characterized by severe loss of dopamine-related neurons in substantia nigra and cytoplasmic inclusions. Although conventional therapeutic agents enable treatments of PD, they suffer from the major problem of transport into the brain by crossing the BBB. Thus, optimizing the route of delivery of the therapeutics and the composition of a suitable drug are major goals in research in this domain. Intranasal administration of therapeutics using NPs like nanoemulsions, lipid NPs and polymeric nanomicelles to the CNS bypassing the BBB have been reported to offer advantage in treating neurodegenerative disorders by avoiding hepatic metabolism and noninvasiveness [8]. Neurotropic virus and human immunodeficiency virus (HIV) induce neurotoxicity and subsequent brain pathologies. HIV-associated neurocognitive disorders (HAND) are reported in more than 50% of AIDS patients, which weakens the connections between neurons affecting synaptic plasticity causing altered gene expression and subsequent loss of dendritic and spine morphology and physiology. Magnetic nanoparticles (MNPs) find importance in targeted drug delivery, release and bioavailability of drugs in noninvasive methods [9].

Quantum dots (QDs) can cross the BBB and enter the brain, but the underlying mechanism of interaction with host CNS is not well studied. Few studies have revealed QDs induced neurotoxicity in CNS leading to injury by OS, upregulation of cytoplasmic Ca^{2+} synthesis and autophagy leading to damage of neural cells, impairments of synaptic transmission and plasticity and other brain functions [10].

5.2 Different types of NPs and nanotoxicity

Copper (Cu) is an essential trace metal that plays important role but is toxic to the body. Copper and copper oxide nanoparticles (CuNPs and CuONPs, respectively) have applications as NPs in drug delivery devices but are highly toxic to the body [11]. They are major agents generating neurotoxic effects mediated by their property to accumulate in the basal ganglia of the CNS and lead to activation of neurotoxic cascades, production of reactive oxygen species (ROS) as well as their pathogenic interaction with intracellular targets in the dopamine neuron. It has also been known that exposure to metals may lead to Parkinson's and is considered an occupational hazard for people from the mining industry, E-waste generation and metal NP manufacturing [12].

Although the applications of TiO_2NPs are increasing in several fields of drug delivery and imaging, the concerns over their hazardous effects are on the rise. With their ability to cross the placental barrier in pregnant mice and cause neurotoxicity in their offsprings, disruption in the anatomical structure of fetal brain and liver in a pregnant mice model with profound toxic effects on the brain and liver cells leading to necrotic effects on neural tissues [13] is an object of major concern.

CeO_2NPs, also termed as nanoceria, are known to display antioxidant properties. Experiments with exposure of nanoceria on neural progenitor cells using the C17.2 murine cell line model reveals that it inhibits neuronal differentiation, interference with cytoskeletal organization and reduced neuron specific β3-tubulin expression, which is the marker for neuronal differentiation, glial fibrillary acidic protein (GFAP) and neuroglia, altered cytoskeletal organization and altered structure of neural growth cones posing to be a developmental neurotoxicity hazard [14].

Manganese (Mn) is also known for its toxic effects affecting the nervous system and is particulartly considered as a occupational hazard to the miners. Mn welding fumes with metal oxide NPs pose toxic effects [15].

Cadmium (Cd) has been reported to have profound neurotoxic and nephrotoxic effects mediated by its property of generation of ROS. Selenium nanoparticles due to their properties of high bioavailability and antioxidant activities have been reported in a study to reduce $CdCl_2$-induced neuro- and nephrotoxicity in rats by removing free radicals, chelating the metal ions, inhibiting apoptosis and altering the cell-protective pathways [16].

TiO_2NPs when administered systematically can accumulate in the brain and induce brain dysfunctioning. However, the underlying mechanisms of neurotoxicity are not known, but indications on generation of OS, apoptosis, inflammatory response, genotoxicity, impairment of cell components, altered distribution of trace elements, signaling pathways, dysregulated neurotransmitters and synaptic plasticity, autophagy and DNA methylation are being highlighted [17]. Metallic NPs including metals and metal oxides contributing to OS, apoptosis and inflammatory response have been reported to contribute to their neurotoxic effects. Some studies have revealed that the neurotoxic effect could be reduced by antioxidant treatment by decreasing the level of ROS, upregulating the activities of antioxidant enzymes, decreasing the proportion of apoptotic cells and suppressing the inflammatory response [18].

Polyamidoamine (PAMAM) dendrimers although find application in brain imaging, they have an adverse effect on neuronal differentiation and cause OS and DNA damage during development of neural cells [19].

5.3 Neurotoxicity in animals models

Most of the studies of nanotoxicology in the animal models are conducted in the nematode *Caenorhabditis elegans* (*C. elegans*). In a study it has been observed that when exposed to AgNPs, life span of *C. elegans* was reduced. Defects in the neumuscular functions, decline of swimming potential with age and increase of uncoordinated movements at dose of ≥10 µg/mL were observed, whereas nano zinc oxide (ZnO) and CeO_2 at concentration of 1–160 µg/mL did not affect longevity or cause neuromuscular defects [20]. Although inhaled TiO_2NPs are reported to be neurotoxic with negative effects on the BBB, male rats on exposure to inhaled TiO_2 nano-aerosol revealed increased BBB permeability in aging rats together with increased cytokines/chemokines in the brain, including interleukin-1β, interferon-y and fractalkine and decreased expression of synaptophysin, a neuronal activity marker, neuroinflammation and decreased expression of neuronal activity marker, which was further exacerbated in the brain of aged animals [21]. Zinc oxide (ZnONPs) in the environment have been known to affect the behavior of male Swiss mice [22]. Male adult Wistar rats subjected to acute and long-term administration of gold nanoparticles (AuNPs) of 10 nm and 30 nm, 70 µg kg^{-1}led to inhibition of energy metabolism caused by OS [23]. Multi-walled carbon nanotubes (MWCNTs) although bear promises for their wide-scale application, they have been reported to be toxic to bivalves. The Manila clam *Ruditapes philippinarum*, a bivalve species, when expososed to different MWCNT concentration revealed OS, higher lipid peroxidation and lower ratio between reduced and oxidized glutathione and neurotoxicity observed by inhibition of cholinesterase activities [24]. Neurotoxicity of ZnONPs was confirmed from a study involving prenatal exposure of ZnO on pregnant Sprague Dawley (SD) rats leading to the zinc accumulation in offspring brains, abnormal neuron ultrastructures, decreased proliferation, higher apoptotic death, imbalanced antioxidant status with impaired learning and memory behavior causing impaired learning and memory capabilities in adulthood, particularly in female rats.

Manufactured NP exposure to marine scallop *Chlamys farreri* caused increased OS, neurotoxicity and histopathological effects of TiO_2NPs at predicted environmentally relevant concentration (1 mg/L), significantly elevated superoxide dismutase (SOD), catalase (CAT) activities and malondialdehyde (MDA) contents. The increased neurotransmitter acetylcholineesterase (AChE) activities reflected neurotoxicity of TiO_2NPs with dysplastic and necrosis alteration in gill and digestive gland and integrated biomarker response revealing strong toxic effects [25]. Zebrafish embryos when exposed to TiO_2NPs for 96 hrs post fertilization, the hatching time of zebrafish was decreased, accompanied by an increase in malformation rate with no mortality, ROS generation and cell death of hypothalamus leading to the formation of Lewy bodies, loss of dopaminergic neurons linked to PD induces neurotoxicity in vivo and in vitro [26]. AgNPs showed testicular/sperm toxicity in males and ovarian and embryonic toxicity in females. Maternal injection of AgNPs delayed physical development and impaired cognitive behavior in offsprings and accumulation in visceral yolk sac when administered at early gestation in mice. Radiolabeled AgNPs revealed accumulation in placenta, breast milk and pre- and postnatal offsprings after injection during late gestation in rats [27]. SiO_2NPs with diameters 20, 50 and 80 nm in vivo zebrafish have been reported to accelerate the hatching time and alterations in the behavior of zebrafish embryos/larvae, but is no other toxic effects have been observed [28]. Chitosan nanoparticles (CSNPs) and the Tween 80 modified CSNPs (TmCSNPs) find importance as vehicles targeting brain. However, they have been reported of their developmental toxicity, leading to decreased hatching rate, increased mortality, malformation, neurobehavioral changes with decreased spontaneous movement in TmCSNP-treated embryos and hyperactive effect in CSNP-treated larvae, inhibition of axonal development of primary and secondary motor neurons, affected muscle structure, disrupted neurobehavior of zebrafish larvae and affected muscle and neuron development in zebrafish embryos [29]. Mediterranean sea urchin *Paracentrotus lividus* revealed accumulation of NPs in coelomocytes and toxicity in sperms [30]. Zebrafish (*Danio rerio*) revealed effects of treatment of different NPs including silver, gold and metal oxide NPs (TiO_2, Al_2O_3, CuO, NiO and ZnO) causing affected hatching success rate, malformation of organs, damage in gills and skin, abnormality in movement, immunotoxicity, genotoxicity or gene expression, neurotoxicity, endocrine system disruption, reproduction toxicity and finally mortality [31].

CuONPs add to the aquatic toxicity causing hepatotoxicity and neuronal differentiation on zebrafish inducing inflammatory response [32]. Toxicity due to oral intake of nano-silver in rats revealed neurotoxic effects resulting in behavioral and cerebral myelin changes [33]. AgNPs lead to release of silver ions causing neurotoxicity in rat cerebral astrocytes, AgNP-mediated increased caspase activities and cell apoptosis, silver ion-mediated cell membrane integrity and cell necrosis, generation of ROS together with activation of downstream signalling pathways including c-Jun N-terminal kinases (JNK) phosphorylation and apaptosis, increased multiple cytokines by astrocytes instigating their involvement in neuroinflammation [34].

N-vinyl-2-pyrrolidone-based NPs (Me-PSs) containing cobalt (Co^{2+}) or zinc (Zn^{2+}) when tested on marsh frog *Pelophylax ridibundus*, except Co^{2+} caused neurotoxicity and overall toxicity of Co- and Zn-containing Me-PSs and their parent compounds (Zn^{2+} and PS) with disrupted hormonal pathways and reduced organic xenobiotic detoxification [35].

Water-soluble cadmium telluride (CdTe) QDs capped with 3-mercaptopropionic acid in *C. elegans* exert neurotoxic effects, behavoril changes due to alterations to body bending, head thrashing, pharyngeal pumping and defecation intervals, impaired learning and memory behavior plasticity, based on chemotaxis or thermotaxis, inhibited transcription of transporters and receptors of glutamate, serotonin and dopamine with excessive ROS generation and nanotoxicity by OS damage [36].

Freshwater benthonic juvenile fish *Prochilodus lineatus* when exposed to TiO_2 lead to increased aggregation, decreased AChE activities in muscles or brain and increased protein oxidative damage in the brain and gills but not in the liver, which is indicative of neurotoxic effects [37].

Neurotoxicity induced by ZnONP in adult and old male C57BL/6J mice of different ages led to overproduction of pro-inflammatory cytokines in the serum, together with increased OS, impaired learning and memory abilities, hippocampal pathological changes in old mice, decreased hippocampal cAMP response element binding (CREB) protein, phosphorylated CREB, synapsin I and suppression of cAMP/CREB signaling [38].

AgNP suspension-induced neurotoxicity was demonstrated on a triple coculture BBB model of rat brain by induction of ultrastructural changes of the microvascular endothelial cells, pericytes and astrocytes with increased BBB permeability and decreased tight junction (TJ) protein ZO-1 with discontinuous TJs between microvascular endothelial cells, severe mitochondrial shrinkage, vacuolations, endoplasmic reticulum expansion and astrocytes. Gene ontology and Kyoto Encyclopedia of Genes and Genomes (KEGG) pathway analysis revealed 23 genes associated with metabolic processes, biosynthetic processes, response to stimuli, cell death, mitogen-activated protein kinases (MAPK) pathway, induced inflammation and apoptosis through modulation of the MAPK pathway or B-cell lymphoma-2 expression or mTOR activity in astrocytes [39] as downstream pathways of AgNP suspension induced neurotoxicity on BB model of rat brain. SiO_2NPs reveal protein aggregation in *C. elegans* and induction of amyloid in intestinal cell nuclei. Serotonergic neural cells revealed SiO_2NP-induced protein aggregation in axons of hermaphrodite specific neurons, where presynaptic accumulation of serotonin occurs [40]. Systemic toxicity has been reported of TiO_2 on earthworm *Pheretima hawayana* [41]. Nuerotoxicity of AgNPs in rats after intragastric administration revealed alterations in the BBB including neuronal shrinkage, cytoplasmic or foot swelling of astrocytes and extra-vascular lymphocytes in AgNP exposure groups with increased cadherin-1 and Claudin-1 and IL-4 together with inflammatory effects [42].

5.4 Neurotoxicity in humans

Platinum (Pt) drugs find applications in treatment of cancers with nephro- and neurotoxic effects on humans [43]. ZnONPs and TiO_2NPs can be transported into the CNS via the taste nerve pathway, which were observed to be significantly deposited in the nerves–brain translocation pathway and induce a certain adverse effect [44].

Manufactured NPs can cross the BBB leading to neurotoxicity of tissues, cytotoxicity in nondifferentiated PC-12 cells exposed to CB (10–100 µg/mL), SWNTs (10–100 µg/mL), C_{60} (100 µg/mL), CdSe (10 µg/mL), CB (500 µg/mL) and dye-doped SiNSs (10 µg/mL). Exposure to higher concentrations (100 µg/mL) of SWNTs, CB and C_{60} increased the formation of SBDP150/145, as well as cell membrane contraction and the formation of cytosolic vacuoles [45]. Particulate matter (PM) combined with meteorological factors affect human health causing cardiopulmonary disease, asthma and heart attacks. When PM was exposed to brain, it could stimulate pro-inflammatory cytokines, induce oxidative stress (OS), inflammation, dysfunctioning of cellular organelles, protein homeostasis, loss of neurons inflammation, ROS, microglial activation, thus affecting the CNS [46].

AgNP exposure of hippocampal tissue resulted in a significant decrease in cell survival in a dose-dependent manner, distributed in the extracellular matrix and were taken into the cytoplasm of the neurons. Larger AgNPs were taken into the neurons via phagocytosis. This study showed that pure AgNPs produced by laser ablation were toxic to the neural tissue [47].

Nickel oxide NPs exposed to neuronal (SH-SY5Y) cells affected the cell viability, induced morphological changes, induced dose-dependent DNA damage and apoptosis with oxidative damage [48].

Anatase TiO_2NPs on the brain of Wistar rats post oral intake downregulated AChE activities, interleukin 6 (IL-6) secretion, expression of GFAP, increased plasmatic IL-6 level, increased cerebral IL-6 level, local inflammation, elevated levels of immunoreactivity to GFAP in rat cerebral cortex, neuroinflammation and neurotoxic hazards to health [49]. Neurotoxicity of silica-coated MNPs containing rhodamine B isothiocyanate dye (MNPs@SiO2(RITC)) in HEK293 cells and SH-SY5Y cells has been reported [50]. TiO_2NPs affect the CNS and brain, the murine microglial (BV-2) cell growth and viability and neurotoxicity-inhibited cell growth, leading to mitochondrial dysfunctions and OS [51]. AgNPs in primary neuronal cultures lead to disrupted calcium homeostasis by glutamatergic NMDA receptors (that controls normal synaptic plasticity and memory function) and imbalance between extracellular and intracellular zinc levels [52], cause neurotoxic effects in human embryonic stem cell (hESC)-derived neural stem/progenitor cells (NPCs) revealing apoptosis and increased OS and dysfunctional neurogenesis [53].

5.4.1 Down stream effects of NP-mediated neurotoxicity

Although not fully recognized, the neurotoxic effects of AgNPs are thought to occur through induction of OS and apoptosis possible attributed changes in the mRNA expression of Bcl-2 and Bax genes in the rat hippocampus in a study of male Wistar rats. AgNP in a dose-dependent manner could induce apoptosis, reduced mRNA level of Bcl-2 in the rat hippocampal cells and increased mRNA level of Bax [54].

Aluminum nanoparticles (AlNPs) show the neurotoxicity on rat brain and isolated mitochondria by inducing ROS generation, lipid peroxidation, protein oxidation, glutathione depletion, mitochondrial dysfunction, gait abnormalities in a dose-dependent manner, OS and decreased mitochondrial membrane potential leading to toxicity effects [55].

Maternal exposure to carbon black nanoparticles (CBNPs) from particulate air pollution induce long-lasting diffuse perivascular abnormalities in pregnant ICR (Institute of Cancer Research) mice when exposed intranasally to CBNPs on gestational days 5 and 9 showing protein denaturation with an increased β-sheet content and decreased α-helix content in proteins, thus accumulating in the perivascular space leading to denaturation of perivascular macrophages and astrocyte activation

[56]. Maternal CBNP exposure induced long-term activation of astrocytes resulting in reactive astrogliosis in the brains of young mice with potentially increased risk of the onset of age-related neurodegenerative diseases causing phenotypic changes in the CNS of the offspring [57].

SiO_2NPs enter the circulatory system and negatively effect BBB in SD rats by induction of tight junction, loss of cytoskeleton arrangement, increased inflammatory response and release of vascular endothelial growth factor (VEGF) of brain microvessel endothelial cells (BMECs), further activating astrocytes to amplify the generation of VEGF and increase the aquaporin-4 expression, which is a water channel integral protein that conducts water through the cell membrane, thereby leading to BBB disruption [58].

Metallic NPs cause cell death and is regulated by certain genes leading to apoptosis, autophagy, necroptosis and pyroptosis involved in brain development, neurodegenerative disorders, psychiatric disorders and brain injury [59].Chitosan-coated levodopa liposomes, an N-deacetylated derivative of chitin, in striatum of rat model revealed decreased abnormal involuntary movement (AIM) [60].

Oral administration of ZnONPs in rats revealed neurotoxicity by elevated serum inflammatory markers including tumor necrosis factor alpha (TNF-α), interleukin 1 (IL-1β), interleukin-6 (IL-6), C-reactive protein (CRP), CAT, glutathione peroxidase (GPx), glutathione reductase (GR) and glutathione (GSH) in rat brains [61] indicative of stress response.

Protein misfolding to amyloid aggregates is the hallmark for neurodegenerative disease. Designed nanozyme and Ceria/Polyoxometalates hybrid (CeONP@POMs) with both proteolytic and SOD activities could efficiently degrade Aβ aggregates and reduce intracellular ROS, promote PC12 cell proliferation and can cross BBB, inhibit Aβ-induced biocompatible BV2 microglial cell activation [62].

PAMAM dendrimers have been reported to be toxic to the CNS causing OS and autophagy in neuronal cells [63].

ZnONPs are known to induce OS, initiate inflammatory pathway, cytotoxicity and neurotoxicity. ZnONP treated to SH-SY5Y human neuroblastoma cell line revealed reduced neuronal viability and altered morphology, increased apoptotic proteins by annexin V and caspase-3/7 activities, necrosis manifested by LOX-mediated ROS production elevation, swelling or loss of cell organelles and rupture of the cytosolic or nuclear membrane and chacteristics of neuronal injury manifested by elevated PI3 kinase and p-Akt/Akt activities induced by ZnONP [64]. Subchronic exposure of male and female wistar rats to diesel exhaust nanoparticles led to overexpression of pro-inflammatory cytokines, amyloid beta 42 (Aβ 42), ROS, H_2O_2, nitrate (NO_3^-), nitrite (NO_2^-) and apurinic/apyrimidinic sites in rat brain leading to increased inflammation, DNA damage and OS [65]. AgNP exposure can transmit them in the hippocampus of adult male Winstar rat revealing increase in the number of dark neurons (DNs), apoptotic cells and dentate gyrus of hippocampus [66]. MWCNTs have been reported to induce apoptosis and oxidative damage in nerve cells or autophagy. When rats are exposed to them, they show neurotoxic effects on hippocampal synaptic plasticity and spatial cognition. Rats subjected to intraperitoneal injection or i.p. when tested by Morris water maze (MWM) test, and the long-term potentiation (LTP), revealed cognitive deficits, histopathological alteration and increased autophagy by alteration in microtubule-associated protein 1A/1B-light chain 3 (LC3). In an autophagy response, a cytosolic form of LC3 (LC3-I) in conjugation with phosphatidylethanolamine forms LC3-phosphatidylethanolamine conjugate (LC3-II) that is recruited to autophago-somal membranes. AgNP treatment has been reported to increase LC3 II/LC3 I ratio, thus promoting autophagy and the expression of Beclin-1 protein indicative of autophagy and apoptosis [67]. Al$_2$O$_3$NP-generated neurotoxicity can lead to neurodegenerative disorders. Exposure of human neuroblastoma SH-SY5Y and mouse hippocampal HT22 cells in vitro and ICR female mice in vivo increased brain aluminium and ROS production, disturbing brain energy homeostasis, leading to hippocampus-dependent memory impairment, induction of AD, neuropathology by enhancing the amyloidogenic pathway of Aβ production, aggregation and implied the progression of neurodegeneration in the cortex and hippocampus [68]. TiO$_2$NPs exposure leads to their absorbance and accumulation in the brain transmitted by the BBB or through the nose-to-brain pathway, leading to dysfunctioning of CNS. They can also affect the brain development of embryo by crossing the placental barrier. It can lead to altered morphology and function of neuronal or glial cells leading to cell necrosis, impaired recognition ability, spatial memory and learning ability, and accumulation in the brain could lead to neurodegeneration [69].

After 24-hr exposure of SHSY-5Y human neuroblastoma cells to ferric oxide nanoparticles activated tyrosine kinase c-Abl with increased levels, OS and neuronal α-synuclein expression, a protein marker of neuronal injury, inhibition of cell-proliferation, significant reductions in the number of active mitochondria and an increase in ROS in neuronal cells were reported [70].

Dopamine (DA) quinone-induced dopaminergic neurotoxicity is involved in neurodegeneration [71]. Neonatal SD rats exposed to AgNPs by intranasal instillation for 14 weeks causes cerebellar ataxia-like symptom in rats, dysfunctioning of motor coordination and impairment of locomotor activity, destruction of cerebellum granular layer with concomitant activation of glial cells and decreasing calcium channel protein (CACNA1A) levels in cerebellum [72].

Inhaled ZnONPs can reach the brain through the olfactory neuronal pathway and affect brain zinc homeostasis in rat C6 glial cells, release ROS and cause nuclear condensation and apoptosis [73].

Embryonic neural stem cells (NSCs) from human and rat fetuses exposed to AgNPs revealed developmental neurotoxicity [74]. TiO$_2$NPs on human neuroblastoma cell line (SH-SY5Y) revealed neurotoxic effects [75] affecting neurite outgrowth of hippocampal neurons [76]. On entering the brain, they can affect mitochondrial membrane potential, cause membrane damage and cell morphology in human cerebral SH-SY5Y and D384 cell lines [77] and affect the CNS, causing OS, mitochondrial damage and apoptosis, reduced glutathione content and major glutathione metabolizing enzymes [78].

NP neurotoxicity of ZnO and SiO_2 on oral, dermal and intravenous administrations exerted toxicological effects on the brain [79]. TiO_2NPs can affect embryonic development, cause retinal neurogenesis, agglomeration and sedimentation on zebrafish embryos post fertilization with delayed embryonic development or retinal neurotoxicity [80]. *C. elegans* reveal phenotype of neurotoxicity of CdTe quantum dots (QDs) on RMEs motor neurons, affect foraging behavior and cause deficits in development of RMEs motor neuron OS [81]. Prolonged exposure of rats to AgNPs affect synapse ultrastructure and specific proteins showing blurred synapse structure and strongly enhanced density of synaptic vesicles clustering in the center of the presynaptic part, disturbed synaptic membrane leading to liberation of synaptic vesicles into neuropil, which testifies for strong synaptic degeneration, decreased levels of the presynaptic proteins synapsin I and synaptophysin and PSD-95 protein indicator of postsynaptic densities and severe synaptic degeneration, mainly in the hippocampal region of brain [82]. Aβ25-35 peptide causes rapid aggregation and high neurotoxicity in AD [83].

TiO_2NPs have been reported to cross the BBB, accumulate in the brain causing glutamate-mediated neurotoxicity with elevated glutamate release and phosphate-activated glutaminase activity and reduction in glutamine and glutamine synthetase in the hippocampus [84]. Targeting Aβ-induced complex neurotoxicity is finding importance in the therapeutic and preventive treatments of AD. AuNPs@POMD-pep (AuNPs: gold nanoparticles, POMD: polyoxometalate with Wells–Dawson structure, pep: peptide), a novel multifunctional Aβ inhibitor, shows synergistic effects in inhibiting Aβ aggregation, dissociating Aβ fibrils and decreasing Aβ-mediated peroxidase activity and Aβ-induced cytotoxicity when crossing the BBB [85]. Nanoparticle-mediated neurotoxicity is recorded in zebrafish brains [86]. PAMAM dendrimers could induce cytotoxicity, neurotoxicity and autophagy [87]. SiO_2NPs have been reported to be potentially neurotoxic leading to impaired morphology of human SK-N-SH and mouse neuro2a cells, with increased number of round cells, diminished dendrite-like processes, decreased cell density and cell viability, induced cellular apoptosis and elevated intracellular ROS in a dose-dependent manner in both cell lines. They are found with increased deposit of intracellular Aβ(1–42) and enhanced phosphorylation of tau at Ser262 and Ser396, markers for AD, upregulation of amyloid precursor protein, downregulation of Aβ-degrading enzyme neprilysin in SiNP-treated cells and decreased phosphorylation of glycogen syntheses kinase-3β at Ser9 (inactive form) in SiNP-treated SK-N-SH cells [88].

Superparamagnetic iron-oxide nanoparticles (SPIONs) coated with cross-linked aminated dextran ($CLIO-NH_2$) on adult fish brain revealed decreased AChE activity and AChE expression in zebrafish brain, with high level of ferric ion in the brain and induction of caspase-8, caspase-9 and Jun genes leading to apoptosis by inhibiting AChE activity [89].

AgNP (20 nm) neurotoxicity on human cerebral (SH-SY5Y and D384) cell lines leads to mitochondrial metabolism changes and cell membrane damage [90].

Polyvinylpyrrolidone-coated silver nanoparticles (PVP-AgNPs) on females of fathead minnow (Pimephales promelas) and onlyAgNPs affected pathways involved in Na^+, K^+ and H^+ homeostasis and OS, but different neurotoxicity pathways [91].

TiO$_2$NPs toxicity affects the hippocampal regions and the learning and memory of offspring of pregnant Wistar rats by reduced cell proliferation [92]. SiO$_2$NPs have been reported to affect the neural behaviors of zebrafish impacting the cognitive neurobehavioral patterns like colour preference and cause potential PD-like behavior. Nanosilica has been reported to act on the retina and dopaminergic (DA) neurons to change colour preference and to cause potential PD-like behavior [93].

AlNPs via inhalation in the environment and the workplace affected the rats by accumulation in olfactory bulb and the brain, overexpression of MAPK activities including Ptprc, P2rx7, Map2k4, Trib3, Trib1 and Fgd4 and activation of ERK1 and p38 MAPK protein expressions in the brain [94].

AgNPs, CuNPs or AuNPs induced release of pro-inflammatory mediators including IL-1β, TNFα and PGE2 affecting the BBB in confluent porcine brain microvessel endothelial cells (pBMECs) [95]. AgNPs contribute to neuronal cell death and decreased mitochondrial potential in primary cultures of rat cerebellar granule cells (CGCs) [96].

CeO$_2$NPs or nanoceria cause neurotoxic effects on serotonin (5-HT), an important neurotransmitter that plays a critical role in various physiological processes and forms a surface adsorbed 5-HT-nanoceria complex [97].

Lipid NPs have been reported to accumulate in the brain parenchyma leading to activation of brain microglia with morphological characteristic of changes of microglial activation in mice, abnormal Ca^{2+} waves in microglia, activation of caspase-1 and IL-1β and neurovascular damage [98].

Chronic administration of engineered NPs from metals, for example, Cu, Ag or Al cause BBB disruption and induce brain pathology in adult rats, BBB breakdown, brain edema formation and neuronal injuries, glial fibrillary acidic protein upregulation and myelin vesiculation in young animals [99].

Lead sulfide NPs affect the neuronal ultrastructure and pathology in hippocampus and show neurotoxicity by means of calcium homeostasis disorder and abnormal calcium transportation [100]. ZnONPs in rat brain lead to alterations in emotional behavior and altered trace elements homeostasis in rat brain [101].

TiO$_2$NPs when administered nasally caused brain injuries, translocated and accumulated in brain that led to OS, overproliferation of all glial cells, tissue necrosis and hippocampal cell apoptosis [102]. AgNPs could cause neurotoxicity in rat CGCs inducing OS, generation of ROS, depletion of antioxidant GSH, increase of intracellular calcium, destruction of the cerebellum granular layer in rats and activation of caspase-3 [103]. A hESC-derived three-dimensional model enabled the study of neurotoxicity [104]. Metal ions play important roles in amyloid aggregation and neurotoxicity, and its chelation finds application in AD treatment. Modified mesoporous silica nanoparticles (MSNs) are biocompatibile, with easy cellular uptake and efficient intracellular release of metal chelators [105]. Aβ fibrils in AD and AuNPs are known to affect amyloids using Aβ as a model system. Bare AuNPs could inhibit Aβ fibrillization to form fragmented fibrils and spherical oligomers

[106]. Nano-CuO has been reported to adversely affect hippocampal CA1 neuron, cause spatial cognition and electrophysiological alterations in rats, histological and biochemical changes in rats' hippocampus, weak learning responses, lowered excitatory postsynaptic potentials slopes increased ROS and MDA in hippocampus while reduced SOD and glutathione peroxidase (GSH-Px) activities and enhanced 4-hydroxynonenal (HNE) and caspase-3 leading to apoptosis in the hippocampus, indicative of neuronal damage, defective oxidation–antioxidation homeostasis and altered animal behaviors [107].

Toxicity of the water-soluble compound, polyhydroxyfullerene (fullerenol), on rat hippocampal neurons induces apoptosis through caspase-dependent pathways and damages cells in a concentration dependant manner due to the reduction–oxidation pathway [108]. Al_2O_3NP-exposed *C. elegans* showed decrease in locomotion behavior with a probable involvement of neurotransmitters [109]. Chronic toxicity on locomotion behavior has been reported from another study in *C. elegans*, with severe stress and OS response, increased ROS production and suppression of ROS defense mechanisms, decreased SOD activity and decreased expression of genes encoding Mn-SODs (sod-2 and sod-3) [110]. CuNPs have been reported to induce the release of proinflammatory mediators that affect the BBB in rat brain microvessel endothelial cells (rBMECs) and their proliferation and/or induce BBB and neurotoxicity at high concentrations [111].

SiO_2NPs caused OS, inflammatory response, increased levels of neurochemicals in the brain, depositions in the striatum, increased levels of lactate dehydrogenase, triggered OS, disturbed cell cycle, induced apoptosis, activated p53-mediated signaling pathway, depleted dopamine in the striatum, downregulated tyrosine hydroxylase protein and caused neurodegenerative diseases [112].

Neurotoxicity of anatase TiO_2NPs on PC12 cells that is an in vitro model of dopaminergic neurons for neurodegenerative diseases was manifested by decreased cell viability, increased lactate dehydrogenase levels, OS, apoptosis, affected cell cycle, upregulated JNK- and p53-mediated signaling pathway [113]. ZnONPs in mouse NSCs indicated to cause toxic effects, revealed by apoptosis [114]. NPs have been reported to inhibit adherence of astrocytes in culture plates and cause cytotoxicity or termination of growth, affect membrane integrity and mitochondrial function [115].

5.5 Safe therapeutic use of NPs in targeting disorders of the nervous system

Neuroprotective lactoferrin-modified nonviral vector has found application in safe brain gene delivery of human glial cell line-derived neurotrophic factor gene [116]. Safe and specific delivery of hyaluronic acid (HA) has been reported by HA conjugated liposomes (HALNPs) in the brain including primary astrocytes, microglia and human glioblastoma cells [117]. Bone marrow stromal cell (BMSC) transplantation and its tracking by noninvasive methods using MRI- and SPION-based labeling agents have been studied as safe methods of BMSC transplanatation in patients and find applications in treatment of stroke [118]. Polymeric and lipid-based nanocarriers with the property of crossing the BBB have been reported to enter the brain capillaries and is being researched for potential applications in safe delivery of drugs and therapeutic molecules in patients with both neurodegenerative and non-neurodegenerative diseases of the CNS [119]. Neurodegenerative diseases are characterized by deposition of misfolded proteins, including PrP(Sc) prions or Aβ in AD, with features of neuronal death and oxidative damage. Nano-PSO, a nanodroplet formulation of pomegranate seed oil (PSO) composed of natural antioxidants, polyunsaturated fatty acid, punicic acid, with neuroprotective effects of reduced lipid oxidation and neuronal loss has been reported to be safe formulations in patients with neurodegenerative disorders [120]. Estrogen (E2), a pleiotropic steroid hormone with anti-inflammatory, anti-apoptotic and neurotrophic properties, has been reported to be safe and effective in treatment of patients with spinal cord injuries (SCI) [121]. Nonviral vectors have been recently researched as safe and effective vehicles for gene and cell therapy in the treatment of brain cancer. Poly(beta-amino ester) polymer has been researched to transfect human adipose-derived mesenchymal stem cells (hAMSCs) and has revealed higher efficacy than Lipofectamine™ 2000 [122]. Japanese encephalitis (JE) is an infectious disease caused by the JE virus (JEV). Biodegradable poly(gamma-glutamic acid) nanoparticles (gamma-PGA-NPs) are used as an adjuvant to JE vaccine that enabled immunize mice to survive a lethal JEV infection, and it is predicted that it may have promising applications in humans as well [123]. Novel neuroprotectant (ZL006)-loaded dual targeted nanocarrier based on liposome (T7&SHp-P-LPs/ZL006) conjugated with T7 peptide (T7) and stroke homing peptide (SHp) has been recently resigned by scientist, which has been successful in crossing the BBB and finds applications in the treatment of stroke [124].

Scientists are testing the efficacy and safety of intravenous administration of autologous mesenchymal stem cells (MSCs) magnetically labeled with the SPION ferumoxides (Feridex) in patients with multiple sclerosis (MS) and amyotrophic lateral sclerosis in open-safety clinical trial with their clinical feasibility and safety [125]. Safe and effective drug delivery nanocarriers to cross BBB in the treatment of CNS by different types of nanomaterials in the brain are being tseted for the delivery of nucleic acid [126]. Nano-amomi paste (nmAP) has been reported to be effective in treatment of children's anorexia [127].

Hydrogels or hydrogels seeded with MSCs are being tested to be implanted in rats with spinal cord injury (SCI) [128]. Iron oxide MNPs, 10 nm in diameter, have been tested to be safe in crossing BBB in normal rats and those with SCI [129].

Solid lipid NPs (SLN, 150 nm) and poly-lactic-co-glycolic NPs, 115 nm, are being tested for safe and effective delivery of therapeutics to human brain cells and offer promises in brain disease therapies [130]. PEGylated SLNs surface modified with axo-glial-glycoprotein, anti-Contactin-2 or anti-Neurofascin antigens have been tested to be efficient transporters of therapeutics in treatment of MS [131]. Methylprednisolone sodium succinate loaded polycaprolactone based NPs have been reported to be effective and safe in treatment of acute SCI in rat model [132]. Synthetic nerve guidance conduits are being applied in treatment of large peripheral nerve injuries. MNPs immobilized growth factors (GFs) including nerve growth factor (NGF) and vascular endothelial growth factor (VEGF), which have been reported to be safe and effective in inducing nerve regeneration and recovery of motor function in nerve injuries [133]. Neuroprotective and anti-inflammatory properties of microRNA-124 (miR-124)-loaded NPs (miR-124 NPs) have found application in adult neurogenesis in mice models in treating post-stoke inflammation [134].

5.6 Green sysnthesis of NPs and application in nervous system disorders

Manganese (Mn) containing metal fumes at workplaces can cause damage in the nervous system including a PD-like syndrome. Green tea has been reported to have the potential to reduce the Mn-induced neurotoxic effects in male Winstar rats [135]. Green synthesis of AgNPs from leaf extract of *Anisomeles indica* by reduction of Ag^+ ions from silver nitrate solution has been reported safe and effective to *Anopheles subpictus*, *Aedes albopictus* and *Culex. tritaeniorhynchus* [136].

5.7 Discussion

Diverse range of metalic and nonmetalic NPs have been reported to cause neurotoxicity in different cell lines, animal models and humans. Recent research across the globe has enabled our understanding of the downstream pathways and molecules that play role in the biological functions that are affected due to NP-mediated neurotoxicity. Various effects of NPs have been discussed in this chapter including neuroinflammation, stress and altered molecular expression and dysfunction of cells and molecules and neurodegeneration.

References

[1]   Wong HL, Wu XY, Bendayan R. Nanotechnological advances for the delivery of CNS therapeutics. Adv Drug Deliv Rev. 2012 May 15;64(7):686–700.
[2]   Pandey A, Malek V, Prabhakar V, Kulkarni YA, Gaikwad AB. Nanoparticles: A neurotoxicological perspective. CNS Neurol Disord Drug Targets. 2015; 14(10):1317–1327.

[3]   Mushtaq G, Khan JA, Joseph E, Kamal MA. Nanoparticles, neurotoxicity and neurodegenerative diseases. Curr Drug Metab. 2015; 16(8):676–684.

[4]   Feng X, Chen A, Zhang Y, Wang J, Shao L, Wei L. Central nervous system toxicity of metallic nanoparticles. Int J Nanomedicine. 2015 Jul 3;10:4321–4340.

[5]   Rueda F, Cruz LJ. Targeting the brain with nanomedicine. Curr Pharm Des. 2017;23(13):1879–1896.

[6]   Valdiglesias V, Fernández-Bertólez N, Kiliç G, Costa C, Costa S, Fraga S, Bessa MJ, Pásaro E, Teixeira JP, Laffon B. Are iron oxide nanoparticles safe? current knowledge and future perspectives. J Trace Elem Med Biol. 2016 Dec.38:53–63.

[7]   Skalska J, Strużyńska L. Toxic effects of silver nanoparticles in mammals–does a risk of neurotoxicity exist? Folia Neuropathol. 2015;53(4):281–300.

[8]   Pathak K, Akhtar N. Nose to brain delivery of nanoformulations for neurotherapeutics in Parkinson's disease: Defining the preclinical, clinical and toxicity issues. 2016 Jun 7;13(8): 1205–1221. doi: 10.2174/1567201813666160607123409.

[9]   Sagar V, Atluri VS, Pilakka-Kanthikeel S, Nair M. Magnetic nanotherapeutics for dysregulated synaptic plasticity during neuroAIDS and drug abuse. Mol Brain. 2016 May 23;9(1):57.

[10]   Wu T, Zhang T, Chen Y, Tang M. Research advances on potential neurotoxicity of quantum dots. J Appl Toxicol. 2016 Mar;36(3):345–351.

[11]   Bulcke F, Dringen R, Scheiber IF. Neurotoxicity of copper. Adv Neurobiol. 2017;18:313–343.

[12]   Caudle WM. Occupational metal exposure and parkinsonism. Adv Neurobiol. 2017.18:143–158.

[13]   Naserzadeh P, Ghanbary F, Ashtari P, Seydi E, Ashtari K, Akbari M. Biocompatibility assessment of titanium dioxide nanoparticles in mice fetoplacental unit. J Biomed Mater Res A. 2018 Feb;106(2):580–589. doi: 10.1002/jbm.a.36221. Epub 2017 Nov 21.

[14]   Gliga AR, Edoff K, Caputo F, Källman T, Blom H, Karlsson HL, Ghibelli L, Traversa E, Ceccatelli S, Fadeel B. Cerium oxide nanoparticles inhibit differentiation of neural stem cells. Sci Rep. 2017 Aug 24;7(1):9284.

[15]   Máté Z, Horváth E, Papp A, Kovács K, Tombácz E, Nesztor D, Szabó T, Szabó A, Paulik E. Neurotoxic effects of subchronic intratracheal n nanoparticle exposure alone and in combination with other welding fume metals in rats. Inhal Toxicol. 2017 Apr;29(5):227–238.

[16]   Sadek KM, Lebda MA, Abouzed TK, Nasr SM, Shoukry M. Neuro- and nephrotoxicity of subchronic cadmium chloride exposure and the potential chemoprotective effects of selenium nanoparticles. Metab Brain Dis. 2017 Oct; 32(5):1659–1673. doi: 10.1007/s11011-017-0053-x. Epub 2017 Jun 28.

[17]   Song B, Zhang Y, Liu J, Feng X, Zhou T, Shao L.Beilstein Unraveling the neurotoxicity of titanium dioxide nanoparticles: Focusing on molecular mechanisms. J Nanotechnol. 2016 Apr 29;7:645–654.

[18]   Song B, Zhang Y, Liu J, Feng X, Zhou T, Shao l. Is neurotoxicity of metallic nanoparticles the cascades of oxidative stress?. Nanoscale. Res Lett. 2016 Dec;11(1):291.

[19]   Zeng Y, Kurokawa Y, Win-Shwe TT, Zeng Q, Hirano S, Zhang Z, Sone H. Effects of PAMAM dendrimers with various surface functional groups and multiple generations on cytotoxicity and neuronal differentiation using human neural progenitor cells. J Toxicol Sci. 2016;41(3):351–370.

[20]   Piechulek A, von Mikecz A. Life span-resolved nanotoxicology enables identification of age-associated neuromuscular vulnerabilities in the nematode Caenorhabditis elegans. Environ Pollut. 2018 Feb;233:1095–1103.

[21]   Disdier C, Chalansonnet M, Gagnaire F, Gaté L, Cosnier F, Devoy J, Saba W, Lund AK, Brun E, Mabondzo A. Brain inflammation, blood brain barrier dysfunction and neuronal synaptophysin decrease after inhalation exposure to titanium dioxide nano-aerosol in aging. Rats.Sci Rep. 2017 Sep 22;7(1):12196.

[22] de Souza JM, Mendes BO, Guimarães ATB, Rodrigues ASL, Chagas TQ, Rocha TL, Malafaia G. Zinc oxide nanoparticles in predicted environmentally relevant concentrations leading to behavioral impairments in male swiss mice. Sci Total Environ. 2017 Sep 19;613–614:653–662.

[23] Ferreira GK, Cardoso E, Vuolo FS, Galant LS, Michels M, Gonçalves CL, Rezin GT, Dal-Pizzol F, Benavides R, Alonso-Núñez G, Andrade VM, Streck EL, da Silva Paula MM. Effect of acute and long-term administration of gold nanoparticles on biochemical parameters in rat brain. Mater Sci Eng C Mater Biol Appl. 2017 Oct 1;79:748–755.

[24] De Marchi L, Neto V, Pretti C, Figueira E, Chiellini F, Soares AMVM, Freitas R. The impacts of emergent pollutants on ruditapes philippinarum: Biochemical responses to carbon nanoparticles exposure. Aquat Toxicol. 2017 Jun;187:38–47.

[25] Xia B, Zhu L, Han Q, Sun X, Chen B, Qu K, Effects of TiO_2 nanoparticles at predicted environmental relevant concentration on the marine scallop chlamys farreri: An integrated biomarker approach. Environ Toxicol Pharmacol. 2017 Mar;50:128–135.

[26] Hu Q, Guo F, Zhao F, Fu Z. Effects of titanium dioxide nanoparticles exposure on parkinsonism in zebrafish larvae and PC12. Chemosphere. 2017 Apr;173:373–379.

[27] Ema M, Okuda H, Gamo M, Honda K. A review of reproductive and developmental toxicity of silver nanoparticles in laboratory animals. Reprod Toxicol. 2017 Jan;67:149–164.

[28] Pham DH, De Roo B, Nguyen XB, Vervaele M, Kecskés A Ny A, Copmans D, Vriens H, Locquet JP, Hoet P, de Witte PA. Use of zebrafish larvae as a multi-endpoint platform to characterize the toxicity profile of silica nanoparticles. Sci Rep. 2016 Nov 22;6:37145.

[29] Yuan Z, Li Y, Hu Y, You J, Higashisaka K, Nagano K, Tsutsumi Y, Gao J. Chitosan nanoparticles and their Tween 80 modified counterparts disrupt the developmental profile of zebrafish embryos. Int J Pharm. 2016 Dec 30; 515 (1–2): 644–656.

[30] Gambardella C, Ferrando S, Gatti AM, Cataldi E, Ramoino P, Aluigi MG, Faimali M, Diaspro A, Falugi C. Review: Morphofunctional and biochemical markers of stress in sea urchin life stages exposed to engineered nanoparticles. Environ Toxicol. 2016 Nov;31(11):1552–1562.

[31] Chakraborty C, Sharma AR, Sharma G, Lee SS. Zebrafish: A complete animal model to enumerate the nanoparticle toxicity. J Nanobiotechnology. 2016 Aug 20;14(1):65.

[32] Sun Y, Zhang G, He Z, Wang Y, Cui J, Li Y. Effects of copper oxide nanoparticles on developing zebrafish embryos and larvae. Int J Nanomedicine. 2016 Mar 7;11:905–918.

[33] Dąbrowska-Bouta B, Zięba M, Orzelska-Górka J, Skalska J, Sulkowski G, Frontczak-Baniewicz M, Talarek S, Listos J, Strużyńska L. Influence of a low dose of silver nanoparticles on cerebral myelin and behavior of adult rats. Toxicology. 2016 Jul;1363–364:29–36.

[34] Sun C, Yin N, Wen R, Liu W, Jia Y, Hu L, Zhou Q, Jiang G Silver nanoparticles induced neurotoxicity through oxidative stress in rat cerebral astrocytes is distinct from the effects of silver ions. Neurotoxicology. 2016 Jan;52:210–221.

[35] Falfushynska H, Gnatyshyna L, Fedoruk O, Sokolova IM, Stoliar O. Endocrine activities and cellular stress responses in the marsh frog pelophylax ridibundus exposed to cobalt, zinc and their organic nanocomplexes. Aquat Toxicol. 2016 Jan;170:62–71.

[36] Wu T, He K, Zhan Q, Ang S, Ying J, Zhang S, Zhang T, Xue Y, Tang M. MPA-capped CdTe quantum dots exposure causes neurotoxic effects in nematode Caenorhabditis elegans by affecting the transporters and receptors of glutamate, serotonin and dopamine at the genetic level, or by increasing ROS, or both. Nanoscale. 2015 Dec 28;7(48):20460–20473.

[37] Miranda RR, Damaso da Silveira AL, de Jesus IP, Grötzner SR, Voigt CL, Campos SX, Garcia JR, Randi MA, Ribeiro CA, Filipak Neto F. Effects of realistic concentrations of TiO_2 and ZnO nanoparticles in Prochilodus lineatus juvenile fish. Environ Sci Pollut Res Int. 2016 Mar;23(6):5179–5188.

[38]   Tian L, Lin B, Wu L, Li K, Liu H, Yan J, Liu X, Xi Z. Neurotoxicity induced by zinc oxide nanoparticles: age-related differences and interaction. Sci Rep. 2015 Nov 3;5:16117.

[39]   Xu L, Dan M, Shao A, Cheng X, Zhang C, Yokel RA, Takemura T, Hanagata N, Niwa M, Watanabe D. Silver nanoparticles induce tight junction disruption and astrocyte neurotoxicity in a rat blood-brain barrier primary triple coculture model. Int J Nanomedicine. 2015 Sep 29;10:6105–6118.

[40]   Scharf A, Gührs KH, von Mikecz A. Anti-amyloid compounds protect from silica nanoparticle-induced neurotoxicity in the nematode C. elegans. Nanotoxicology. 2016;10(4):426–435.

[41]   Khalil AM. Neurotoxicity and biochemical responses in the earthworm Pheretima hawayana exposed to TiO$_2$NPs. Ecotoxicol Environ Saf. 2015 Dec;122:455–461.

[42]   Xu L, Shao A, Zhao Y, Wang Z, Zhang C, Sun Y, Deng J, Chou LL. Neurotoxicity of silver nanoparticles in rat brain after intragastric exposure. J Nanosci Nanotechnol. 2015 Jun;15(6):4215–4223.

[43]   Sun Y, Shi TY, Zhou LY, Zhou YY, Sun BW, Liu XP. Folate-decorated and NIR-activated nanoparticles based on platinum(IV) prodrugs for targeted therapy of ovarian cancer. J Microencapsul. 2017 Nov;34(7):675–686.

[44]   Aijie C, Huimin L, Jia L, Lingling O, Limin W, Junrong W, Xuan L, Xue H, Longquan S. Central neurotoxicity induced by the instillation of ZnO and TiO$_2$ nanoparticles through the taste nerve pathway.Nanomedicine (Lond). 2017 Oct;12(20):2453–2470.

[45]   Larner SF, Wang J, Goodman J, O'Donoghue Altman MB, Xin M, Wang KKW. In Vitro Neurotoxicity Resulting from Exposure of Cultured Neural Cells to Several Types of Nanoparticles. Journal of Cell Death. 2017;10:1179670717694523.

[46]   Wang Y, Xiong L, Tang M. Toxicity of inhaled particulate matter on the central nervous system: neuroinflammation, neuropsychological effects and neurodegenerative disease. J Appl Toxicol. 2017 Jun;37(6):644–667.

[47]   Kursungoz C, Taş ST, Sargon MF, Sara Y, Ortaç B. Toxicity of internalized laser generated pure silver nanoparticles to the isolated rat hippocampus cells. Toxicol Ind Health. 2017 Jul;33(7):555–563.

[48]   Abudayyak M, Guzel E, Özhan G. Nickel oxide nanoparticles are highly toxic to SH-SY5Y neuronal cells. Neurochem Int. 2017 Sep;108:7–14.

[49]   Grissa I, Guezguez S, Ezzi L, Chakroun S, Sallem A, Kerkeni E, Elghoul J, El Mir L, Mehdi M, Cheikh HB, Haouas Z. The effect of titanium dioxide nanoparticles on neuroinflammation response in rat brain. Environ Sci Pollut Res Int. 2016 Oct;23(20):20205–20213.

[50]   Phukan G, Shin TH, Shim JS, Paik MJ, Lee JK, Choi S, Kim YM, Kang SH, Kim HS, Kang Y, Lee SH, Mouradian MM, Lee G. Silica-coated magnetic nanoparticles impair proteasome activity and increase the formation of cytoplasmic inclusion bodies in vitro. Sci Rep. 2016 Jul 5;6:29095.

[51]   Rihane N, Nury T, M'rad I, El Mir L, Sakly M, Amara S, Lizard G. Microglial cells (BV-2) internalize titanium dioxide (TiO$_2$) nanoparticles: Toxicity and cellular responses. Environ Sci Pollut Res Int. 2016 May;23(10):9690–9699.

[52]   Ziemińska E, Strużyńska L. Zinc modulates nanosilver-induced toxicity in primary neuronal cultures. Neurotox Res. 2016 Feb;29(2):325–343.

[53]   Oh JH, Son MY, Choi MS, Kim S, Choi AY, Lee HA, Kim KS, Kim J, Song CW, Yoon S. Integrative analysis of genes and miRNA alterations in human embryonic stem cells-derived neural cells after exposure to silver nanoparticles. Toxicol Appl Pharmacol. 2016 May 15; 299:8–23.

[54]   Ghooshchian M, Khodarahmi P, Tafvizi F. Apoptosis-mediated neurotoxicity and altered gene expression induced by silver nanoparticles. Toxicol Ind Health. 2017 Oct;33(10):757–764.

[55]   Mirshafa A, Nazari M, Jahani D, Shaki F. Size-dependent neurotoxicity of aluminum oxide particles: A comparison between nano – and micrometer size on the basis of mitochondrial oxidative damage. Biol Trace Elem Res. 2017 Aug 30.

[56] Onoda A, Kawasaki T, Tsukiyama K, Takeda K, Umezawa M. Perivascular accumulation of β-sheet-rich proteins in offspring brain following maternal exposure to carbon black nanoparticles. Front Cell Neurosci. 2017 Mar 31; 11:92.

[57] Onoda A, Takeda K, Umezawa M. Dose-dependent induction of astrocyte activation and reactive astrogliosis in mouse brain following maternal exposure to carbon black nanoparticle. Part Fibre Toxicol. 2017 Feb 2;14(1):4.

[58] Liu X, Sui B, Sun J. Blood–brain barrier dysfunction induced by silica NPs in vitro and in vivo: Involvement of oxidative stress and rho-kinase/JNK signaling pathways. Biomaterials. 2017 Mar;121:64–82.

[59] Song B, Zhou T, Liu J, Shao L. Involvement of programmed cell death in neurotoxicity of metallic nanoparticles: Recent advances and future perspectives. Nanoscale Res Lett. 2016 Dec;11(1):484.

[60] Cao X, Hou D, Wang L, Li S, Sun S, Ping Q, Xu Y. Effects and molecular mechanism of chitosan-coated levodopa nanoliposomes on behavior of dyskinesia rats.Biol Res. 2016 Jul 4;49(1):32.

[61] Ansar S, Abudawood M, Hamed SS, Aleem MM. Exposure to zinc oxide nanoparticles induces neurotoxicity and proinflammatory response: Amelioration by hesperidin. Biol Trace Elem Res. 2017 Feb;175(2):360–366.

[62] Guan Y, Li M, Dong K, Gao N, Ren J, Zheng Y, Qu X. Ceria/POMs hybrid nanoparticles as a mimicking metallopeptidase for treatment of neurotoxicity of amyloid-β peptide. Biomaterials. 2016 Aug;98:92–102.

[63] Li Y, Zhu H, Wang S, Qian X, Fan J, Wang Z, Song P, Zhang X, Lu W, Ju D.Interplay of oxidative stress and autophagy in PAMAM dendrimers-induced neuronal cell death. Theranostics. 2015 Oct 8;5(12):1363–1377.

[64] Kim JH, Jeong MS, Kim DY, Her S, Wie MB. Zinc oxide nanoparticles induce lipoxygenase-mediated apoptosis and necrosis in human neuroblastoma SH-SY5Y cells. Neurochem Int. 2015 Nov;90:204–214.

[65] Durga M, Devasena T, Rajasekar A.Determination of LC50 and sub-chronic neurotoxicity of diesel exhaust nanoparticles. Environ Toxicol Pharmacol. 2015 Sep;40(2):615–625.

[66] Bagheri-Abassi F, Alavi H, Mohammadipour A, Motejaded F, Ebrahimzadeh-Bideskan A.The effect of silver nanoparticles on apoptosis and dark neuron production in rat hippocampus. Iran J Basic Med Sci. 2015 Jul;18(7):644–648.

[67] Gao J, Zhang X, Yu M, Ren G, Yang Z. Cognitive deficits induced by multi-walled carbon nanotubes via the autophagic pathway. Toxicology. 2015 Nov 4;337:21–29.

[68] Shah SA, Yoon GH, Ahmad A, Ullah F, Ul Amin F, Kim MO. Nanoscale-alumina induces oxidative stress and accelerates amyloid beta (Aβ) production in ICR female mice. Nanoscale. 2015 Oct 7; 7(37):15225–15237.

[69] Song B, Liu J, Feng X, Wei L, Shao L. A review on potential neurotoxicity of titanium dioxide nanoparticles. Nanoscale Res Lett. 2015 Dec; 10(1):1042.

[70] Imam SZ, Lantz-McPeak SM, Cuevas E, Rosas-Hernandez H, Liachenko S, Zhang Y, Sarkar S, Ramu J, Robinson BL, Jones Y, Gough B, Paule MG, Ali SF, Binienda ZK. Iron oxide anoparticles induce dopaminergic damage: In vitro pathways and in vivo imaging reveals mechanism of neuronal damage. Mol Neurobiol. 2015 Oct; 52(2):913–926.

[71] Ma W, Liu HT, Long YT. Monitoring dopamine quinone-induced dopaminergic neurotoxicityusing dopamine functionalized quantum dots. ACS Appl Mater Interfaces. 2015 Jul; 87(26): 14352–14358.

[72] Yin N, Zhang Y, Yun Z, Liu Q, Qu G, Zhou Q, Hu L, Jiang G. Silver nanoparticle exposure induces rat motor dysfunction through decrease in expression of calcium channel protein in cerebellum. Toxicol Lett. 2015 Sep 2; 237(2):112–120.

[73] Sruthi S, Mohanan PV. Investigation on cellular interactions of astrocytes with zinc oxide nanoparticles using rat C6 cell lines. Colloids Surf B Biointerfaces. 2015 Sep; 1133:1–11.

[74] Liu F, Mahmood M, Xu Y, Watanabe F, Biris AS, Hansen DK, Inselman A, Casciano D, Patterson TA, Paule MG, Slikker W Jr, Wang C. Effects of silver nanoparticles on human and rat embryonic neural stem cells. Front Neurosci. 2015 Apr; 8(9):115.

[75] Mao Z, Xu B, Ji X, Zhou K, Zhang X, Chen M, Han X, Tang Q, Wang X, Xia Y. Titanium dioxide nanoparticles alter cellular morphology via disturbing the microtubule dynamics. Nanoscale. 2015 May 14;7(18):8466–8475.

[76] Hong F, Sheng L, Ze Y, Hong J, Zhou Y, Wang L, Liu D, Yu X, Xu B, Zhao X, Ze X. Suppression of neurite outgrowth of primary cultured hippocampal neurons is involved in impairment of glutamate metabolism and NMDA receptor function caused by nanoparticulate TiO2. Biomaterials. 2015 Jun;53:76–85.

[77] Coccini T, Grandi S, Lonati D, Locatelli C, De Simone U. Comparative cellular toxicity of titanium dioxide nanoparticles on human astrocyte and neuronal cells after acute and prolonged exposure. Neurotoxicology. 2015 May;48:77–89.

[78] Nalika N, Parvez S. Mitochondrial dysfunction in titanium dioxide nanoparticle-induced neurotoxicity. Toxicol Mech Methods. 2015;25(5):355–363.

[79] Shim KH, Jeong KH, Bae SO, Kang MO, Maeng EH, Choi CS, Kim YR, Hulme J, Lee EK, Kim MK, An SS. Assessment of ZnO and SiO_2 nanoparticle permeability through and toxicity to the blood–brain barrier using Evans blue and TEM. Int J Nanomedicine. 2014 Dec 15; 9 Suppl 2:225–233.

[80] Wang YJ, He ZZ, Fang YW, Xu Y, Chen YN, Wang GQ, Yang YQ, Yang Z, Li YH. Effect of titanium dioxide nanoparticles on zebrafish embryos and developing retina. Int J Ophthalmol. 2014 Dec 18;7(6):917–923.

[81] Zhao Y, Wang X, Wu Q, Li Y, Wang D. Translocation and neurotoxicity of CdTe quantum dots in RMEs motor neurons in nematode Caenorhabditis elegans. J Hazard Mater. 2015;283:480–489.

[82]. Skalska J, Frontczak-Baniewicz M, Strużyńska L. Synaptic degeneration in rat brain after prolonged oral exposure to silver nanoparticles. Neurotoxicology. 2015 Jan;46:145–154.

[83] Peng J, Weng J, Ren L, Sun LP. Interactions between gold nanoparticles and amyloid β25-35 peptide. IET Nanobiotechnol. 2014 Dec;8(4):295–303.

[84] Ze X, Su M, Zhao X, Jiang H, Hong J, Yu X, Liu D, Xu B, Sheng L, Zhou Q, Zhou J, Cui J, Li K, Wang L, Ze Y, Hong F. TiO_2 nanoparticle-induced neurotoxicity may be involved in dysfunction of glutamate metabolism and its receptor expression in mice. Environ Toxicol. 2016 Jun;31(6):655–662.

[85] Gao N, Sun H, Dong K, Ren J, Qu X. Gold-nanoparticle-based multifunctional amyloid-β inhibitor against Alzheimer's disease. Chemistry. 2015 Jan 7;21(2):829–835.

[86] Sheng L, Wang L, Su M, Zhao X, Hu R, Yu X, Hong J, Liu D, Xu B, Zhu Y, Wang H, Hong F. Mechanism of TiO_2 nanoparticle-induced neurotoxicity in zebrafish (Danio rerio). Environ Toxicol. 2016 Feb;31(2):163–175.

[87] Wang S, Li Y, Fan J, Wang Z, Zeng X, Sun Y, Song P, Ju D. The role of autophagy in the neurotoxicity of cationic PAMAM dendrimers. Biomaterials. 2014 Aug;35(26):7588–7597.

[88] Yang X, He C, Li J, Chen H, Ma Q, Sui X, Tian S, Ying M, Zhang Q, Luo Y, Zhuang Z, Liu J. Uptake of silica nanoparticles: Neurotoxicity and Alzheimer-like pathology in human SK-N-SH and mouse neuro2a neuroblastoma cells. Toxicol Lett. 2014 Aug 17;229(1):240–249.

[89] de Oliveira GM, Kist LW, Pereira TC, Bortolotto JW, Paquete FL, de Oliveira EM, Leite CE, Bonan CD, de Souza Basso NR, Papaleo RM, Bogo MR. Transient modulation of acetylcholinesterase activity caused by exposure to dextran-coated iron oxide nanoparticles in brain of adult zebrafish. Comp Biochem Physiol C Toxicol Pharmacol. 2014 May;162:77–84.

[90]   Coccini T, Manzo L, Bellotti V, De Simone U. Assessment of cellular responses after short- and long-term exposure to silver nanoparticles in human neuroblastoma (SH-SY5Y) and astrocytoma (D384) cells. ScientificWorldJournal. 2014 Feb 13:259765.

[91]   Garcia-Reyero N, Kennedy AJ, Escalon BL, Habib T, Laird JG, Rawat A, Wiseman S, Hecker M, Denslow N, Steevens JA, Perkins EJ. Differential effects and potential adverse outcomes of ionic silver and silver nanoparticles in vivo and in vitro. Environ Sci Technol. 2014 Apr 15;48(8):4546–4555.

[92]   Mohammadipour A, Fazel A, Haghir H, Motejaded F, Rafatpanah H, Zabihi H, Hosseini M[5], Bideskan AE. Maternal exposure to titanium dioxide nanoparticles during pregnancy; impaired memory and decreased hippocampal cell proliferation in rat offspring. Environ Toxicol Pharmacol. 2014 Mar; 37(2):617–625.

[93]   Li X, Liu B, Li XL, Li YX, Sun MZ, Chen DY, Zhao X, Feng XZ. SiO_2 nanoparticles change colour preference and cause parkinson's-like behaviour in zebrafish. Sci Rep. 2014 Jan 22;4:3810.

[94]   Kwon JT, Seo GB, Jo, Lee M, Kim HM, Shim I, Lee BW, Yoon BI, Kim P, Choi K. Aluminum nanoparticles induce ERK and p38MAPK activation in rat brain. Toxicol Res. 2013 Sep;29(3):181–185.

[95]   Trickler WJ, Lantz-McPeak SM, Robinson BL, Paule MG, Slikker W Jr, Biris AS, Schlager JJ, Hussain SM, Kanungo J, Gonzalez C, Ali SF. Porcine brain microvessel endothelial cells show pro-inflammatory response to the size and composition of metallic nanoparticles. Drug Metab Rev. 2014 May;46(2):224–231.

[96]   Ziemińska E, Stafiej A, Strużyńska L. The role of the glutamatergic NMDA receptor in nanosilver-evoked neurotoxicity in primary cultures of cerebellar granule cells. Toxicology. 2014 Jan 6; 315:38–48.

[97]   Ozel RE, Hayat A, Wallace KN, Andreescu S. Effect of cerium oxide nanoparticles on intestinal serotonin in zebrafish. RSC Adv. 2013 Sep 21;3(35):15298–15309.

[98]   Huang JY, Lu YM, Wang H, Liu J, Liao MH, Hong LJ, Tao RR, Ahmed MM, Liu P, Liu SS, Fukunaga K, Du YZ, Han F. The effect of lipid nanoparticle PEGylation on neuroinflammatory response in mouse brain. Biomaterials. 2013 Oct;34(32):7960–7970.

[99]   Sharma A, Muresanu DF, Patnaik R, Sharma HS. Size- and age-dependent neurotoxicity of engineered metal nanoparticles in rats. Mol Neurobiol. 2013 Oct; 48(2):386–396.

[100]   Cao Y, Liu H, Li Q, Wang Q, Zhang W, Chen Y, Wang D, Cai Y. Effect of lead sulfide nanoparticles exposure on calcium homeostasis in rat hippocampus neurons. J Inorg Biochem. 2013 Sep;126:70–75.

[101]   Amara S, Slama IB, Omri K, El Ghoul J, El Mir L, Rhouma KB, Abdelmelek H, Sakly M. Effects of nanoparticle zinc oxide on emotional behavior and trace elements homeostasis in rat brain. Toxicol Ind Health. 2015 Dec;31(12):1202–1209.

[102]   Ze Y, Hu R, Wang X, Sang X, Ze X, Li B, Su J, Wang Y, Guan N, Zhao X, Gui S, Zhu L, Cheng Z, Cheng J, Sheng L, Sun Q, Wang L, Hong F. Neurotoxicity and gene-expressed profile in brain-injured mice caused by exposure to titanium dioxide nanoparticles. J Biomed Mater Res A. 2014 Feb;102(2):470–478.

[103]   Yin N, Liu Q, Liu J, He B, Cui L, Li Z, Yun Z, Qu G, Liu S, Zhou Q, Jiang G. Silver nanoparticle exposure attenuates the viability of rat cerebellum granule cells through apoptosis coupled to oxidative stress. Small. 2013 May 27; 9(9–10): 1831–1841.

[104]   Hoelting L, Scheinhardt B, Bondarenko O, Schildknecht S, Kapitza M, Tanavde V, Tan B, Lee QY, Mecking S, Leist M, Kadereit S. A 3-dimensional human embryonic stem cell (hESC)-derived model to detect developmental neurotoxicity of nanoparticles. Arch Toxicol. 2013 Apr;87(4):721–733.

[105] Geng J, Li M, Wu L, Chen C, Qu X. Mesoporous silica nanoparticle-based H_2O_2 responsive controlled-release system used for alzheimer's disease treatment. Adv Healthc Mater. 2012 May;1(3):332–336.

[106] Liao YH, Chang YJ, Yoshiike Y, Chang YC, Chen YR. Negatively charged gold nanoparticles inhibit Alzheimer's amyloid-β fibrillization, induce fibril dissociation, and mitigate neurotoxicity. Small. 2012 Dec 7;8(23):3631–3639.

[107] An L, Liu S, Yang Z, Zhang T. Cognitive impairment in rats induced by nano-CuO and its possible mechanisms. Toxicol Lett. 2012 Sep 3; 213(2):220–227.

[108] Zha YY, Yang B, Tang ML, Guo QC, Chen JT, Wen LP, Wang M. Concentration-dependent effects of fullerenol on cultured hippocampal neuron viability. Int J Nanomedicine. 2012;7:3099–3109.

[109] Li Y, Yu S, Wu Q, Tang M, Wang D. Transmissions of serotonin, dopamine, and glutamate are required for the formation of neurotoxicity from Al_2O_3-NPs in nematode Caenorhabditis elegans. Nanotoxicology. 2013 Aug;7(5):1004–1013.

[110] Li Y, Yu S, Wu Q, Tang M, Pu Y, Wang D. Chronic Al_2O_3-nanoparticle exposure causes neurotoxic effects on locomotion behaviors by inducing severe ROS production and disruption of ROS defense mechanisms in nematode Caenorhabditis elegans. J Hazard Mater. 2012 Jun 15; 219–220:221–230.

[111] Trickler WJ, Lantz SM, Schrand AM, Robinson BL, Newport GD, Schlager JJ, Paule MG, Slikker W, Biris AS, Hussain SM, Ali SF. Effects of copper nanoparticles on rat cerebral microvessel endothelial cells. Nanomedicine (Lond). 2012 Jun;7(6):835–846.

[112] Wu J, Wang C, Sun J, Xue Y. Neurotoxicity of silica nanoparticles: Brain localization and dopaminergic neurons damage pathways. ACS Nano. 2011 Jun 28;5(6):4476–4489.

[113] Wu J, Sun J, Xue Y. Involvement of JNK and P53 activation in G2/M cell cycle arrest and apoptosis induced by titanium dioxide nanoparticles in neuron cells. Toxicol Lett. 2010 Dec 15;199(3):269–276.

[114] Deng X, Luan Q, Chen W, Wang Y, Wu M, Zhang H, Jiao Z. Nanosized zinc oxide particles induce neural stem cell apoptosis. Nanotechnology. 2009 Mar 18; 20(11):115101.

[115] Au C, Mutkus L, Dobson A, Riffle J, Lalli J, Aschner M. Effects of nanoparticles on the adhesion and cell viability on astrocytes. Biol Trace Elem Res. 2007 Winter;120(1–3):248–56.

[116] Huang R, Han L, Li J, Ren F, Ke W, Jiang C, Pei Y. Neuroprotection in a 6-hydroxydopamine-lesioned Parkinson model using lactoferrin-modified nanoparticles. J Gene Med. 2009 Sep; 11(9):754–763.

[117] Hayward SL, Wilson CL, Hyaluronic acid-conjugated liposome nanoparticles for targeted delivery to CD44 overexpressing glioblastoma cells. Kidambi S Oncotarget. 2016 Jun 7;7(23):34158–34171.

[118] Tan C, Shichinohe H, Abumiya T, Nakayama N, Kazumata K, Hokari M, Hamauchi S, Houkin K. Short-, middle- and long-term safety of superparamagnetic iron oxide-labeled allogeneic bone marrow stromal cell transplantation in rat model of lacunar infarction. Neuropathology. 2015 Jun;35(3):197–208.

[119] Haque S1, Md S, Alam MI, Sahni JK, Ali J, Baboota S. Nanostructure-based drug delivery systems for brain targeting. Drug Dev Ind Pharm. 2012 Apr; 38(4):387–411.

[120] Mizrahi M, Friedman-Levi Y, Larush L, Frid K, Binyamin O, Dori D, Fainstein N, Ovadia H, Ben-Hur T, Magdassi S, Gabizon R. Pomegranate seed oil nanoemulsions for the prevention and treatment of neurodegenerative diseases: The case of genetic CJD. Nanomedicine. 2014 Aug; 10(6):1353–1363.

[121] Cox A, Varma A, Barry J, Vertegel A, Banik N. Nanoparticle estrogen in rat spinal cord injury elicits rapid anti-inflammatory effects in plasma, cerebrospinal fluid, and tissue. J Neurotrauma. 2015 Sep 15;32(18):1413–1421.

[122] Mangraviti A, Tzeng SY, Gullotti D, Kozielski KL, Kim JE, Seng M, Abbadi S, Schiapparelli P, Sarabia-Estrada R, Vescovi A, Brem H, Olivi A, Tyler B, Green JJ, Quinones-Hinojosa A. Non-virally engineered human adipose mesenchymal stem cells produce BMP4, target brain tumors, and extend survival. Biomaterials. 2016 Sep; 100:53–66.

[123] Okamoto S, Yoshii H, Ishikawa T, Akagi T, Akashi M, Takahashi M, Yamanishi K, Mori Y. Single dose of inactivated Japanese encephalitis vaccine with poly(gamma-glutamic acid) nanoparticles provides effective protection from Japanese encephalitis virus. Vaccine. 2008 Jan 30;26(5):589–594.

[124] Zhao Y, Jiang Y, Lv W, Wang Z, Lv L, Wang B, Liu X, Liu Y, Hu Q, Sun W, Xu Q, Xin H, Gu Z. Dual targeted nanocarrier for brain ischemic stroke treatment. J Control Release. 2016 Jul 10;233:64–71.

[125] Karussis D, Karageorgiou C, Vaknin-Dembinsky A, Gowda-Kurkalli B, Gomori JM, Kassis I, Bulte JW, Petrou P, Ben-Hur T, Abramsky O, Slavin S. Safety and immunological effects of mesenchymal stem cell transplantation in patients with multiple sclerosis and amyotrophic lateral sclerosis. Arch Neurol. 2010 Oct;67(10):1187–1194.

[126] Wang D, Wu LP. Nanomaterials for delivery of nucleic acid to the central nervous system(CNS). Mater Sci Eng C Mater Biol Appl. 2017 Jan 1; 70(Pt 2):1039–1046.

[127] Wu M, Li Z, Yu JE, Lu WW, Ni JX, Xia YL. Multi-centered clinical study on effects of nano-amomi paste in treating children's anorexia. Chin J Integr Med. 2007 Mar;13(1):55–58.

[128] Syková E, Jendelová P, Urdzíková L, Lesný P, Hejcl A. Bone marrow stem cells and polymer hydrogels–two strategies for spinal cord injury repair. Cell Mol Neurobiol. 2006 Oct-Nov; 26(7–8):1113–1129.

[129] Menon PK, Sharma A, Lafuente JV, Muresanu DF, Aguilar ZP, Wang YA, Patnaik R, Mössler H, Sharma HS. Intravenous administration of functionalized magnetic iron oxide nanoparticles does not induce CNS injury in the rat: Influence of spinal cord trauma and cerebrolysin treatment. Int Rev Neurobiol. 2017;137:47–63.

[130] Gomes MJ, Fernandes C, Martins S, Borges F, Sarmento B. Tailoring lipid and polymeric nanoparticles as siRNA carriers towards the blood-brain barrier - from targeting to safe administration. J Neuroimmune Pharmacol. 2017 Mar;12(1):107–119.

[131] Gandomi N, Varshochian R, Atyabi F, Ghahremani MH, Sharifzadeh M, Amini M, Dinarvand R. Solid lipid nanoparticles surface modified with anti-contactin-2 or anti-Neurofascin for brain-targeted delivery of medicines. Pharm Dev Technol. 2017 May;22(3):426–435.

[132] Karabey-Akyurek Y, Gurcay AG, Gurcan O, Turkoglu OF, Yabanoglu-Ciftci S, Eroglu H, Sargon MF, Bilensoy E, Oner L. Localized delivery of methylprednisolone sodium succinate with polymeric nanoparticles in experimental injured spinal cord model. Pharm Dev Technol. 2017 Dec;22(8):972–981.

[133] Giannaccini M, Calatayud MP, Poggetti A, Corbianco S, Novelli M, Paoli M, Battistini P, Castagna M, Dente L, Parchi P, Lisanti M, Cavallini G, Junquera C, Goya GF, Raffa V. Magnetic nanoparticles for efficient delivery of growth factors: Stimulation of peripheral nerve regeneration. Adv Healthc Mater. 2017 Apr;6(7).

[134] Saraiva C, Talhada D, Rai A, Ferreira R, Ferreira L, Bernardino L, Ruscher K. MicroRNA-124-loaded nanoparticles increase survival and neuronal differentiation of neural stem cells in vitro but do not contribute to stroke outcome in vivo. PLoS One. 2018 Mar 1;13(3):e0193609.

[135] Sárközi K, Papp A, Horváth E, Máté Z, Hermesz E, Kozma G, Zomborszki ZP, Kálomista I, Galbács G, Szabó A. Protective effect of green tea against neuro-functional alterations in rats treated with MnO_2 nanoparticles. J Sci Food Agric. 2017 Apr;97(6):1717–1724.

[136] Govindarajan M, Rajeswary M, Veerakumar K, Muthukumaran U, Hoti SL, Benelli G. Green synthesis and characterization of silver nanoparticles fabricated using Anisomeles indica: Mosquitocidal potential against malaria, dengue and Japanese encephalitis vectors. Exp Parasitol. 2016 Feb;161:40–47.

6 Health effects of inhaled nanomaterials

Abstract: Inhaled nanoparticles (NPs) from the environment or those that are released from the industrial and manufacturing sources are a risk factor for lung and cardiovascular diseases.The detection and quantification of risk in a dose-response manner find importance in determining hazardous nature of NPs. In this chapter, we outline the different NPs and different effects of inhaled NPs on human health.

Keywords: titanium dioxide nanoparticles, MWCNTs, SWCNTs, ENMs, AgNPs, NPS, P53, P21, Bax, Bcl2, cleaved caspase-3, CVD.

6.1 Introduction

Various applications of nanotechnology are utilized in day-to-day life. The past records have shown tremendous applications of nanotechnology in our day-to-day life including applications in medicine, agriculture, information technology and mobile phones. Engineered nanoparticles (NPs) with the desired property have led to overcome the drawbacks of conventional technology.

While applications of nanotechnology are tremendous, they causemany health hazards and their synthesis and handling may lead to unintentional exposure to workers, and thus their exposure in the environment and in various industrial applications have become the leading cause of global concern. The inhalation of NPs has been observed in the easiest routes of entry of NPs through the nose to the human body, thus gaining access to cells and tissues through circulation. Based on the diversity of their chemical nature, the solubility, surface properties and biodegradability properties are either degraded or accumulated in cells and tissues.

Inhaled NPs pose a major threat to health showing symptoms of oxidative stress, inflammation, cytotoxicity, genotoxicity and fibrosis that might lead to primary or secondary pulmonary disorders and are reported to be major cause of lung disorders (inclusive of lung cancer). Inhaled NPs when circulating through blood and body fluids are reported to show systemic cytotoxicity and associated disorders of the body and adversely affect the cardiovascular system. When they cross the blood–brain barrier, they have been reported to have neurotoxic effects. They cause detrimental effects on the developing fetus when they cross the placental barrier in pregnant female animal models.

The air quality is an important domain for understanding the risk associated with the exposure of particulate matter (PM) in air to the human body. In the past decade, significant deterioration of the air quality has been detected and the presence of PM in the air from various anthropogenic sources resulting in the rise of lung disorders that are lethal is alarming.

https://doi.org/10.1515/9783110579093-006

The World Health Organization (WHO) in 2012 has recorded death of around 7 million people due to air pollution, thus confirming the role of air pollution, in various environmental health hazards that cause cardiovascular disorders, cancer and respiratory diseases ranging from acute respiratory infections to chronic obstructive pulmonary diseases (COPD) and lung cancer [1]. PM composed of mixture of particles and air pollutants have been reported to be associated with increased morbidity and mortality of cardiovascular and pulmonary diseases. Ultrafine particles (UFPs) known to adsorb high concentrations of toxic air pollutants are easily inhaled into the lungs and are most harmful to health [2]. WHO reported in 2012 that low- and middle-income countries in the South-East Asia and Western Pacific regions had the largest air pollution-related burden with a total 3.3 million deaths linked to indoor air pollution and 2.6 million deaths related to outdoor air pollution.Thus, it is very important in the current scenario to understand the contributions of NPs to air pollution and their role in causing different pulmonary and other disorders (Table 6.1). In this chapter, we discuss different NPs and their effects on human from in vitro studies in human and animal models.

Table 6.1: Air pollution and diseases[$].

Outdoor air pollution-caused deaths – breakdown by disease:	Indoor air pollution-caused deaths – breakdown by disease:
40% – ischemic heart disease	34% – stroke
40% – stroke	26% – ischemic heart disease
11% – chronic obstructive pulmonary disease (COPD)	22% – COPD
6% – lung cancer	12% – acute lower respiratory infections in children
3% – acute lower respiratory infections in children	6% – lung cancer

[$] [1]

Amongst the different components of air pollution (Table 6.2), we focus attention on various constituents of PM since it is a mixture of chemicals constituting organic and inorganic chemicals in both solid and liquid states that remain suspended in air and

Table 6.2: Major contributors to air pollution:
From http://www.who.int/mediacentre/news/releases/2014/air-pollution/en/ [1].

Particulate matter
Ozone (O_3)
Nitrogen dioxide (NO_2)
Sulfur dioxide (SO_2)

are also known to be the major component of air pollutant that pose health hazards to humans globally.

WHO has confirmed that sulfate, nitrates, ammonia, sodium chloride, CB, mineral dust and water are the most common constituents of PM. Molecules with diameter smaller than or 10 microns ($\leq PM_{10}$), with their lung penetration property pose the most common health hazards and is a major contributor of pulmonary disorders. Chronic exposure poses risk to develop cardiovascular and respiratory diseases, as well as lung cancer leading to increased mortality or morbidity. In the next portion, we will discuss NP, PM and its contribution to deterioration of air quality.

6.2 NPs, particulate matter and air pollution

Ultrafine nanoparticles (UFPs) with less than 100 nm diameter (PM10) have been reported to be toxic to the body causing inflammation, manifestation by release of inflammatory mediators, activation of signaling cascades, altered gene transcription, oxidative stress by production of reactive oxygen species, calcium signaling leading to respiratory and cardiovascular disorders are known. Epidemiological studies have indicated the effect of particulate air pollutants and UFPs can affect both mortality in adults and children. Carbon, polystyrene (PS) beads and titanium dioxide (TiO_2) have been detected as major components of PM in air pollution. Engineered NPs varying in size and surface properties that remain largely uncharacterized are also a new level of health hazard [3] and exposure of UFPs to workers pose an occupational health hazard. Therefore, we discuss the impact of toxicity of PM NPs on health in the following sections.

6.3 Effect of airborne NPs on health

Several studies on human cell lines and animal models have confirmed the effect of different types of NPs on human health (Figure 6.1).

6.3.1 Ultrafine particles and particulate matter

NP and air pollution particles may lead to systemic disorders in cardiovascular system, pulmonary system, GI tract, uterine system and pregnancy and fetal development that remain a major unexplored domain and future scope of research [4]. Recently, studies in rodents have shown that airborne PM, ultrafine NPs, engineered NPs gaining access to the system by air pathways through mouth or nose and their movement though the olfactory epithelium by the circulation has been shown to be one of the major environmental factors that are potentially involved in the

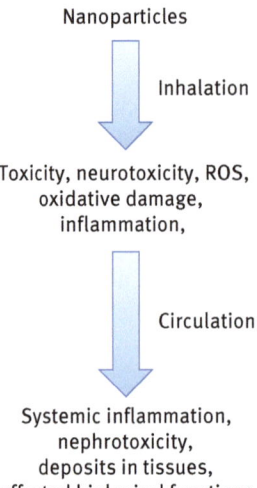

Nanoparticles

Inhalation

Toxicity, neurotoxicity, ROS,
oxidative damage,
inflammation,

Circulation

Systemic inflammation,
nephrotoxicity,
deposits in tissues,
affected biological functions   **Figure 6.1:** Impact of NPs exposed by routes of inhalation.

pathogenesis of Alzheimer's disease (AD) and Parkinson's disease (PD) that affect the oxidative stress and neuroinflammation leading to neurodegeneration and cognitive impairment [5]. UFPs, engineered NPs deposited on the respiratory epithelial tract are thought to cross the air–blood barrier via the expansive alveolar region into the systemic circulation to reach different organs such as myocardium, liver, pancreas, kidney and spleen [6].

6.3.2 Carbon-based NPs

Inhalation studies, intratracheal instillation (IT) and pharyngeal aspiration studies of carbon-based NPs with increased surface area leading to tissue aggregation and or dispersion reveal pulmonary inflammation and injury, neutrophil and macrophage infiltration in the alveolar and/or interstitial space, release of chemokines, lactate dehydrogenase (LDH) and alkaline phosphatase (ALP) release in the bronchoalveolar lavage fluid (BALF), the expression of oxidative stress-related genes, presence of granulomatous lesion and pulmonary fibrosis neutrophil inflammation or granulomatous formation in the lung and persistent inflammation in the lung [7].

6.3.3 Carbon black NPs

Carbon black (CB) NPs find importance as a constituent in the rubber manufacturing industry. CB exposure to the human body through routes of inhalation is known to generate reactive oxygen and nitrogen species, including nitric oxide that causes damage to the inflammed and adjoining tissues; DNA damage that leads to malignant lung tumor formation; chronic inflammation in respiratory system that leads to metastasis and carcinogenesis in experimental animal models. In a study on murine macrophage cell line RAW 264.7 and human epithelial lung cells A549, exposure to CB is reported to form 8-nitroguanine (8-nitroG), leading to nitrative DNA damage, tissue inflammation and oxidative damage to tissues probably involving the activation of clathrin-mediated endocytosis pathway [8]. The exposure to NPs to human tissues by different routes, including the route through skin may lead to immunomodulatory effects involving the protein–corona interaction between NPs and biological matter. The exposure through the skin may lead to allergic responses [9].

6.3.4 Carbon nanohorns

Carbon nanohorns (CNH) are synthetic NPs that pose a potential occupational risk to employees who are associated with the manufacturing. It is known to be taken up by human tissues on inhalation and exerts its toxic effects on interaction with blood and other tissues. When exposed to human epithelial cell lines of respiratory systems, it is reported to cause loss in cell viability and metabolic activity. Although the bronchial epithelial cells were the most robust in nature, CNH exposure could lead to reduced mitochondrial activity that is indicative of low metabolic activity and overall cytotoxic effects [10]. Inhaled NPs are reported to be difficult to be separated from the sites of deposition, and thus pose difficulty in cleaning mechanism by macrophage and is known to enter the circulatory system via lymph and blood and enter into the circulatory and nervous systems and different tissues and organs, including the brain [11]. C_{60} fullerene (C_{60}), or buckminster fullerene, with 60 carbon atoms, of 1 nm diameter leads to respiratory toxicity and immunotoxicity when exposed to B6C3F1/N mice and Wistar Han rats by inhalation and inflammatory responses manifested by histiocytic infiltration, macrophage pigmentation, chronic inflammation, increased expression of monocyte chemoattractant protein (MCP)-1 (rats) and macrophage inflammatory protein (MIP)-1α (mice) in BALF [12].

6.3.5 Carbon nanotubes

Carbon nanotubes (CNTs) in aerosols entering through nose are known to be transported through the airways and observed to be deposited in human respiratory tract and lower airways enforcing its role as an occupational hazard [13]. Both single-walled carbon nanotubes (SWCNT) and multiwalled carbon nanotubes (MWCNT) have been reported for their deposition in the respiratory tract [14] posing pulmonary hazard. SWCNT could induce pulmonary fibrosis through ROS production, oxidative stress, collagen expression and transforming growth factor-beta (TGF-β) production and that the size of the fibers is correlated with the intensity of toxicity [15]. Studies from mass spectrometry (MS)-based oxidative lipidomic analysis of mice exposed to SWCNT through inhalation revealed pulmonary phospholipid peroxidation reactions, occurrence of anionic phospholipids, cardiolipin, phosphatidylserine and phosphatidylinositol and oxygenation of polyunsaturated fatty acid (PUFA) together with the formation of apoptotic cells [16].

MWCNT have been known for causing lung toxicity, pulmonary inflammation and pulmonary fibrosis initiated by transition from epithelial to mesenchymal tissues also termed as epithelial mesenchymal transition (EMT) together with granuloma formation, symptoms of airway injury and malignant cancer [17–18]. Short-time exposure to normal human bronchial epithelial cells, MWCNT could generate cytotoxic effects, whereas their long-term exposure is affected by cell proliferation [17]. MWCNT in human bronchial epithelial cells could activate the TGF-β/Smad signaling pathway, thereby leading to pulmonary fibrosis. Studies in C57BL/6 mice and human BEAS-2B cell line revealed that MWCNT affected transcription regulation of SNAIL-1 by activation of TGF-β, upregulation of nuclear β-catenin, repression of E-cadherin, thus leading to MWCNT-induced EMT [18]. MWCNTs exposure has been reported to affect the liver pathogenesis revealed by the development of nonalcoholic steatohepatitis (NASH)-like phenotype, inflammation, hepatic steatosis and fibrosis, IL-6 and plasminogen activator inhibitor (PAI)-1upregulation, overexpression of NF-κBp65 and impaired cholesterol homeostasis and suppression of peroxisome proliferator-activated receptor gamma (PPARy) in mice liver [19].

MWCNT when exposed to epithelium affect the cocultured endothelial cell (EC) barrier leading to increased ROS, actin rearrangement, loss of cell surface VE-cadherin, increase in endothelial angiogenic ability, secretion of vascular endothelial growth factor A (VEGFA), soluble intercellular adhesion molecule-1 (sICAM-1) and circulating vascular adhesion molecule-1 (sVCAM-1), intracellular phospho-NF-κB, phospho-Stat3 and phospho-p38 MAPK-induced multiple changes to the underlying endothelium through cell signaling mediators derived from MWCNT-exposed epithelial cells [20]. MWCNT when exposed to small airway epithelial cells (SAEC) penetrate into the epithelial lining leading to inflammation and progressive fibrosis, altered transcription profiles, ROS production, protein phosphotyrosine and phosphothreonine levels and cell migratory behavior leading to lung damage, carcinogenesis and tumor progression [21]. MWCNTs in suspensions of IT and inhalation studies in rats have been reported to induce pulmonary inflammation, granulomatous lesion and persistent neutrophil infiltration [22]. MWCNTs exposure has been reported to cause pulmonary secretion of acute phase proteins (APPs), including serum amyloid A1/2 (SAA1/2) and SAA3 in blood plasma hepatic Saa1 and pulmonary Saa3 mRNA levels [23] and lead to granuloma formation, fibrosis and cancer in C57BL/6 mice manifested by eosinophil infiltration, giant cell formation, mucus production and IL-13 expression in lungs [24].

MWCNTs exposure by inhalation by male and female F344 rats revealed increased carcinomas and adenomas in males without pleural mesothelioma formation. Epithelial hyperplasia, granulomatous change, localized fibrosis and altered BALF indicated their carcinogenic potential [25]. Inhaled MWCNT has been studied that led to inflammatory and rapid fibrotic response, increased LDH activity and polymorphonuclear cells (PMNs) or polymorphonuclear leukocytes infiltration that are indicative of pathological changes in the lung toward fibrosis [26]. MWCNT exposure is known to cause inflammation and disease by altering gene expression-like promoter methylation of inflammatory genes, interferon gamma and tumor necrosis factor alpha (IFN-y and TNF-α) correlating with cytokine production, methylation of thy-1 involved in tissue fibrosis correlated with collagen deposition and DNA hypomethylation in the lung and blood led to disease development. Global DNA methylation has been identified as an important biomarker of the disease in response to exposure by MWCNTs [27]. Inhaled MWCNT has been reported to be carcinogenic in promoting cell growth with the existing DNA damage and the development of lung bronchioloalveolar adenomas and lung adenocarcinomas in mouse models [28].

Double-walled carbon nanotubes (DWCNT) exposed by pharyngeal aspiration in C57Bl/6 mice has been reported to induce pulmonary toxicity revealing pathophysiology of inflammation and injury, rise in LDH activity, albumin levels and decrease in the integrity of the blood–gas barrier in the lung and significant alveolitis and fibrosis [29].

6.3.6 Titanium dioxide NPs

Cosmetics and sunscreens represent the major cause of dermal exposure to titanium dioxide nanoparticles (TiO$_2$ NPs). Using a probabilistic approach nano-TiO$_2$ powder has been reported to be the major cause of risk to human [30] health that affects the routes of inhalation.TiO$_2$-treated cell lines representative of alveolocapillary barrier revealed ROS generation and apoptosis of macrophage-like THP-1 and HPMEC-ST1.6R microvascular cells due to redox changes. Genotoxic potential of TiO$_2$ NPs was manifested by activation of yH2AX, activation of DNA repair proteins, DNA damage, cell stress, cell cycle arrest and activation of heat shock protein 27 (HSP27) and Stress-activated protein kinase/c-Jun NH2-terminal kinase (SAPK/JNK) pathways [31]. Exposure of TiO$_2$ NPS to rats through sprays and inhalation route revealed depositions in lungs and whole body and histopathological changes in lungs [32]. Protein corona formation has been observed on NPs in serum, even for inhaled NPs. Experimental studies with exposure of human BALF from patients with pulmonary alveolar proteinosis (PAP) have revealed protein corona formation on pulmonary surfactant (PS) and on TiO$_2$ particles under high resolution MS studies and their probable interaction with lipids and pulmonary surfactant (PS), leading to chronic airway diseases such as asthma and Chronic obstructive pulmonary disease (COPD), and leading to increased susceptibility toward other respiratory diseases [33]. Neurological disorders such as dementia, AD are projected to rise till 2040 [33, 1].

Female Sprague Dawley (SD) rats exposed to nano-TiO$_2$ aerosols revealed uterine microvascular sensitivity [34]. Chronic overload exposure of TiO$_2$NP by inhalation or by skin or oral routes in rats has been reported to lead to pathophysiology of lung cancer and noncancerous chronic respiratory effects in workers. Experimental dermal studies indicate a lack of penetration of particles beyond the epidermis with no consequent health risks [35]. TiO$_2$ and engineered NPs on primary vascular ECs led to increased cellular oxidative stress and DNA binding of NF-κB, increased phosphorylation of Akt (serine/threonine-protein kinase), ERK (kinase), c-Jun N-terminal kinase (JNK) and p38 (mitogen-activated protein kinases) with overexpression of VCAM-1 (vascular cell adhesion molecule-1) and MCP-[1, 36] are indicative of the fact that NPs affected cellular signaling processes.

TiO$_2$NPs-induced cytotoxicity leading to apoptosis manifested by apoptotic bodies altered expression of P53 (tumor suppressor protein), P21 (cyclin-dependent kinase inhibitor [CKI]), Bax (apoptotic regulator protein), Bcl2 (apoptotic regulator protein) and cleaved caspase-3 proteins as markers of apoptosis, oxidative stress by release of ROS and genotoxicity in the human lung cancer cell line, A549 [37]. Biodistribution of TiO$_2$ NPs from lungs to healthy young adults and in elderly rats exposed to TiO$_2$ nanostructured aerosol [38] has been reported.

6.3.7 Zinc oxide NPs

Exposure of ZnO nanowires (NW) to porcine PS or pulmonary surfactant decreased their dissolution at acidic pH through the formation of a phospholipid corona with increased uptake of ZnONWs within TT1 cells, thus increasing their toxicity and their increased dispersion in serum [39]. Inhaled NPs have been reported to move to the olfactory bulb in rats. ZnONP both coated and uncoated on primary human cells cultured from the olfactory mucosa containing the multipotent cells revealed stress, inflammatory response and apoptosis, thus indicating that the inhaled NPs are dangerous to stem cells of olfactory system [40]. Inhaled ZnONPs have been reported of high deposition in lung alveoli and adjoining regions. Intratracheal administration of ZnONPs in female BALB/c mice, BEAS-2B and adenocarcinoma A549 cells using molecular imaging (SPECT and CT) has been reported to cause increased LDH in BALF and 8-hydroxy-2′-deoxyguanosine (8-OHdG), caspase-3 and the p63 tumor marker in lung tissues [41]. Inhaled NPs have been reported to have high-deposition rates in the alveolar region of the lungs affecting the PS [41].

6.3.8 Iron NPs

Airborne water insoluble iron nanorods chemically composed to Fe$_2$O$_3$ NPs on exposure through inhalation in rats revealed pulmonary accumulation of nanorods, penetration into the alveolar system and active endocytosis, thereby leading to alveolocytes and myelin sheath damage. These NPs were reported to dissolve in cell-free BALF supernatant [42]. Of the several routes of NPs into the body, including oral uptake, skin exposure and inhalation, large absorption area and the relatively high translocation rate, mechanisms to cross mucus and epithelial barrier and or gastrointestinal tract play important role in transportation of NPs and are also areas of interaction of NPs in biological systems causing health effects [43]. Zerovalent iron (nZVI) NPs, when exposed to a alveolar-capillary coculture model has been reported to induce both pulmonary and cardiovascular toxicity, even without passing through the epithelial barrier into the endothelium, but caused oxidative and inflammatory responses in both epithelial and ECs. Increased oxidized α1-antitrypsin and low-density lipoprotein in the coculture model increased the risk of COPD and atherosclerosis [44].

6.3.9 Cerium dioxide

Cerium dioxide (CeO_2) and engineered $nCeO_2$ NPs with enhanced desired properties are largely uncharacterized but finds its applications in our day-to-day lives in fuel additives, catalyst and polish; however, it suffers from potential toxic effects. In a rat model, $nCeO_2$ Nps has been reported to deposit in the lungs leading to induction of fibroblastic nodules, thus initiating fibrosis, increased cell proliferation and increased production of collagen I [45]. Time kinetic study of effect on inhaled cobalt and CeO_2 NPs in alveolar A549 and bronchial BEAS-2B epithelial cells led to downregulated gene transcription [46]. CeO_2 and $BaSO_4$ NP frequently found in industrial and consumer products and exposure to humans and other organisms are related with toxic effects revealing tissue inflammation with increasing severity and post-exposure persistency for CeO_2, increased genotoxicity and cell proliferation [47]. CeO_2 NPs revealed the acute effect of IT leading to oxidative stress and thrombosis in mice with significant increase of neutrophils into the BAL fluid, increased TNFα, decreased antioxidant catalase activity, macrophages and neutrophils infiltration, rise in plasma levels of C reactive protein (CRP), TNFα, fibrinogen and Plasminogen activator inhibitor-1 (PAI-1) and promotes thrombosis in vivo [48].

6.3.10 Uncharacterized NPs

A study involving inhalation of NPs liberated from electrical discharge machine, which is not characterized by revealed NPs deposited in the nasal cavity, upper respiratory tract and laryngeal region and enhancement of airway dosages. [49]. Laser printer generated exposure of engineered nanomaterials (ENMs) from toners, which are largely uncharacterized pose a major health hazard and is a great cause of concern. These printer emitted particles (PEPs) have been reported to adversely affect the pulmonary tissues in a nine-week old male *Balb/c* mice exposed to various doses of PEPs by IT, revealing pulmonary immune response manifested by influx of neutrophils and macrophages, inflammation, upregulation of Ccl5 (Rantes), Nos1 and Ucp2 genes, involved in the repair of oxidative damage and immune response against foreign bodies and regulation of DNA methylation patterns (Dnmt3a) and expression of transposable element (TE) LINE-1 [50]. Exposure to sprays containing uncharacterized NPs capable of surface adsorption and exposure to the human body through inhalation raises concern. It is reported that it leads to reduced macrophage mobility and chronic inflammation that may then lead to diseases including cancer [51].

NPs from toner could lead to chronic inflammation and fibrosis in animal models; however, their detailed toxic effects are yet to be characterized. In the alveolar-capillary coculture model with Human Small Airway Epithelial Cells (SAEC) and human microvascular endothelial cells (HMVEC) exposure to toner NPs have been observed to cause morphological alterations of actin remodeling and gap formations within the endothelial monolayer, oxidative stress revealed by ROS generation and angiogenesis, IL-6 and MCP-1 [52]. Inhalation of uncharacterized matter in diesel exhaust (DE) is a potential threat to pulmonary disorder [53]. NP-rich diesel exhaust (NR-DE) has been reported to affect the testicular steroidogenesis [54]. Welders fumes with uncharacterized NPs but predominant presence of manganese (Mn), nickel (Ni), chromium (Cr) and hexavalent chromium (Cr(VI) are reported to cause severe adverse health outcomes [55].

6.3.11 Silver NPs

Inhaled silver NPs (AgNPs) have been shown to express acute toxicity to pulmonary system revealing shift in particle surface area to particle mass as suitable dose metric using monolayer lung epithelial cells [56]. AgNP exposure has been observed to reveal systemic defects in the body, depositions in the liver, gut-associated lymphoid tissue (GALT) and brain of mice, with decreased cellularity in spleen follicles, altered cell number and populations, reduced splenic GSH:GSSG [glutathione(GSH) to oxidized glutathione (GSSG)] ratio, overexpression of oxidative stress-responsive gene Hmox1 in the hippocampus [57]. AgNPs with varying sizes and coatings affect pulmonary responses at different time points when exposed to male SD rats were intratracheally instilled with AgNPs leading to elevated PMNs in BALF with inflammation [58]. Silver nanowires (Ag NWs) have large-scale applications in electronic items. Male, SD rats with intratracheal instillation of Ag NWs revealed a strong eosinophil influx, formation of Langhans and foreign body giant cells, epithelial sloughing in the terminal bronchioles (TB) and alveolar cellular exudates, granuloma formation and macrophage infiltration indicative of pulmonary toxicity [59]. AgNP-exposed rats by inhalation reveal toxicity affecting the kidney gene with overexpression of 104 genes, of which 96 of them showed altered expression in male SD rat kidneys [60]. Inhaled engineered AgNPs reach and cross mouse placental barrier and reach maternal tissues, placenta and fetal tissues and cause adverse effects to fetus when exposed to pregnant mouse females [61].

6.3.12 Silica NPs

Corona formation is very important to understand the safety associated with the clearance of NPs, and therefore it is important in the designing of inhaled nanomedicines. NPs deposited in the lungs, transited through respiratory tract lining fluid (RTLF) and interaction with the RTLF lining acquire a biomolecular corona that are indicative of the presence of innate immunity proteins, including surfactant protein A, napsin A and complement (C1q and C3) proteins, revealing opsonization and initiation of phagocytosis and removal from lungs. Label-free snapshot proteomics have enabled the study of semiquantitative profiles of corona proteins formed around SiO_2 and poly(vinyl) acetate (PVAc) NPs in RTLF that finds applications in drug delivery vehicles [62]. Synergistic effects of toxicity was observed in lung adenocarcinoma A549 cells when exposed to SiO_2 NPs, UFPs and lead acetate as air pollutants lead to oxidative stress, ROS generation, lipid peroxidation, reduced glutathione content and superoxide dismutase (SOD) and glutathione peroxidase activities and DNA damage [2]. Coexposure of silica NPs (nano-SiO_2) and methyl mercury (MeHg) on lung adenocarcinoma cells (A549) led to the generation of ROS, lipid peroxidation and reduced SOD activity and glutathione peroxidase (GSH-px) indicative of oxidative stress lead to oxidative DNA damage and cellular apoptosis [63].

6.3.13 Other NPs

Aluminosilicate NPs have been reported to affect PS formation and toxicity is reported to be NP load dependent and pose a risk to the workplace [64]. Cellulose nanomaterials although have advantages as composites in medicine, pose a pulmonary hazard by generation of OH radicals causing stress to the host [65].Inhaled tungsten (IV) oxide nanoparticles (WO_3 NPs) lead to toxicity in Golden Syrian hamsters; they are detected by altered cell numbers, LDH activity, ALP activity, total protein content, TNF-α and High mobility group box 1 (HMGB1) levels in BLF fluids, expression of cathepsin B, Thioredoxin-interacting protein (TXNIP), NLR family pyrin domain containing 3 (NLRP3), Apoptosis-associated speck-like protein containing a CARD (ASC), IL-1β and caspase-1, deposition and their translocation in airway epithelia, macrophages and interstitial areas of alveolar spaces indicative of inflammasome, morphological changes, lung injury and cytotoxic pathway [66]. Although cobalt monoxide (CoO) and lanthanum oxide (La_2O_3) NPs are both transition and rare earth metal oxide NPs that show differences in toxic effects to lungs when exposed through inhalation in mice. La_2O_3 NPs are accumulated in the tracheobronchial lymph nodes and formed urchin-shaped lanthanum(III) phosphate ($LaPO_4$) structures in phagolysosomal fluid, whereas CoO NPs led to increased LDH in BALF indicative of acute pulmonary toxicity by CoO while chronic toxicity by La_2O_3 NPs [67]. La_2O_3 NP inhaled by male SD rats reveals transportation of NP by blood to the entire system, BALF and oxidative stress in lung tissues, increased cells such as neutrophils and macrophages in BALF, increased LDH activity, albumin, NO and TNF-alpha [68]. Copper oxide nanoparticles (CuO NPs) inhaled by rats reveal symptoms of lung inflammation and cytotoxicity, alveolitis, bronchiolitis, vacuolation of the respiratory epithelium and emphysema in the lung. Inhaled nickel nanoparticles (NiNPs) are reported to cause damage to the brain and lungs when exposed to C57BL/6J mice by nicotinamide adenine dinucleotide phosphate (NADPH) oxidase-mediated ROS generation in activated microglia [69] and cause DNA damage. ENMs inhalation leads to toxicity in the lungs and subsequent translocation. ENMs composed of CeO_2, ZnO and SiO_2-coated CeO_2 and ZnO on Calu-3 lung epithelial cells reveal translocation of ENMs by transcellular pathways without affecting monolayer integrity or disrupting tight junctions [70].

6.4 Pathogenesis

Different studies in animal and human models for in vivo and in vitro effects of toxicity by exposure to NPs through inhalation lead to disorders ranging from pulmonary disorders to systemic disorders. Many studies have tried to understand the downstream pathway of pathogenesis in NP-induced disorders. Symptoms of oxidation, stress, inflammation reactions, role of inflammatory cells, activation of cytotoxic pathways, DNA methylation, DNA damages and failure of DNA repair mechanisms are reported to be the major effects of NP-induced toxicity in host cells. Pulmonary inflammation induced by NPs in animal models involved influx of neutrophils and alveolar macrophages, induction of cytokine-neutrophil chemoattractant (CINC) family, macrophage inflammatory protein-1α and oxidant stress-related genes such as heme oxygenase-1 (HO-1), failure of DNA repair and accumulation of mutations that may lead to malignancy, generation of free radicals leading to induced signaling molecules in immune responses are reported to activate Th2 immune responses such as activation of eosinophil and induction of IgE [71]. Inhalation of the aerosolized product has been reported to damage the PS system with increased airway resistance in the mice and decreased expiratory flow rate, thus becoming toxic to the pulmonary system [72]. Immune response has been reported to follow chronic pulmonary accumulation of Fe–NPs and Fe_2O_3NPs. FeNPs introduced by IT in mice were observed in lungs and particle-laden macrophages in the BAL fluid with increased neutrophil and lymphocytes, apoptotic cells, LDH level and Th1 inflammatory response in lungs with increased chemokines, including GM-CSF, MCP-1 and MIP-1 and overexpression of antigen presentation proteins, including CD80, CD86 and MHC class II on antigen-presenting cells (APC) in BAL fluid [73].

NPs affect the CV system reported by the damage of cellular functions of vascular EC, inflammation, oxidative stress, autonomic dysregulation impacted integrity, thereby leading to impaired circulation, altered heart rate and electrical activity via catecholamine release, increased susceptibility to ischemia/reperfusion injury and modified baroreceptor control of cardiac function, promoting myocardial infarction, hypertension, cardiac arrhythmias and thrombosis [74].

Cadmium oxide (CdO) NPs known for their nephrotoxic effects pose a major health hazard when exposed to pregnant mice. Inhaled CdONP on pregnant CD-1 mice affected maternal and neonatal renal function revealing kidney injury, including kidney injury molecule-1 (Kim-1) and neutrophil gelatinase-associated lipocalin (NGAL), and increase in neonatal Kim-1 [75]. ENMs from toner in SAECs, macrophages (THP-1 cells) and lymphoblasts (TK6 cells) have been reported to cause membrane damage with increased ROS production, and increased proinflammatory cytokines and differences in methylation patterns leading to pulmonary disorders [76].

Exposure to metallic and metal oxide NPs such as TiO_2, CeO_2 and AgNP through nonsurgical IT in pregnant female mice affected the lung development of the offspring together with decreased placental efficiency, VEGF-α and matrix metalloproteinase 9 (MMP-9) and fibroblast growth factor-18 (FGF-18), irrespective of the chemical properties of the NPs that are indicative of the toxic effect of NP transmitted to offsprings through inhalation [77]. The adult female C57BL/6 mice exposed to TiO_2 NPs via IT revealed their translocation to heart and liver through blood together leading to activation of complement cascade and inflammatory processes in heart and complement factor 3 in blood, indicating innate immune response for opsonization and clearance. The liver showed acute phase response [78]. CNT is reported to induce granuloma formation in abdominal cavity and subpleural fibrosis and allergic disease such as asthma in mice. The inhalation of CNTs is reported to cause eosinophilia with mucus hypersecretion, airway hyperresponsiveness (AHR) and the expression of Th2 cytokines and innate immune responses mediated by mast cells and alveolar macrophages in regulating inflammation [79].

MWCNT exposure by whole-body inhalation in male and female rats revealed that IL-1β, IL-6, IL-10 MIP-1α mRNA in female rats are significantly higher than in control cells. In males, IL-1β, IL-6 and TNFα were overexpressed but both rats revealed lowered IL2 expression. Systemic inflammation was found to occur due to inflammatory cytokines in splenic macrophages [80].

Endothelial progenitor cells (EPCs) when exposed to inhaled Ni–NPs through inhalation by the whole body led to increase in vascular inflammation, ROS generation, altered vasomotor tone and atherosclerosis in experimental mouse model, EPCs, circulating endothelial progenitor cells (CEPCs), circulating endothelial cells (CECs) and endothelial microparticles (EMPs) revealed endothelial damage [81].

Inhaled ZnONP by mice led to dose-dependent responses with acute inflammation at bronchioloalveolar junctions of lungs with increased metallothionein expression in epithelial cells of brochioloalveolar junction. ZnONP increased activities of LDH, glutamate oxaloacetate transaminase, glutamate pyruvate transaminase and creatine phosphokinase in blood. In human bronchial epithelial cells, ZnONP induced IL8 and activation of LR4 [82]. MWCNT inhalation has been reported to cause inflammatory and progressive fibrotic response with lung inflammation and fibrosis that affect signaling cascades, inflammation and fibrosis, VEGFa and C–C chemokine 2 (ccl2) [83]. Quantum dots (QDs) of CdSe/ZnS NPs when exposed to mice by inhalation show translocation from the olfactory tract to the brain [84].

6.5 Recent developments in tests for the detection of pulmonary toxicity

One of the major limitations in detecting NP toxicity is the availability of suitable models to assess long-term effects of inhaled NPs. Inhalation and IT tests are the most common methods for the estimation of pulmonary toxicity and inflammation [85]. Nano aerosol chamber is designed for in vitro toxicity (NACIVT) of inhaled NP, including silver (Ag) and carbon (C) NP on normal and compromised airway epithelia [86]. A computational method integrated with fluid flow and calculation of particle dispersion is being developed to estimate the occupational exposure of TiO_2 [87].

6.6 Discussion

Characterization, quantification and biological monitoring with imaging methods on the reactions or systemic disposition or transport to different organs after exposure to NPs is an important domain that needs further research and remains the scope of future research [88–89]. More research should be taken up to understand the underlying mechanism of cytotoxicity caused by NPs in pulmonary and CV disorders. More methods for detecting minute levels of toxicity induced by NPs in tissues need to be researched [90]. The point of concern is exposure of human to uncharacterized components of air pollution as NPs, the effects of which are yet to be known. It is therefore the need of the hour for an awareness among masses, research information to percolate more from laboratory benches to the masses for the health of the mankind.

References

[1]   http://www.who.int/mediacentre/news/releases/2014/air-pollution/en/.
[2]   Lu CF, Yuan XY, Li LZ, Zhou W, Zhao J, Wang YM, Peng SQ. Combined exposure to nano-silica and lead induced potentiation of oxidative stress and DNA damage in human lung epithelial cells. Ecotoxicol Environ Saf. 2015. Dec;122:537–544.

[3]   Stone V, Johnston H, Clift MJ. Air pollution, ultrafine and nanoparticle toxicology: cellular and molecular interactions. IEEE Trans Nanobioscience. 2007. Dec;6(4):331–340.

[4]   Hougaard KS, Campagnolo L, Chavatte-Palmer P, Tarrade A, Rousseau-Ralliard D, Valentino S, Park MV, de Jong WH, Wolterink G, Piersma AH, Ross BL, Hutchison GR, Hansen JS, Vogel U, Jackson P, Slama R, Pietroiusti A, Cassee FR. A perspective on the developmental toxicity of inhaled nanoparticles. Reprod Toxicol. 2015. Aug 15;56:118–140.

[5]   Heusinkveld HJ, Wahle T, Campbell A, Westerink RHS, Tran L, Johnston H, Stone V, Cassee FR, Schins RPF. Neurodegenerative and neurological disorders by small inhaled particles. Neurotoxicol. 2016 Sep;56:94–106.

[6]   Yacobi NR, Fazllolahi F, Kim YH, Sipos A, Borok Z, Kim KJ, Crandall ED. Nanomaterial interactions with and trafficking across the lung alveolar epithelial barrier: implications for health effects of air-pollution particles. Air Qual Atmos Health. 2011. Mar 1;4(1):65–78.

[7]   Morimoto Y, Horie M, Kobayashi N, Shinohara N, Shimada M. Inhalation toxicity assessment of carbon-based nanoparticles. Acc Chem Res. 2013 Mar 19;46(3):770–778.

[8]   Hiraku Y, Nishikawa Y, Ma N, Afroz T, Mizobuchi K, Ishiyama R, Matsunaga Y, Ichinose T, Kawanishi S, Murata M. Nitrative DNA damage induced by carbon-black nanoparticles in macrophages and lung epithelial cells.

[9]   Yoshioka Y, Kuroda E, Hirai T, Tsutsumi Y, Ishii KJ. Allergic responses Induced by the immuno-modulatory effects of nanomaterials upon skin exposure. Front Immunol. 2017 Feb 16;8:169.

[10]   Schramm F, Lange M, Hoppmann P, Heutelbeck A. Cytotoxicity of carbon nanohorns in different human cells of the respiratory system.J. Toxicol Environ Health A. 2016; 79 (22–23): 1085–1093.

[11.]   Viswanath B, Kim S. Influence of nanotoxicity on human health and environment: The alternative strategies. Rev Environ Contam Toxicol. 2017;242:61–104.

[12]   Sayers BC, Germolec DR, Walker NJ, Shipkowski KA, Stout MD, Cesta MF, Roycroft JH, White KL, Baker GL, Dill JA, Smith MJ. Respiratory toxicity and immunotoxicity evaluations of microparticle and nanoparticle C_{60} fullerene aggregates in mice and rats following nose-only inhalation for 13 weeks. Nanotoxicol. 2016 Dec;10(10):1458–1468.

[13]   Su WC, Cheng YS. Carbon nanotubes size classification, characterization and nasal airway deposition. Inhal Toxicol. 2014 Dec;26(14):843–852.

[14]   Sturm R. Nanotubes in the human respiratory tract - Deposition modeling. Z Med Phys. 2015 Jun;25(2):135–145.

[15]   Manke A, Luanpitpong S, Dong C, Wang L, He X, Battelli L, Derk R, Stueckle TA, Porter DW, Sager T, Gou H, Dinu CZ, Wu N, Mercer RR, Rojanasakul Y.Effect of fiber length on carbon nanotube-induced fibrogenesis. Int J Mol Sci. 2014 Apr 29;15(5):7444–7461.

[16]   Tyurina YY, Kisin ER, Murray A, Tyurin VA, Kapralova VI, Sparvero LJ, Amoscato AA, Samhan-Arias AK, Swedin L, Lahesmaa R, Fadeel B, Shvedova AA, Kagan VE. Global phospho-lipidomics analysis reveals selective pulmonary peroxidation profiles upon inhalation of single-walled carbon nanotubes. ACS Nano. 2011 Sep 27;5(9):7342–7353.

[17]   Phuyal S, Kasem M, Rubio L, Karlsson HL, Marcos R, Skaug V, Zienolddiny S., Effects on human bronchial epithelial cells following low-dose chronic exposure to nanomaterials: A 6-month transformation study. Toxicol In Vitro. 2017 Oct;44:230–240.

[18]   Polimeni M, Gulino GR, Gazzano E, Kopecka J, Marucco A, Fenoglio I, Cesano F, Campagnolo L, Magrini A, Pietroiusti A, Ghigo D, Aldieri E.Multi-walled carbon nanotubes directly induce epithelial-mesenchymal transition in human bronchial epithelial cells via the TGF-β-mediated Akt/GSK-3β/SNAIL-1 signalling pathway. Part Fibre Toxicol. 2016 Jun 1;13(1):27.

[19]   Kim JE, Lee S, Lee AY, Seo HW, Chae C, Cho MH. Intratracheal exposure to multi-walled carbon nanotubes induces a nonalcoholic steatohepatitis-like phenotype in C57BL/6J mice. Nanotoxicol. 2015;9(5):613–623.

[20] Snyder-Talkington BN, Schwegler-Berry D, Castranova V, Qian Y, Guo NL. Multi-walled carbon nanotubes induce human microvascular endothelial cellular effects in an alveolar-capillary co-culture with small airway epithelial cells. Part Fibre Toxicol. 2013 Aug 1;10:35.

[21] Snyder-Talkington BN, Pacurari M, Dong C, Leonard SS, Schwegler-Berry D, Castranova V, Qian Y, Guo NL.Systematic analysis of multiwalled carbon nanotube-induced cellular signaling and gene expression in human small airway epithelial cells. Toxicol Sci. 2013 May;133(1):79–89.

[22] Morimoto Y, Hirohashi M, Ogami A, Oyabu T, Myojo T, Todoroki M, Yamamoto M, Hashiba M, Mizuguchi Y, Lee BW, Kuroda E, Shimada M, Wang WN, Yamamoto K, Fujita K, Endoh S, Uchida K, Kobayashi N, Mizuno K, Inada M, Tao H, Nakazato T, Nakanishi J, Tanaka I. Pulmonary toxicity of well-dispersed multi-wall carbon nanotubes following inhalation and intratracheal instillation. Nanotoxicol. 2012 Sep;6(6):587–599.

[23] Poulsen SS, Knudsen KB, Jackson P, Weydahl IE, Saber AT, Wallin H, Vogel U.Multi-walled carbon nanotube-physicochemical properties predict the systemic acute phase response following pulmonary exposure in mice. PLoS One. 2017 Apr 5;12(4):e0174167.

[24] Kinaret P, Ilves M, Fortino V, Rydman E, Karisola P, Lähde A, Koivisto J, Jokiniemi J, Wolff H, Savolainen K, Greco D, Alenius H. Inhalation and oropharyngeal aspiration exposure to rod-like carbon nanotubes induce similar airway inflammation and biological responses in mouse lungs. ACS Nano. 2017 Jan 24;11(1):291–303.

[25] Kasai T, Umeda Y, Ohnishi M, Mine T, Kondo H, Takeuchi T, Matsumoto M, Fukushima S.Lung carcinogenicity of inhaled multi-walled carbon nanotube in rats. Part Fibre Toxicol. 2016 Oct 13; 13(1):53.

[26] Snyder-Talkington BN, Dong C, Porter DW, Ducatman B, Wolfarth MG, Andrew M, Battelli L, Raese R, Castranova V, Guo NL, Qian Y. Multiwalled carbon nanotube-induced pulmonary inflammatory and fibrotic responses and genomic changes following aspiration exposure in mice: A 1-year postexposure study. J Toxicol Environ Health A. 2016;79(8):352–366.

[27] Brown TA, Lee JW, Holian A, Porter V, Fredriksen H, Kim M, Cho YH.Alterations in DNA methylation corresponding with lung inflammation and as a biomarker for disease development after MWCNT exposure. Nanotoxicol. 2016;10(4):453–461.

[28] Sargent LM, Porter DW, Staska LM, Hubbs AF, Lowry DT, Battelli L, Siegrist KJ, Kashon ML, Mercer RR, Bauer AK, Chen BT, Salisbury JL, Frazer D, McKinney W, Andrew M, Tsuruoka S, Endo M, Fluharty KL, Castranova V, Reynolds SH. Promotion of lung adenocarcinoma following inhalation exposure to multi-walled carbon nanotubes. Part Fibre Toxicol. 2014 Jan 9;11:3.

[29] Sager TM, Wolfarth MW, Battelli LA, Leonard SS, Andrew M, Steinbach T, Endo M, Tsuruoka S, Porter DW, Castranova V. Investigation of the pulmonary bioactivity of double-walled carbon nanotubes. J Toxicol Environ Health A. 2013;76(15):922–936.

[30] Tsang MP, Hristozov D, Zabeo A, Koivisto AJ, Jensen ACØ, Jensen KA, Pang C, Marcomini A, Sonnemann G. Probabilistic risk assessment of emerging materials: case study of titanium dioxide nanoparticles. Nanotoxicol. 2017 May;11(4):558–568.

[31] Hanot-Roy M, Tubeuf E, Guilbert A, Bado-Nilles A, Vigneron P, Trouiller B, Braun A, Lacroix G. Oxidative stress pathways involved in cytotoxicity and genotoxicity of titanium dioxide (TiO2) nanoparticles on cells constitutive of alveolo-capillary barrier in vitro. Toxicol In Vitro. 2016 Jun;33:125–135.

[32] Oyabu T, Morimoto Y, Izumi H, Yoshiura Y, Tomonaga T, Lee BW, Okada T, Myojo T, Shimada M, Kubo M, Yamamoto K, Kawaguchi K, Sasaki T.Comparison between whole-body inhalation and nose-only inhalationon the deposition and health effects of nanoparticles.

[33] Whitwell H, Mackay RM, Elgy C, Morgan C, Griffiths M, Clark H, Skipp P, Madsen J. Nanoparticles in the lung and their protein corona: the few proteins that. count.Nanotoxicol. 2016 Nov;10(9):1385–1394.

[34] Stapleton PA, McBride CR, Yi J, Nurkiewicz TR. Uterine microvascular sensitivity to nanomaterial inhalation: An in vivo assessment. Toxicol Appl Pharmacol. 2015 Nov 1; 288(3):420–428.

[35] Warheit DB, Donner EM. Risk assessment strategies for nanoscale and fine-sized titanium dioxide particles: Recognizing hazard and exposure issues. Food Chem Toxicol. 2015 Nov;85:138–147.

[36] Han SG, Newsome B, Hennig B.Titanium dioxide nanoparticles increase inflammatory responses in vascular endothelial cells.

[37] Srivastava RK, Rahman Q, Kashyap MP, Singh AK, Jain G, Jahan S, Lohani M, Lantow M, Pant AB. Nano-titanium dioxide induces genotoxicity and apoptosis in human lung cancer cell line, A549. Hum Exp Toxicol. 2013 Feb;32(2):153–166.

[38] Gaté L, Disdier C, Cosnier F, Gagnaire F, Devoy J, Saba W, Brun E, Chalansonnet M, Mabondzo A. Biopersistence and translocation to extrapulmonary organs of titanium dioxide nanoparticles after subacute inhalation exposure to aerosol in adult and elderly rats. Toxicol Lett. 2017 Jan 4;265:61–69.

[39] Theodorou IG, Ruenraroengsak P, Gow A, Schwander S, Zhang JJ, Chung KF, Tetley TD, Ryan MP, Porter AE. Effect of pulmonary surfactant on the dissolution, stability and uptake of zinc oxide nanowires by human respiratory epithelial cellsNanotoxicol.2016 Nov;10(9):1351–1362.

[40] Osmond-McLeod MJ, Osmond RI, Oytam Y, McCall MJ, Feltis B, Mackay-Sim A, Wood SA, Cook AL.Surface coatings of ZnO nanoparticles mitigate differentially a host of transcriptional, protein and signalling responses in primary human olfactory cells. Part Fibre Toxicol. 2013 Oct 21;10(1):54.

[41] Chuang HC, Chuang KJ, Chen JK, Hua HE, Shen YL, Liao WN, Lee CH, Pan CH, Chen KY, Lee KY, Hsiao TC, Cheng TJ. Pulmonary pathobiology induced by zinc oxide nanoparticles in mice: A 24-hour and 28-day follow-up study. Toxicol Appl Pharmacol. 2017 Jul 15;327:13–22.

[42] Sutunkova MP, Katsnelson BA, Privalova LI, Gurvich VB, Konysheva LK, Shur VY, Shishkina EV, Minigalieva IA, Solovjeva SN, Grebenkina SV, Zubarev IV. On the contribution of the phagocytosis and the solubilization to the iron oxide nanoparticles retention in and elimination from lungs under long-term inhalation exposure. Toxicol, 2016 Jul; 1: 363–364:19–28.

[43] Fröhlich E, Roblegg E. Oral uptake of nanoparticles: human relevance and the role of in vitro systems. Arch Toxicol. 2016 Oct;90(10):2297–2314.

[44] Sun Z, Yang L, Chen KF, Chen GW, Peng YP, Chen JK, Suo G, Yu J, Wang WC, Lin CH. Nano zerovalent iron particles induce pulmonary and cardiovascular toxicity in an in vitro human co-culture model. Nanotoxicol. 2016 Sep;10(7):881–890.

[45] Davidson DC, Derk R, He X, Stueckle TA, Cohen J, Pirela SV, Demokritou P, Rojanasakul Y, Wang L. Direct stimulation of human fibroblasts by nCeO2 in vitro is attenuated with an amorphous silica coating. Part Fibre Toxicol. 2016 May;413(1):23.

[46] Verstraelen S, Remy S, Casals E, De Boever P, Witters H, Gatti A, Puntes V, Nelissen I. Gene expressionprofiles reveal distinct immunological responses of cobalt and cerium dioxide nanoparticles in two in vitro lung epithelial cell models. Toxicol Lett. 2014 Aug 4;228(3):157–169.

[47] Schwotzer D, Ernst H, Schaudien D, Kock H, Pohlmann G, Dasenbrock C, Creutzenberg O. Effects from a 90-day inhalation toxicity study with cerium oxide and barium sulfate nanoparticles in rats. Part Fibre Toxicol. 2017 Jul 12;14(1):23.

[48] Nemmar A, Al-Salam S, Beegam S, Yuvaraju P, Ali BH.The acute pulmonary and thrombotic effects of cerium oxide nanoparticles after intratracheal instillation in mice. Int J Nanomedicine. 2017 Apr 10;12:2913–2922.

[49] Tian L, Shang Y, Chen R, Bai R, Chen C, Inthavong K, Tu J. A combined experimental and numerical study on upper airway dosimetry of inhaled nanoparticles from an electrical discharge machine shop. Part Fibre Toxicol. 2017 Jul 12;14(1):24.

[50] Pirela SV, Lu X, Miousse I, Sisler JD, Qian Y, Guo N, Koturbash I, Castranova V, Thomas T, Godleski J, Demokritou P. Effects of intratracheally instilled laser printer-emitted engineered nanoparticles in a mouse model: A case study of toxicological implications from nanomaterials released during consumer use. NanoImpact. 2016 Jan;1:1–8.

[51] Riebeling C, Luch A, Götz ME. Comparative modeling of exposure to airborne nanoparticles released by consumer spray products. Nanotoxicol. 2016;10(3):343–351.

[52] Sisler JD, Pirela SV, Friend S, Farcas M, Schwegler-Berry D, Shvedova A, Castranova V, Demokritou P, Qian Y. Small airway epithelial cells exposure to printer-emitted engineered nanoparticles induces cellular effects on human microvascular endothelial cells in an alveolar-capillary co-culture model. Nanotoxicol. 2015;9(6):769–779.

[53] Mills NL, Miller MR, Lucking AJ, Beveridge J, Flint L, Boere AJ, Fokkens PH, Boon NA, Sandstrom T, Blomberg A, Duffin R, Donaldson K, Hadoke PW, Cassee FR, Newby DE. Combustion-derived nanoparticulate induces the adverse vascular effects of diesel exhaust inhalation. Eur Heart J. 2011 Nov;32(21):2660–2671.

[54] Yamagishi N, Ito Y, Ramdhan DH, Yanagiba Y, Hayashi Y, Wang D, Li CM, Taneda S, Suzuki AK, Taya K, Watanabe G, Kamijima M, Nakajima T. Effect of nanoparticle-rich diesel exhaust on testicular and hippocampus steroidogenesis in male rats. Inhal Toxicol. 2012 Jul;24(8):459–467.

[55] Cena LG, Keane MJ, Chisholm WP, Stone S, Harper M, Chen BT. A novel method for assessing respiratory deposition of welding fume nanoparticles. J Occup Environ Hyg. 2014;11(12):771–780.

[56] Braakhuis HM, Giannakou C, Peijnenburg WJ, Vermeulen J, van Loveren H, Park MV. Simple in vitro models can predict pulmonary toxicity of silver nanoparticles. Nanotoxicol. 2016 Aug;10(6):770–779.

[57] Davenport LL, Hsieh H, Eppert BL, Carreira VS, Krishan M, Ingle T, Howard PC, Williams MT, Vorhees CV, Genter MB. Systemic and behavioral effects of intranasal administration of silver nanoparticles. Neurotoxicol Teratol. 2015 Sep-Oct;51:68–76.

[58] Silva RM, Anderson DS, Franzi LM, Peake JL, Edwards PC, Van Winkle LS, Pinkerton KE. Pulmonary effects of silver nanoparticle size, coating, and dose over time upon intratracheal instillation. Toxicol Sci. 2015 Mar;144(1):151–162.

[59] Silva RM, Xu J, Saiki C, Anderson DS, Franzi LM, Vulpe CD, Gilbert B, Van Winkle LS, Pinkerton KE. Short versus long silver nanowires: a comparison of in vivo pulmonary effects post instillation. Part Fibre Toxicol. 2014 Oct 8;11:52.

[60] Dong MS, Choi JY, Sung JH, Kim JS, Song KS, Ryu HR, Lee JH, Bang IS, An K, Park HM, Song NW, Yu IJ. Gene expression profiling of kidneys from Sprague-Dawley rats following 12-week inhalation exposure to silver nanoparticles. Toxicol Mech Methods. 2013 Jul;23(6):437–448.

[61] Campagnolo L, Massimiani M, Vecchione L, Piccirilli D, Toschi N, Magrini A, Bonanno E, Scimeca M, Castagnozzi L, Buonanno G, Stabile L, Cubadda F, Aureli F, Fokkens PH, Kreyling WG, Cassee FR, Pietroiusti A. Silver nanoparticles inhaled during pregnancy reach and affect the placenta and the foetus. Nanotoxicol. 2017 Jun;11(5):687–698.

[62] Kumar A, Bicer EM, Morgan AB, Pfeffer PE, Monopoli M, Dawson KA, Eriksson J, Edwards K, Lynham S, Arno M, Behndig AF, Blomberg A, Somers G, Hassall D, Dailey LA, Forbes B, Mudway IS. Enrichment of immunoregulatory proteins in the biomolecular corona of nanoparticles within human respiratory tract lining fluid. Nanomedicine. 2016 May;12(4):1033–1043.

[63]   Yu Y, Duan J, Li Y, Yu Y, Jin M, Li C, Wang Y. Sun Z Combined toxicity of amorphous silica
       nanoparticles and methylmercury to human lung epithelial cells. Ecotoxicol Environ Saf. 2015
       Feb;112:144–152.
[64]   Kondej D, Sosnowski TR. Alteration of biophysical activity of pulmonary surfactant by
       aluminosilicate nanoparticles. Inhal Toxicol. 2013 Feb;25(2):77–83.
[65]   Stefaniak AB, Seehra MS, Fix NR, Leonard SS. Lung biodurability and free radical production of
       cellulose nanomaterials. Inhal Toxicol. 2014 Oct;26(12):733–749.
[66]   Prajapati MV, Adebolu OO, Morrow BM, Cerreta JM. Original Research: Evaluation of
       pulmonary response to inhaledtungsten (IV) oxide nanoparticles in golden Syrian hamsters.
       Exp Biol Med (Maywood). 2017 Jan;242(1):29–44.
[67]   Sisler JD, Li R, McKinney W, Mercer RR, Ji Z, Xia T, Wang X, Shaffer J, Orandle M, Mihalchik AL,
       Battelli L, Chen BT, Wolfarth M, Andrew ME, Schwegler-Berry D, Porter DW, Castranova V, Nel
       A, Qian Y.Differentialpulmonary effects of CoO and La2O3 metal oxide nanoparticle responses
       during aerosolized inhalation in mice.Part Fibre Toxicol.2016 Aug 15;13(1):42.
[68]   Shin SH, Lim CH, Kim YS, Lee YH, Kim SH, Kim JC. Twenty-eight-day repeated inhalation
       toxicity study of nano-sized lanthanum oxide in male sprague-dawley rats. Environ Toxicol.
       2017 Apr;32(4):1226–1240.
[69]   van Berlo D, Hullmann M, Wessels A, Scherbart AM, Cassee FR, Gerlofs-Nijland ME, Albrecht
       C, Schins RP.Investigation of the effects of short-term inhalation of carbon nanoparticles on
       brains and lungs of c57bl/6j and p47(phox-/-) mice. Neurotoxicol. 2014 Jul;43:65–72.
[70]   Cohen JM, Derk R, Wang L, Godleski J, Kobzik L, Brain J, Demokritou P. Tracking translocation
       of industrially relevant engineered nanomaterials (ENMs) across alveolar epithelial
       monolayers in vitro. Nanotoxicol. 2014 Aug;8 Suppl 1:216–225.
[71]   Morimoto Y, Izumi H, Kuroda E. Significance of persistent inflammation in respiratory
       disorders induced by nanoparticles. J Immunol Res. 2014;2014:962871.
[72]   Larsen ST, Dallot C, Larsen SW, Rose F, Poulsen SS, Nørgaard AW, Hansen JS, Sørli JB, Nielsen
       GD, Foged C.Mechanism of action of lung damage caused by a nanofilm spray product. Toxicol
       Sci. 2014 Aug 1;140(2):436–444.
[73]   Park EJ, Oh SY, Lee SJ, Lee K, Kim Y, Lee BS, Kim JS.Chronic pulmonary accumulation of
       iron oxide nanoparticles induced Th1-type immune response stimulating the function of
       antigen-presenting cells. Environ Res. 2015 Nov;143(Pt A):138–147.
[74]   Mann EE, Thompson LC, Shannahan JH, Wingard CJ. Changes in cardiopulmonary function induced
       by nanoparticles. Wiley Interdiscip Rev Nanomed Nanobiotechnol. 2012 Nov-Dec;4(6):691–702.
[75]   J Toxicol Environ Health A. 2015;78(12):711–24. Blum JL[1], Edwards JR, Prozialeck WC, Xiong JQ,
       Zelikoff JT. Effects of maternal exposure to Cadmium oxide nanoparticles during pregnancy on
       maternal and offspring kidney injury markers using a murine model.
[76]   Pirela SV, Miousse IR, Lu X, Castranova V, Thomas T, Qian Y, Bello D, Kobzik L, Koturbash
       I, Demokritou P. Effects of laser printer-emitted engineered nanoparticles on cytotoxicity,
       chemokine expression, reactive oxygen species, DNA methylation, and DNA Damage:
       A comprehensive in vitro analysis in human small airway epithelial cells, macrophages, and
       lymphoblasts. Environ Health Perspect. 2016 Feb;124(2):210–219.
[77]   Paul E, Franco-Montoya ML, Paineau E, Angeletti B, Vibhushan S, Ridoux A, Tiendrebeogo A,
       Salome M, Hesse B, Vantelon D, Rose J, Canouï-Poitrine F, Boczkowski J, Lanone S, Delacourt
       C, Pairon JC. Pulmonary exposure to metallic nanomaterials during pregnancy irreversibly
       impairs lung development of the offspring. Nanotoxicol. 2017 May;11(4):484–495.
[78]   Husain M, Wu D, Saber AT, Decan N, Jacobsen NR, Williams A, Yauk CL, Wallin H, Vogel U,
       Halappanavar S.Intratracheally instilled titanium dioxide nanoparticles translocate to heart
       and liver and activate complement cascade in the heart of C57BL/6 mice. Nanotoxicol.
       2015;9(8):1013–1022.

[79]   Rydman EM, Ilves M, Koivisto AJ, Kinaret PA, Fortino V, Savinko TS, Lehto MT, Pulkkinen V, Vippola M, Hämeri KJ, Matikainen S, Wolff H, Savolainen KM, Greco D, Alenius H. Inhalation of rod-like carbon nanotubes causes unconventional allergic airway inflammation. Part Fibre Toxicol. 2014 Oct 16;11:48.

[80]   Kido T, Tsunoda M, Kasai T, Sasaki T, Umeda Y, Senoh H, Yanagisawa H, Asakura M, Aizawa Y, Fukushima S.The increases in relative mRNA expressions of inflammatory cytokines and chemokines in splenic macrophages from rats exposed to multi-walled carbon nanotubes by whole-body inhalation for 13 weeks. Inhal Toxicol. 2014 Oct;26(12):750–758.

[81]   Liberda EN, Cuevas AK, Qu Q, Chen LC. The acute exposure effects of inhaled nickel nanoparticles on murine endothelial progenitor cells. Inhal Toxicol. 2014 Aug;26(10):588–597.

[82]   Chen JK, Ho CC, Chang H, Lin JF, Yang CS, Tsai MH, Tsai HT, Lin P.Particulate nature of inhaled zinc oxide nanoparticles determines systemic effects and mechanisms of pulmonary inflammation in mice. Nanotoxicol. 2015 Feb;9(1):43–53.

[83]   Snyder-Talkington BN, Dymacek J, Porter DW, Wolfarth MG, Mercer RR, Pacurari M, Denvir J, Castranova V, Qian Y, Guo NL. System-based identification of toxicity pathways associated with multi-walled carbon nanotube-induced pathological responses. Toxicol Appl Pharmacol. 2013 Oct 15;272(2):476–489.

[84]   Hopkins LE, Patchin ES, Chiu PL, Brandenberger C, Smiley-Jewell S, Pinkerton KE. Nose-to-brain transport of aerosolised quantum dots following acute exposure. Nanotoxicol. 2014 Dec8(8):885–893.

[85]   Morimoto Y, Izumi H, Yoshiura Y, Fujisawa Y Fujita K Significance of Inratracheal Instillation Tests for the Screening of Pulmonary Toxicity of Nanomaterials. J UOEH. 2017;39(2):123–132.

[86]   Pilou M, Vaquero-Moralejo C, Jaén M, Lopez De Ipiña Peña J.Evaluating adverse effects of inhaled nanoparticles by realistic in vitro technology. Nanomaterials (Basel). 2017 Feb 22;7(2). pii E49.

[87]   Geiser M, Jeannet N, Fierz M, Burtscher H., Neofytou P, Housiadas C. Modeling of occupational exposure to accidentally released manufactured nanomaterials in a production facility and calculation of internal doses by inhalation. Int J Occup Environ Health. 2016 Jul;22(3):249–258.

[88]   Rinaldo M, Andujar P, Lacourt A, Martinon L, Canal Raffin M, Dumortier P, Pairon JC, Brochard P. Perspectives in biological monitoring of inhaled nanosized particles. Ann Occup Hyg. 2015 Jul;59(6):669–680.

[89]   Brandi N. Snyder-Talkington, Yong Qian, Vincent Castranova, Nancy L. Guo. New perspectives for *in vitro* risk assessment of multi-walled carbon nanotubes: Application of coculture and bioinformatics. J Toxicol Environ Health B Crit Rev. 2012;15(7): 468–492.

[90]   Landsiedel R, Ma-Hock L, Haussmann HJ, van Ravenzwaay B, Kayser M, Wiench K. Inhalation studies for the safety assessment of nanomaterials: status quo and the way forward. Wiley Interdiscip Rev Nanomed Nanobiotechnol. 2012 Jul-Aug;4(4):399–413.

7 Nanoparticles and genotoxicity

Abstract: Nanotechnology has wide scale applications in industry, medicine, our day-to-day lives and commercial sectors. However, many nanoparticles (NPs) have caused chromosomal aberrations, DNA, oxidative DNA damage and mutations and mediated cytotoxicity and genotoxic effects. Recently, it has been observed that NPs cause epigenetic changes. We discuss in this chapter the effect of different NPs mediating genotoxic effects. Green synthesis of nanoparticles have been able to reduce the genotoxic effects caused by NPs. In this chapter, we discuss the different genotoxic effects of different types of NPs detected by biochemical assays. We also discuss the strategies used by scientists to overcome genotoxicity of NPs.

Future research needs to focuss on designing the NPs without genotoxic properties.

Keywords: Formamidopyrimidine DNA glycosylase, Reactive oxygen species (ROS), Engineered nanoparticles (ENPs), ECs, Ultrafine particles (UFPs), endothelial cells (ECs), genotoxity

7.1 Introduction

Together with cytotoxicity and genotoxicity induced by NPs (Figure 7.1), epigenetic changes that include those heritable gene expression changes that do not reveal alteration in the underlying DNA coding sequence, including alterations in DNA methylation patterns, histone modifications and miRNA expressions are associated with NP-mediated toxicity [1,2].

Ultrafine particles (UFPs d < 100 nm) generated from combustion of logwood and pellet revealed genotoxic effects on A549 cells (human lung carcinoma cells) [3]. NPs affect the human umbilical vein endothelial cells (HUVECs) used as a model to study endothelial cells (ECs) that form the lining of blood vessels in vitro. HUVECs have been reported to uptake NPs by endocytosis and transport them by exocytosis and paracellular pathways. NPs could cause cytotoxicity, genotoxicity, endothelial nitric oxide synthase 3 (NOS3) or eNOS expression and EC activation, oxidative stress, inflammatory response and dysfunctioning of organelles in HUVEC. [4].

Engineered nanoparticles (ENPs) have been known to induce genotoxicity, cytotoxicity and carcinogenicity by reactive oxygen species (ROS) production, toxic ion release from soluble ENPs and binding with mitotic spindle or its components have been reported to cause an increase in oxidative stress, thus affecting the cell cycle checkpoint and causing inhibition of antioxidant defense system [5,6]. Several methods including single-cell gel electrophoresis, or comet assay, cytokinesis-block micronucleus assay and detection of oxidative stress parameters including detection of ROS, lipid peroxidation and glutathione depletion find application in detecting genotoxicity induced by ENPs [7].

https://doi.org/10.1515/9783110579093-007

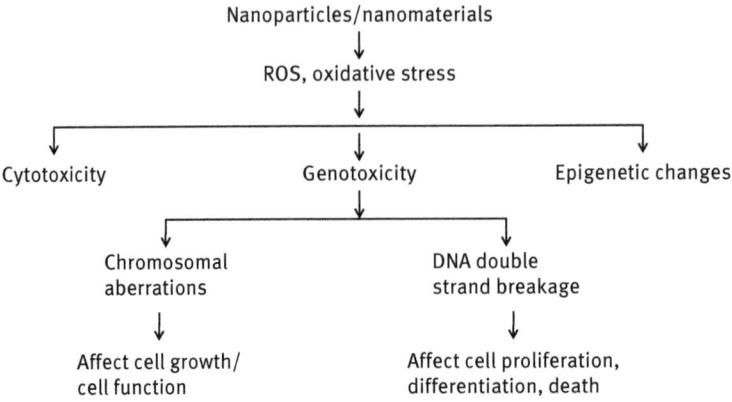

Figure 7.1: NPs and genotoxicity.

The current day research has revealed genotoxicty mediated by a diverse metal, polymeric, engineered and magnetic NPs [8,9]. In parallel, there are developments in the advancement of NP genotoxicity testing [10]. Comet assay finds wide scale application in detection of DNA damage [11]; it was first described by Ostling and Johansonin the OECD Guideline 489 for the in vivo mammalian comet assay and finds large-scale application in detecting NP-mediated genotoxicity [12]. In vitro cytokinesis-block micronucleus (CBMN) assay detected genotoxic effects of 20 nm nanosilver in human liver carcinoma cell line HepG2 and colon cancer cell line $_{Caco-2}$that are human epithelial cell line are used as a model of the intestinal epithelial barrier. [13].

Different genotoxic effects mediated by NPs warrant their usage prior to their complete investigation. Green synthesis of NPs, that is, synthesis of NPs from natural products has been reported with reduced genotoxic effects. In this chapter, we discuss the (i) genetoxic effects of NPs and (ii) different strategies used to reduce their genotoxicity and enhance their bioapplicability.

7.2 Titanium dioxide NPs

Titanium dioxide nanoparticles (TiO_2NPs) could increase the genotoxicity of As(III) in human–hamster hybrid (A_L) cells by partially inhibiting endocytosis pathway inhibitors together with release of ROS [14]. However, green synthesis Ginkgo biloba leaves polyprenol (GBP) in an oil-in-water (o/w) type nanoemulsion (NE) revealed low cytotoxic and genotoxic effects possibly by inhibiting Ca^{2+}-ATPase and Ca^{2+}/Mg^{2+}-ATPase and intracellular Ca^{2+} release into extracellular spaces [15]. Inhalation of nanomaterials (NMs) or engineered surface functionalized NMs reveal cytotoxicity and genotoxicity in EpiAirway™ 3D human bronchial models in nMs and sur 15 nm SiO_2 (four variants), 10 nm ZrO_2 [16] and nanosilver [17], ZnONM-110, TiO_2 NM-105, $BaSO_4$ NM-220 and AlOOH NMs and their genotoxicity is controlled by the core material composition [18]. TiO_2NPs and zinc oxide NPs (ZnONPs) have been exploited maximally in cancer eradication [19]. Curcumin-loaded TiO_2NPs (CTNPs) by coprecipitation method have been reported to be with improved bioavailability of curcumin and site-specific drug delivery with no genotoxicity [20]. TiO_2NP has been reported for its toxic cytotoxicity, genotoxicity and epigenetic properties effect alone or in its form of surface functionalization coated with silica or citrate, and P25 in human lung epithelial cell line A549 wherein coating by citrate increased their cytotoxicity and genotoxicity, also exerting epigenotoxic effects with reduction of LINE-1 methylation levels [21]. TiO_2NPs and ZnONPs when tested on human nasal mucosa (monolayer, air–liquid interface and mini organ culture) revealed nontoxic effects of TiO_2NPs but ZnONPs revealed cyto- and genotoxicity and was enhanced by TiO_2NPs. But adsorption of dissolved zinc ions onto TiO_2NPs acted as a major antagonistic mechanism, thus proving to be a safe alternative to the sole use of ZnONPs [22]. TiO_2NPs on mammalian lung fibroblast cells (V-79) revealed ultra-structural changes such as swollen mitochondria and nuclear membrane disruption, induction of indirect mutagenic and genotoxic responses by free radical generation with increase in 6-thioguanine-resistant (6TGR), Hypoxanthine-guanine phosphoribosyltransferase (HGPRT) and mutant frequency that is an indicator of potential carcinogenicity stress in lungs has been reported [23]. TiO_2 P25 NP has been reported to induce genotoxic, inflammation, cytotoxicity and oxidative stress in lungs [25]. Engineered TiO_2 NPs caused genotoxicity manifested by chromosomal aberrations [25]. TiO_2NPs including titanium carbide, titanium carbonitride, titanium (II) oxide, titanium (III) oxide, titanium (IV) oxide, titanium nitride and titanium silicon oxide on human alveolar epithelial (HPAEpiC) and pharynx (HPPC) cell lines revealed cytotoxic and genotoxic effects with genetic alterations in DNA controlling damage or repair, oxidative stress and apoptosis [26]. TiO_2NPs lead to genotoxicity causing DNA damage mainly resulting from oxidized nucleotides in liver and lung tissues with impaired DNA damage, interrupted liver metabolic homeostasis, oxidative stress, inflammatory responses and lung apoptosis [27]. TiO_2NPs have been reported to show inflammatory response, the acute phase response (APR) and the genotoxic effect after single intratracheal instillation. [28]. However, TiO_2NPs have not been able to show genotoxicity on mouse liver or bone marror when injected in multiple intravenous injections in male gpt Delta C57BL/6J mice [29].

　　TiO_2NPs exposed to male Swiss Webster mice orally caused apoptotic DNA lysis, point mutation of Presenilin 1 gene at exon 5, mutagenesis in brain tissue probably leading to Alzheimer's disease (AD, [30]). Inhaled TiO_2NPs cause oxidative stress on cell lines causing ROS generation, activation of γH2AX, DNA damage, cell cycle arrest and HSP27 and SAPK/JNK proteins were reported as biomarkers to intracellular stress [31]. TiO_2NPs on A549 alveolar epithelial cells reveal genotoxic effects by causing DNA damage, oxidative stress and increased 53BP1 foci counts [32]. TiO_2NPs in A549 cells induce DNA damage with increased micronucleus frequency, increased oxidative stress and ROS generation and DNA breaks and cell cycle arrest in G2/M phase [33].

Different crystal phases of TiO$_2$NP (anatase, rutile and anatase: rutile mixture; 20–26 nm) revealed genotoxic effects, cell cycle arrest in the S-phase indicative of epigenetic changes [34]. TiO$_2$NM on BEAS-2B cells cultured under serum-free conditions led to DNA damage, occurrence of oxidized bases [35]. TiO$_2$ and AgNPs were reported to have genotoxic and cytotoxic effects on rat bone marrow cells when injected intraveneously. [36]. TiO$_2$NPS affected lymphocytes by causing membrane damage, mitochondrial function, metabolic activity and lysosomal membrane stability [37]. Polyacrylate-coated nano-TiO$_2$ product could not cause DNA damage. [38]. TiO$_2$NPs can cause apoptosis, oxidative stress, ROS production and genotoxicity in A549 cell line, P53, P21 upregulation and downregulation of Bcl-2, cleaved caspase-3, reduced catalase and glutathione activity and apoptotic bodies and micronucleus formation. [39]. TiO$_2$NPs induce oxidative DNA damage, increase in the micronucleus frequency, reduced glutathione levels, increased lipid peroxidation and ROS, p53, BAX, Cyto-c, Apaf-1, caspase-9 and caspase-3 increased expression and Bcl-2 decreased expression and apoptosis by caspase-dependent pathway [40].

7.3 Iron oxide NPs

Iron oxide NPs (IONPs) less than 100 nm in diameter, although with promising therapy and imaging, pose a threat to workers and are reported to have genotoxic potential in human, animal and cell culture-based studies [41]. Genotoxic potential of iron oxide (Fe$_2$O$_3$) NPs and ionic form were investigated on orally fed transheterozygous larvae of *Drosophila melanogaster* (*D. melanogaster)* that led to dose-dependent genotoxicity revealed by chromosomal abnormalities [42].

Superparamagnetic iron-oxide NPs (SPIONs) covalently coated with a rhamnose derivative when targeted to glioblastoma tumor cell lines revealed biocompatibility and find applications in the treatment of magnetic fluid hyperthermia (MFH) in malignant astrocytis tumor glioblastoma [43].

Of TiO$_2$NPs (20 nm), iron oxide (8 nm) both uncoated (U-Fe$_3$O$_4$) and oleic acid coated iron oxide NPs (OC-Fe$_3$O$_4$), rhodamine-labeled amorphous silica 25 (Fl-25 SiO$_2$) and 50 nm (Fl-50 SiO) and polylactic glycolic acid polyethylene oxide polymeric NPs tested on diverse cell types including primary cells and those derived from human lymphocytes and lymphoblastoid TK6 cells, vascular and central nervous system (CNS) including human endothelial human cerebral ECs, liver including rat hepatocytes and Kupffer cells, kidney cells including monkey Cos-1 and human embryonic kidney cells 293 (HEK293) cells, lung cells including human bronchial epithelial cell line 16HBE14o widely used to study barrier function of the airway epithelium, respiratory ion transport and cystic fibrosis transmembrane conductance regulator (CFTR) and placenta including human BeWo b30 celline which is a malignant gestational choriocarcinoma of the fetal placenta finding importance in study of placental barrier, revealed

OC-Fe_3O_4 and TiO_2 NPs to be genotoxic [44]. IONPs have been known for inducing cytotoxicity, genotoxicity, developmental toxicity and neurotoxicity [45], causing oxidative stress, DNA starnd damage and apoptosis in human breast cancer cells (MCF-7, [46]). TiO_2NPs have been reported to cause inflammation, pulmonary damage, fibrosis and lung tumors with carcinogenic effects and genotoxity by oxidative stress [47]. Saline dissolved Fe_2O_3 and SiO_2 NPs orally administered to female Winster rats caused oxidative stress, increased membrane lipid peroxidation and release of ROS with decreased intracellular glutathione affecting cell signalling and inflammatory responses, genotoxicity, affect cell proliferation and induce DNA damage in a dose-dependent manner [48]. Oleate-coated iron oxide (OC-Fe_3O_4) NPs were observed to be cytotoxic that caused DNA damage, thereby proving to be genotoxic on human lymphoblastoid TK6 cells and in primary human blood cells. [49]. Fe_3O_4 (IONPs) caused cytotoxicity, oxidative stress and genotoxicity in two human cell lines; skin epithelial A431 and lung epithelial A549, reduced glutathione and ROS production, lipid peroxidation, DNA damage, increased caspase-3 and caspase-9 transcription. [50].

7.4 Nano silver and AgNPs

Nano silver although known for their antimicrobial properties have been reported about their genotoxicity and mutagenicity. Cadmium chloride (CdCl2) has been reported to enhance genotoxic effects of nanosilver in mice liver, kidney and brain tissues indicated by increased apoptosis, DNA damage, higher rates of mutation in presenilin-1 and p53 overexpression, congested blood vessels and the infiltration of leukocytes in the liver, kidney and brain with increased ROS generation [16]. AgNPs lead to release of Ag+ ions but were reported to be genotoxic after oral digestion in an in vitro gastrointestinal model. Citrate-coated AgNPs induced chromosomal damage in bone marrow and oxidative DNA damage and accumulated in digestive juices indicating that AgNPs coatings could add to their genotoxic effects, which might lead to cancer [51]. Cytotoxicity, genotoxicity and antibacterial activity of poly(vinyl alcohol)-coated silver NPs (AgNPs-PVA) and farnesol (FAR) revealed their less cytotoxic and genotoxic potential and shows promises as agents in final endodontic irrigation protocols [52].Genotoxic effects of AgNP on human breast cell lines (MCF-10A, MCF-7 and MDB-MB-231), manifested by DNA damage, increased the level of oxidative DNA damage that could be reduced by combining with aluminium salts, butylparaben or di-n-butylphthalate [53]. Surface functionalization of AgNPs affects cytotoxicity, genotoxicity and mechanism of cellular uptake by macropinocytosis and clathrin-mediated endocytosis in mammalian liver cells [54]. Ag and Au NPs size, surface functionalization and concentration determine distribution and toxicity outcomes in murine animals deposited on mononuclear phagocyte system (MPS) such as the liver and spleen with AgNPs induced alteration of gene expression in oxidative stress, apoptosis and ion transport [55]. Exposure to AgNPs has been reported to activate GADD45α gene, an important stress sensor, when treated on HepG2 and A549 cell lines, which is an adenocarcinomic human alveolar basal epithelial cell line containing the GADD45α promoter-driven luciferase reporter in a dose dependant manner, through their uptake and subsequent release of ions together with cytotoxicity, cell cycle arrest of Sub G1 and G2/M phase, Olive tail moment, micronuclei frequency, and the cellular Ag content. [56]. AgNPs lead to DNA damage and mitotic disorders in human keratinocytes (HaCaT cell line), micronuclei (MNi) induction [57]. Argovit™ AgNPs tested on cancer cell lines including HeLa and breast (MDA-MB-231 and MCF7) cancer revealed release of ROS, cytotoxic effect in a time and dose-dependent manner [58]. AgNPs nanotoxicity is expressed by DNA damage responsive pathway activated by GADD45a gene after 24 h of AgNPs exposure in vitro cell models, an immortal HepG2 and lung epithelial A549 cells [59]. AgNPs including silver acetate and silver nitrate and ionic silver when tested on mouse lymphoma L5718Y cells lead to cytotoxity, mutations of the thymidine kinase (Tk) gene, induction of micronuclei in a concentration, size and composition dependent manner. The mutations induced by AgNPs and those by ionic silver were different [60]. AgNPs leads to Ag^+ formation exposed to human thymidine kinase heterozygote **cell** line TK6 cell line led to cytotoxicity and genotoxicity induced oxidative stress and ROS [61]. In invitro models, AgNPs caused apoptosis, necrosis and DNA strand breaks [62]. AgNPs has been reported of cytotoxic and genotoxic effects in Chinese hamster ovary CHO-K1 cells that are an epithelial cell line obtained from Chinese hamster ovary and CHO-XRS5 cell lines [63] that is X-ray sensitive, CHO-K1 mutant cell line obtained from CHO-K1 cells by treating them with ethyl methanesulphonate and then growing in agar. AgNPs have been reported to cause genotoxicity revealed by mutagenicity, clastogenicity and inducing DNA damage [64]. Silver NPs including c-AgNPs and b-AgNPs and graphene oxide (GO) nanosheets have been reported to cause DNA melting on T-lymphocyte cell lines and primary lymphocytes [65]. AgNPs expressed genotoxicity, dose dependent chromosomal aberrations and heaptotoxicity on mature female albino rats [66]. AgNPs exerted genotoxicity, oxidative stress depletion of GSH and induction of ROS, lipid peroxidation (LPO), SOD and catalase increased expression of proinflammatory cytokines interleukin-1β (IL-1β) and interleukin-6 (IL-6) and DNA damage in human lung epithelial (A549) cells [67]. Cytotoxic and genotoxic effects of silver-NPs exerted on primary Syrian hamster embryo (SHE) cells have been reported to cause cell cycle arrest in GO/G1 phase, inhibition of DNA replication and cell proliferation and apoptosis [68]. It has been reported that 40–59 nm AgNPs cause genotoxic effects [69] in human lymphocytes plants including *Allium cepa* and *Nicotiana tabacum* and animal (Swiss albino male mice) with impairment of nuclear DNA [70]. They have also revealed genotoxic effects and oxidative stress in immortalized, human lung epithelial cell line BEAS-2B cells finding importance as model for understanding of pulmonary epithelial function [71]. AgNPs affected cytotoxicity, inflammation, genotoxicity and developmental toxicity in L929 mouse fibroblast cell line [72].

7.5 Zinc oxide (ZnO) NPs (ZnONPs)

ZnONPs although with tremendous application as drug carriers suffer from the lack of safety and effects of genotoxicity. When orally administered to Swiss mice, chromosomal aberration (CA) was observed by the parameters mentioned in the Economic Co-operation and Development guidelines: observation of DNA damage, increased ROS, aberrant sperm morphology with reduced sperm count and motility in a dose dependent manner [73]. Concanavalin A (Con A) immobilized on the hexagonal ZnONPs has been reported to better the hemagglutination activity but without any cytotoxicity or genotoxicity [74]. ZnONPs exerted size dependent genotoxicity and cytotoxicity mediated by ROS generation and oxidative stress on human peripheral blood lymphocytes (PBL). Vitamin C or Quercetin treatment is reported to cirumvent such toxic effects. [75]. ZnONP and Zn ion can lead to cyto- and genotoxic effects in vitro when tested on Madin-Darby canine kidney (MDCK) cells with ZnO NPs with more DNA and chromosomal damaging potential [76]. ZnONP on human peripheral blood mononuclear cells (PBMCs) *in vitro* and in Swiss albino male mice *in vivo* revealed for cyto-genotoxicity and oxidative damage with significant decrease in mitochondrial membrane potential, ROS generation, leading to apoptosis, G0/G1 cell cycle arrest, chromosome aberrations and micronuclei formation. In liver cells genotoxicity together with decrease in inhibition of antioxidant enzymes were noted. [77]. ZnNP in presence and absence of bovine serum albumin (BSA; 0.06%) in human bronchial epithelial BEAS-2B cells were cytotoxic and genotoxic. ZnONPs treated with human colon carcinoma cells (LoVo) revealed their uptake by either passive diffusion or endocytosis or both, based on their agglomeration state, in contact with acid pH of lysosomes- altered organelles structure leading to the release of Zn^{2+} ion, ROS at the mitochondrial and nuclear level, inducing severe DNA damage indicative of toxicity by both ZnONP and Zn^{2+} ion [78]. ZnO particles treated on WIL2-NS human lymphoblastoid cells (26 nm, 78 nm, and 147 nm) revealed genotoxicity, micronucleus formation by medium and large-sized particles, while necrosis enhanced by smaller sized particles [79]. ZnONP has revealed genotoxic effects on human lymphoblastoid cell line [80].

ZnONPs TiO$_2$NPs led to genotoxic effects on human colon carcinoma cells (CaCo2) caused by oxidative stress and DNA damage and affect DNA damage repair pathways [81]. Silver (Ag), cerium dioxide (CeO$_2$) and titanium dioxide (TiO$_2$) NP exposure on normal untransformed human fibroblasts (GM07492) revealed cytotoxicity and genotoxicity by crossing the plasma membrane, accumulated in endocytic vesicles. Both Ag and CeO$_2$ NPs increased DNA fragmentation, oxidative damage and induced apoptotsis while AgNPs caused overexpression of GADD45α and γH2AX protein phosphorylation. CeO$_2$ NPs revealed transient genotoxicity while AgNPs were most cyto-genotoxic [82]. ZnONPs have been proved to have nephrotoxic, cytotoxic and genotoxic potential on rat kidney epithelial cells (NRK-52E) causing DNA damage [83]. ZnONP, 15–18 nm in diameter targeted on A549 cell line led to increased cytoplasmic Zn^{+2} and nuclear accumulation, reduced cellular viability, caused breaks in the DNA double strand and enhanced ROS levels [84]. ZnO and TiO$_2$NPs hemolytic activity, cytotoxicity to human RBCs with increased SOD, CAT and lipid peroxidation together with oxidative stress, [85] ZnONPs and alluminium oxide NPs (Al2O3NPs) have revealed genotoxic effects on human peripheral blood lymphocytes and concentration-dependent increase of DNA single-strand breaks [86]. The copper-zinc alloy NPs (Cu-Zn ANPs) could cause in vitro cytotoxicity and genotoxicity on human lung epithelial cells (BEAS-2B, [87]). ZnONPs caused both cytotoxic and genotoxic effects on human SHSY5Y neuronal cells leading to apoptosis, cell cycle alterations, genotoxicity, including micronuclei production, H2AX phosphorylation and DNA damage, [88]. CdSe-ZnS QDs could cause genotoxic effects in Drosophila melanogaster post oral intake with increased ROS, apoptosis [89]. ZnONPs has been reported to cause cytotoxic and genotoxic effects on primary human epidermal keratinocytes (HEK, [90]), human liver cells (HepG2) With DNA damage, ROS, oxidative stress DNA damage and cytotoxicity [91].

7.6 Carbon NPs

The C$_{60}$ fullerene and cisplatin in aqueous solution with the comet assay on human resting lymphocytes and lymphocytes after blast transformation and cytotoxicity fromindicated no induction of DNA strand breaks in normal and transformed cells and reduces the fraction of necrotic cells [92]. Multi-walled carbon nanotubes (MWCNTs) after ingestion in rat caused DNA damage, in their macrophages [93]. Coal and coal ash showed cytotoxicity and genotoxicity on V79 cell line revealed by apoptosis, necrosis, chromosomal instability, nucleoplasmic bridges, nuclear buds, and micronucleus (MN) formation. [94]. Nanodiamonds (ND)s assessed in *Acheta domesticus* lead to oxidative stress and heat shock proteins, DNA damage [95]. Negative (carboxyl), neutral (hexadecylamine; HDA) or positive (amine) polymer coating Quantum dots treated to human lymphoblastoid TK6 cells revealed overall oxidative stress and genotoxic effects depending on its surface chemistry. While carboxyl-QD revealed smallest agglomerate size and greatest cellular uptake and increased cytotoxicity and genotoxicity, amine-QD led to minor cellular damage and HDA-QD caused both cell death and genotoxicity while not showing cellular internalisation and free cadmium release [96]. Pristine graphene, reduced graphene oxide, graphene oxide, graphite, when treated to a a human primary glioblastoma cell line U87 cell line has been reported to cause DNA fragmentation [97]. Lipoamphiphile-coated CdSe/ZnS Quantum Dots (QDs) has revealed genotoxic effects in brain, liver, kidneys, lungs and testicles in rats [98]. QDs with CdSe/ZnS core-shell encapsulated by a natural fusogenic lipid (1,2-di-oleoyl-sn-glycero-3-phosphocholine (DOPC)) functionalized by a nucleolipid N-[5'-(2',3'-di-oleoyl) uridine]-N',N',N'-trimethylammoniumtosylate (DOTAU) lead to genotoxic properties [99].

7.7 Cobalt NPs (CoNPs)

CoNPs released from metal-on-metal implants have been associated with generation of oxidative stress, cytotoxicity and genotoxicity in NRK rat kidney cells that increased ROS, reduced cell viability, increase in the ratio of Bcl-2-associated X/Bcl-2, upregulation of cleaved caspase-3, indicative of apoptosis and genotoxicity revealed by DNA damage, increases in the tail DNA % and olive tail moment with DNA-double strand breaks and melatonin is reported to reverse such toxic effects. [100]. Cobalt oxide (Co_3O_4) NPs cell lines of liver, HepG2 hepatocellular carcinoma cells; lung, A549 lung carcinoma cells; gastrointestinal, CaCo-2 colorectal adenocarcinoma cells; and nervous system, SH-SY5Y neuroblastoma cells, although taken up equally by all cell-types showed differences in sensitivity . No cell death was observed in HepG2, Caco-2, or SH-SY5Y cells; only A549 cells showed cytotoxicity at relatively high exposure concentrations. DNA damage or apoptosis only in A549 cell lines, induction of cellular oxidative damage in all cell types except CaCo-2, high concentrations of Co_3O_4 NPs affected the pulmonary system but were unlikely to affect the liver, nervous system, or gastrointestinal system. [101]. Poorly soluble cobalt (II, III) oxide particles (Co3O4P) cause cytotoxic and genotoxic effects upon internalisation into lysosomal intracellular compartments, by DNA damage on BEAS-2B human bronchial epithelial cells and formation of γ-H2Ax foci [102]. Co3O4 NPs on bronchial (BEAS-2B) cells caused membrane damage, direct/oxidative DNA damage, release of inflammatory cytokines interleukin (IL)-6, IL-8 and tumor necrosis factor-alpha (TNF-α). In A549 cells oxidative DNA damage were detected without cytokine release. [103]. Cobalt-, nickel- and copper-based NPs have been reported to be genotoxic from different studies [104].

7.8 Polymeric NPs

Poly-lactic co-glycolic acid (PLGA) NPs were synthesized and three pollutants (benzo(a)pyrene, naphthalene and di-ethyl-hexyl-phthalate) were adsorbed on the surface of the NPs and tested on human airway epithelial cell line revealed vectorization by NPs decreases the cyto and geno toxicity of the adsorbed pollutants [105]. Inclusion of the polymer Eudragit (EUD) into poly(methylmethacrylate) (PMMA) nanoparticles (NP) resulted in a novel nanocarrier (PMMA-EUD) with an improved biomedical and safety performance. [106]. Poly(anhydride) NPs (PANPs) finds importance in oral drug delivery, as they are capable of developing intense adhesive interactions within the gut mucosa. Gantrez® AN 119-NP (GN-NP) and Gantrez® AN 119, covered with mannosamine (GN-MA-NP), induced DNA damage and thymidine kinase ($TK^{+/-}$) mutations in L5178Y $TK^{+/-}$ mouse lymphoma cells but had no effects of DNA strand breaks or oxidative damage [107]. Poly(anhydride) NPs protecting the drug can reduce the drug's toxicity and enhance their permeability and bioadhesion to specific target cells [108]. Quinapyramine sulfate loaded-sodium alginate nanoparticles (QS-NPs) on *Trypanosoma evansi (T. evansi)*, a causative agent of trypanosomosis on Vero cell line derived from kidney epithelial cells isolated from an African green monkey (*Chlorocebus* sp) and Hela cell lines and revealed their safe use as compared to QS alone [109]. On the contrary curcumin loaded phosphatidylserine (PS)-coated chitosan (CS) NPs revealed safe applications as drug delivery tools when applied on human embryonic kidney cells (HEK 293) [110]. Poly(lactic-co-glycolic acid) (PLGA) NPs PLGA(+) NPs were reported of its cytotoxicity. [111]. PLGA-PEO (poly-lactic-co-glycolic acid-polyethylene oxide copolymer) NPs was however reported free of cytotoxicity and genotoxicity when tested on TK6 cells [112]. Different nanopartciles including Poly(ethylene glycol) polymers, neutral and cationic liposomes, micelles, poly(amindo amine) and poly(propyleneimine) dendrimers, quantum dots, mesoporous silica, and supermagnetic iron oxide (SPIO) NPs revealed that dendrimers, cationic liposomes, and SPIO NPs caused micronuclei formation thereby inducing genotoxicity while negatively charged and neutral nanocarriers did not reveal genotoxic effects [113].

7.9 Palladium NPs (PdNPs)

PdNPs has been reported to cause apoptosis, cytotoxicity, and DNA damage on human skin malignant melanoma (A375) cell line. [114].

7.10 Magnetic NPs

Magnetic NPs followed by oxidative polymerization of dopamine, forming a polydopamine (PDA) shell indicated low cyto and genotoxicity of thus promises as biocompatibe and low genotoxicity with biomedical application [115]. Mutagenicity of magnetic iron oxide NPs IONPs depends on their particle size and surface coating [116].

7.11 Alumina NPs (AlNPs)

Alumina NPs (AlNPs) were reported to be genotoxic in mice mediated by the ROS which enter cells and induce DNA damage, micronucleus in bone marrow, sperm deformation in a time-, dose- and size-dependent manner. [117]. Al2O3NPs and SiO_2NPs revealed DNA damage to cultured murine macrophage RAW264 cells [118].

7.12 Silica NPs

BASF Levasil® silica NPs used for 3D reconstructed skin micronucleus (RSMN) assay on monocultured human B cells (TK6) revealed genotoxicity, cell death [119] in 2D models but not in 3D models. Mesoporous silica nanoparticles (MSNs) expressed oxidative damage, mitotic chromosomal aberrations in repair-impaied cells as compared to cells with functional DNA damage repair mechanism [120]. SiO_2NPs caused DNA strand breaks and show oxidative DNA base modifications [121]. MSNs expressed genotoxic effects on human embryonic kidney 293 (HEK293) by altering upregulation of 579 genes and down-regulation of 1263 genes indicative of carcinogenesis [122]. Synthetic amorphous silica (SAS) NPs can cause genotoxicity and oxidative DNA damage in vivo in Male Sprague Dawley rats after oral exposure [123]. Synthetic amorphous silica NMs (NM-200, NM-201, NM-202, and NM-203), rats treated with intratracheal instillations increased lung inflam-mation but no DNA damage [124]. Colloidal SiO_2 NPs caused cytotoxicity, genotoxicity, apoptosis, oxidative stress and proinflammatory cytokine IL-8 release from human intes-tinal Caco-2 cell line. [125]. Silicate (SiO4) and Silica (SiO_2) caused DNA damage [126]. MSNs on HT29 human intestine cell line revealed cytotoxic and genotoxic effects [127].

7.13 Gold NPs

Quantum dots (QDs), gold NPs (AuNPs), and polystyrene-cored NPs (PSNPs), in differ-ent organsims including *Saccharomyces cerevisiae*, fresh- (F) and saltwater (S) micro-algae including *Raphidocelis subcapitata* (F), *Scenedesmus obliquus* (F) and *Chlorella* spp. (F), and *Phaeodactylum tricornutum* (S) and *Daphnia magna*, and *Xenopus laevis* are reported to induce genotoxicity and oxidative stress that can drastically change according to the coating employed. [128]. The surface and chiral structure could affect toxicity. AuNPs with different surface chirality and structures like poly(acryloyl-l(d)-va-line (l(d)-PAV) chiral molecules anchored to Au nanocube (AuNCs) and nanooctahedras (AuNOs), respectively. The l-PAV capped AuNCs and AuNOs revealed more cytotoxicity to A549 cells than the D-PAV coated ones, and the PAV-AuNOs had larger cytotoxicity than PAV-AuNCs mediated by release of ROS. The structure and surface chirality at the nanoscale can influence cytotoxicity and genotoxicity. [129]. Both citrate-stabilized gold NPs (AuNPs) and AuNPs in Chinese hamster ovary (CHO) cell line caused DNA damage [130]. AuNPs exerted genotoxic effects genotoxicity in a dose-dependent and size depen-dant manner with cell cycle arrest in G1 phase, DNA damage, and ROS production. [131]. AuNps and capped with either sodium citrate or polyamidoamine dendrimers (PAMAM) revealed cytotoxic and genotoxic effects when targetted on HepG2 and peripheral blood mononuclear cells (PBMC) [132]. Uncoated spherical gold NP on BALB/c 3T3 fibroblast cell line has been revealed to show more toxicity as compared to hyaluronic acid (HA) coated spherical gold NP. [133].

7.14 Nickel oxide (NiO) NPs

Nickel oxide (NiO) NPs is toxic on NRK-52E kidney epithelial cells by inducing loss of cell viability, cytotoxicity, apoptotic/necrotic events, morphological changes [134]. Nickel metal (Ni) and nickel oxide (NiO) NPs lead to release of ROS, cellular uptake, cytotoxicity and genotoxicity. [135]. Metallic nickel and nickel-based NPs are known to be genotoxic and potentially carcinogenic [136].

7.15 Cobalt-NPs

Cobalt ions (Co^{2+}) are less cytotoxic compared to Co-NPs when tested on BRL-3A cells leading to release of ROS, Bax/Bcl-2 mRNA expression, IL-8 mRNA expression and DNA damage, cell membrane damage, immune inflammation [137]

7.16 Pesticides

Nanopermethrin (NP), a pesticide with oil-in-water (o-w) emulsion prepared by rapid evaporation caused cell death and cyto and genotoxicity whn exposed to human peripheral erythrocyte/lymphocyte [138].

7.17 Copper NPs

CuO to rat bone marrow mesenchymal stem cells (MSCs) showed dose-dependent and time-dependent toxicity and genotoxicty to MSCs [139]. CuO NP induced significant amounts of DNA strand breaks in HeLa S3 cells, whereas CuO NP, CuO MP and CuCl2 increased H_2O_2-induced DNA damages and reduced H_2O_2-induced poly(ADP-ribosyl) ation, catalysed by poly(ADP-ribose)polymerase-1 (PARP-1) and copper accumulated in cytoplasm and nucleus of A549 cells [140]. CuONPs caused DNA damage, cytotoxicity, genotoxicity by oxidative stress in human lung epithelial (A549) cells [141]. CuONPs, 50 nm could cause cell death in the human skin epidermal (HaCaT) cells in dose- and time-dependent manner with reduced glutathione and increased lipid peroxidation, catalase, SOD production and upregulation of caspase-3 activity indicative of apoptosis and cause DNA damage by oxidative stress. [142]. CuO NPs in murine macrophages RAW 264.7 and in peripheral whole blood exerted both cytotoxic and genotoxic effects [143]. CuO NPs caused genotoxic effects, DNA fragmentation, on mouse neuroblastoma cell line Neuro-2A, DNA methylation, chromosomal damage, lipid peroxidation and micronucleus formation. [144]. CuO NPswere reported to be more toxic to human lung epithelial (A549) as compared to CuO bulk particles (BPs) causing mitochondrial depolarization, ROS generation, p38 and p53 altered expression and DNA damage [145].

7.18 Chromium NPs

Chromium oxide (Cr2O3)-NPs oral treatment in Wistar rats have been reported to cause DNA damage, depositions in body and Cr excretion through faeces. [146].

7.19 Indium tin oxide (ITO) NP

Indium tin oxide (ITO) NM revealed genotoxic effects on human peripheral lymphocytes [147].

7.20 Solid NPs

Galbanic acid (GBA), a sesquiterpene coumarin with medicinal anticancer properties when loaded onto solid lipid nanoparticles (GBA-SLNs) when targeted on A549 cells proved promising tools to GBA delivery [148].

7.21 Tungsten NPs

Tungsten carbide-cobalt (WC-Co) NPs caused genotoxicity and oxidative stress in L5178Y mouse lymphoma cells and primary human lymphocytes [149]. Tungsten carbide-cobalt (WC-Co)-NPs caused ROS generation, y-H2Ax foci indicative of genotoxicity with cell mortality, DNA double-strand breaks, and cell cycle arrest in human renal (Caki-1) and liver (Hep3B) cell lines [150].

7.22 Lithium titanate (Li₂TiO₃) NPs

Lithium titanate (Li₂TiO₃) NPs (LTT NPs; <100 nm) was free of genotoxic effects on human peripheral lymphocytes [151].

7.23 Molybdenum NPs (MoNPs)

The cytotoxicity, oxidative stress, and genotoxicity by G2/M arrest and DNA damage has been reported to be induced by molybdenum NPs (Mo-NPs) in mouse skin fibroblast cells (L929, [152]).

7.24 CeO$_2$ NPs

CeO$_2$-NPs, has been observed to induce genotoxicity, bioaccumulation in female albino Wistar rats [153]. CeO$_2$NP caused clastogenic effects on Primary human dermal fibroblasts [154]. Ceria NPs (CeO$_2$NPs) exposure to A549, CaCo2 and HepG2 cell lines proved to be toxic. [155].

7.25 Nanocrystals

Nanocrystals have been reported of its genotoxic and immunotoxic potential on human bronchial epithelial BEAS 2B cells. [156].

7.26 Yttrium oxide (Y$_2$O$_3$) NPs

Yttrium oxide (Y2O3) NPs can affect cell viability, morphology, integrity and ROS generation, DNA damage, apoptosis on human embryonic kidney (HEK293) cells [157].

7.27 Ni NPs

NiNPs affect cell viability, damage cell membrane, integrity, cytotoxicity, genotoxicity, ROS generation, caspase 3 activation in human breast carcinoma (MCF-7) cells [158].

7.28 Magnetite nanopartcles

Engineered magnetite NPs (MNPs) could cause oxidative stress, ROS generation and genotoxicity in A549 human lung adenocarcinoma epithelial cells and HEL 12469 human embryonic lung fibroblasts [159].

7.29 Phosphorus containing dendrimers (CPDs)

Cationic phosphorus containing dendrimers (CPDs) revealed less G2/M cell increased S phase cells with DNA cross links, cell attachment loss, disrupted cell membrane and nucleus condensation when targetted on human mononuclear blood cells, A549 human cancer cells and human gingival fibroblast cellines (HGFs). [160].

7.30 Dolomite

Dolomite has been reported to exert cytotoxic, genotoxic effects and imflammation in human lung epithelial cells A(549) affecting cell viability, membrane damage, gluta-thione, ROS, lipid peroxidation (LPO), micronucleus (MN) and release of proinflam-matory cytokines including tumor necrosis factor-α (TNF-α), interleukin-1β (IL-1β) and interleukin-6 (IL-6, [161]).

7.31 SPIONs

Surface modification of SPIONs covered with -O$^-$ groups, when modified with hydroxyl (−OH), carboxylic (−COOH), and amine (-NH$_2$) groups- by surfaces coated with tetra-ethyl orthosilicate (TEOS), (3-aminopropyl)trimethoxysilane (APTMS), TEOS-APTMS, or citrate caused changes in cell cytotoxicity and genotoxicity [162]. Ultrafine super-paramagnetic iron oxide nanoparticles (USPION) can cause DNA damage [163].

7.32 Discussion

Nanotechnology has given rise to several applications to industry, medicine and commercial applications. However their applications is restricted because of their genotoxic effects. Many NPs have been observed to cause chromosomal aberra-tions, DNA, oxidative DNA damage and mutations and mediate cytotoxicity and genotoxic effects [164]. Green sysnthesis [165] involves the biosynthesis of biocom-patible NPs with less or reduced or no genetotoxic effects and cytotoxic effects are being designed by scientists to circumvent the associated genotoxicity of nanopart-ciels and thus improving the different biomedical applications of nanopartciels in targeting. Plant mediated or natutal product mediated synthesis of NPs is a green chemistry approach and holds greater promises to alleiavate the problems associ-ated with NPs.

References

[1]   Wong BSE, Hu Q, Baeg GH. Epigenetic modulations in nanoparticle-mediated toxicity. Food Chem Toxicol. 2017. Nov;109(Pt 1):746 -752.

[2]   Pattan G, Kaul G. Health hazards associated with nanomaterials. Toxicol Ind Health. 2014. Jul;30(6):499–519.

[3]   Marabini L, Ozgen S, Turacchi S, Aminti S, Arnaboldi F, Lonati G, Fermo P, Corbella L, Valli G, Bernardoni V, Dell'Acqua M, Vecchi R, Becagli S, Caruso D, Corrado GL, Marinovich M. Ultrafine particles (UFPs) from domestic wood stoves: genotoxicity in human lung carcinoma A549 cells. Mutat Res. 2017Aug;820:39–46.

[4]   Cao Y, Gong Y, Liu L, Zhou Y, Fang X, Zhang C, Li Y, Li J. The use of human umbilical vein endothelial cells (HUVECs) as an in vitro model to assess the toxicity of nanoparticles to endothelium: a review.J Appl Toxicol. 2017 Dec;37(12):1359–1369.

[5]   Kumar A, Dhawan A. Genotoxic and carcinogenic potential of engineered nanoparticles: an update. Arch Toxicol. 2013. Nov;87(11):1883–1900.

[6]   Magdolenova Z, Collins A, Kumar A, Dhawan A, Stone V, Dusinska M. A Review Of In Vitro and In Vivo Studies With Engineered Nanoparticles. Nanotoxicology. Mech genotoxicity.2014. May;8(3):233–278.

[7]   Kumar A, Sharma V, Dhawan A. Methods for detection of oxidative stress and genotoxicity of engineered nanoparticles. Methods Mol Biol. 2013 1028:231–246.

[8]   Golbamaki N, Rasulev B, Cassano A, Marchese Robinson RL, Benfenati E, Leszczynski J, Cronin MT. Genotoxicity of metal oxide nanomaterials: review of recent data and discussion of possible mechanisms. Nanoscale. 2015 Feb 14;7(6):2154–2198.

[9]   Kwon JY, Koedrith P, Seo YR. Current investigations into the genotoxicity of zinc oxide and silica nanoparticles in mammalian models in vitro and in vivo: carcinogenic/genotoxic potential, relevant mechanisms and biomarkers, artifacts, and limitations. Int J Nanomedicine. 2014 Dec 15;9 Suppl 2:271–286.

[10] Evans SJ, Clift MJ, Singh N, de Oliveira Mallia J, Burgum M, Wills JW, Wilkinson TS, Jenkins GJ, Doak SH. Critical review of the current and future challenges associated with advanced in vitro systems towards the study of nanoparticle (secondary) genotoxicity. Mutagenesis. 2017 Jan;32(1):233–241.

[11] Karlsson HL, Di Bucchianico S, Collins AR, Dusinska M. Can the comet assay be used reliably to detect nanoparticle-induced genotoxicity?. Environ Mol Mutagen. 2015. Mar;56(2):82–96.

[12] Glei M, Schneider T, Schlörmann W. Comet assay: an essential tool in toxicological research. Arch Toxicol. 2016. Oct;90(10):2315–2336.

[13] Sahu SC, Roy S, Zheng J, Ihrie J. Contribution of ionic silver to genotoxic potential of nanosilver in human liver HepG2 and colon Caco2 cells evaluated by the cytokinesis-block micronucleus assay. J Appl Toxicol. 2016. Apr36(4):532–542.

[14] Wang X, Liu Y, Wang J, Nie Y, Chen S, Hei TK, Deng Z, Wu L, Zhao G, Xu A. Amplification of arsenic genotoxicity by TiO_2 nanoparticles in mammalian cells: new insights from physico-chemical interactions and mitochondria. Nanotoxicology. 2017 Oct 18:1–18.

[15] Tao R, Wang C, Zhang C, Li W, Zhou H, Chen H, Ye J. Characterization, Cytotoxicity, and Genotoxicity of TiO_2 and Folate-Coupled Chitosan Nanoparticles Loading Polyprenol-Based Nanoemulsion. Biol Trace Elem Res. 2017. Oct 9.

[16] Mohamed HRH. Estimating the modulatory effect of cadmium chloride on the genotoxicity and mutagenicity of silver nanoparticles in mice. Cell Mol Biol (Noisy-le-grand). 2017. Sep 3063(9):132–143.

[17] Kornberg TG, Stueckle TA, Antonini JA, Rojanasakul Y, Castranova V, Yang Y, Wang L. Potential toxicity and underlying mechanisms associated with pulmonary exposure to iron oxide

nanoparticles: Conflicting literature and unclear risk. Nanomaterials (Basel). 2017. Oct 67(10). pii: E307.

[18] Haase A, Dommershausen N, Schulz M, Landsiedel R, Reichardt P, Krause BC, Tentschert J, Luch A. Genotoxicity testing of different surface-functionalized SiO$_2$, ZrO$_2$ and silver nanomaterials in 3D human bronchial models. Arch Toxicol. 2017. Jun 22.

[19] Bogdan J, Pławińska-Czarnak J, Zarzyńska J. Nanoparticles of titanium and Zinc oxides as novel agents in tumor treatment: a aeview. Nanoscale Res Lett. 2017. Dec;12(1):225.

[20] Sherin S, Sheeja S, Sudha Devi R, Balachandran S, Soumya RS, Abraham A. In vitro and in vivo pharmacokinetics and toxicity evaluation of curcumin incorporated titanium dioxide nanoparticles for biomedical applications. Chem Biol Interact. 2017. Sep;25275:35–46.

[21] Stoccoro A, Di Bucchianico S, Coppedè F, Ponti J, Uboldi C, Blosi M, Delpivo C, Ortelli S, Costa AL. Migliore L. Multiple endpoints to evaluate pristine and remediated titanium dioxide nanoparticles genotoxicity in lung epithelial A549 cells. Toxicol Lett. 2017. Jul 5;276:48–61.

[22] Hackenberg S, Scherzed A, Zapp A, Radeloff K, Ginzkey C, Gehrke T, Ickrath P, Kleinsasser N. Genotoxic effects of zinc oxide nanoparticles in nasal mucosa cells are antagonized by titanium dioxide nanoparticles. Mutat Res Genet Toxicol Environ Mutagen. 2017 Apr;816–817:32–37.

[23] Jain AK, Senapati VA, Singh D, Dubey K, Maurya R, Pandey AK. Impact of anatase titanium dioxide nanoparticles on mutagenic and genotoxic response in Chinese hamster lung fibroblast cells (V-79): The role of cellular uptake. Food Chem Toxicol. 2017. Jul;105:127–139.

[24] Relier C, Dubreuil M, Lozano Garcìa O, Cordelli E, Mejia J, Eleuteri P, Robidel F, Loret T, Pacchierotti F, Lucas S, Lacroix G, Trouiller B.Study of TiO$_2$ P25 nanoparticles genotoxicity on lung, blood, and liver cells in lung overload and non-overload conditions after repeated respiratory exposure in rats. Toxicol Sci. 2017. Apr 1;156(2):527–537.

[25] Patel S, Patel P, Bakshi SR. Titanium dioxide nanoparticles: an in vitro study of DNA binding, chromosome aberration assay, and comet assay. Cytotechnology. 2017. Apr;69(2):245–263.

[26] Aydın E, Türkez H, Hacımüftüoğlu F, Tatar A, Geyikoğlu F.Molecular genetic and biochemical responses in human airway epithelial cell cultures exposed to titanium nanoparticles in vitro. J Biomed Mater Res A. 2017. Jul;105(7):2056–2064.

[27] Li Y, Yan J, Ding W, Chen Y, Pack LM, Chen T. Genotoxicity and gene expression analyses of liver and lung tissues of mice treated with titanium dioxide nanoparticles. Mutagenesis. 2017. Jan;32(1):33–46.

[28] Wallin H, Kyjovska ZO, Poulsen SS, Jacobsen NR, Saber AT, Bengtson S, Jackson P, Vogel U.Surface modification does not influence the genotoxic and inflammatory effects of TiO$_2$ nanoparticles after pulmonary exposure by instillation in mice. Mutagenesis. 2017. Jan;32(1):47–57.

[29] Suzuki T, Miura N, Hojo R, Yanagiba Y, Suda M, Hasegawa T, Miyagawa M, Wang RS. Genotoxicity assessment of intravenously injected titanium dioxide nanoparticles in gpt delta transgenic mice. Mutat Res Genet Toxicol Environ Mutagen. 2016. May;802:30–37.

[30] Genotoxicity studies of titanium dioxide nanoparticles (TiO$_2$NPs) in the brain of mice, Mohamed HR, Hussien NA. Scientifica (Cairo).,.2016;2016:6710840.

[31] Hanot-Roy M, Tubeuf E, Guilbert A, Bado-Nilles A, Vigneron P, Trouiller B, Braun A, Lacroix G. Oxidative stress pathways involved in cytotoxicity and genotoxicity of titanium dioxide (TiO$_2$) nanoparticles on cells constitutive of alveolo-capillary barrier in vitro. Toxicol In Vitro. 2016. Jun;33:125–135.

[32] Armand L, Tarantini A, Beal D, Biola-Clier M, Bobyk L, Sorieul S, Pernet-Gallay K, Marie-Desvergne C, Lynch I, Herlin-Boime N, Carriere M. Long-term exposure of A549 cells to titanium dioxide nanoparticlesinduces DNA damage and sensitizes cells towards genotoxic agents. Nanotoxicol. 2016. Sep;10(7):913–923.

[33] Kansara K, Patel P, Shah D, Shukla RK, Singh S, Kumar A, Dhawan A. TiO$_2$ nanoparticles induce DNA double strand breaks and cell cycle arrest in human alveolar cells. Environ Mol Mutagen. 2015 Mar;56(2):204–217.

[34] Ghosh M, Öner D, Duca RC, Cokic SM, Seys S, Kerkhofs S, Van Landuyt K, Hoet P, Godderis L.Cyto-genotoxic and DNA methylation changes induced by different crystal phases of TiO$_2$-np in bronchial epithelial (16-HBE) cells. Mutat Res. 2017. Feb;796:1–12.

[35] Di Bucchianico S, Cappellini F, Le Bihanic F, Zhang Y, Dreij K, Karlsson HL.Genotoxicity of TiO$_2$ nanoparticles assessed by mini-gel comet assay and micronucleus scoring with flow cytometry. Mutagenesis. 2017. Jan;32(1):127–137.

[36] Dobrzyńska MM, Gajowik A, Radzikowska J, Lankoff A, Dušinská M, Kruszewski M. Genotoxicity of silver and titanium dioxide nanoparticles in bone marrow cells of rats in vivo. Toxicol. 2014. Jan ;6315:86–91.

[37] Ghosh M, Chakraborty A, Mukherjee A. Cytotoxic, genotoxic and the hemolytic effect of titanium dioxide (TiO$_2$) nanoparticles on human erythrocyte and lymphocyte cells in vitro. J Appl Toxicol. 2013. Oct;33(10):1097–1110.

[38] Hamzeh M, Sunahara GI. In vitro cytotoxicity and genotoxicity studies of titanium dioxide (TiO$_2$) nanoparticles in Chinese hamster lung fibroblast cells. Toxicol In Vitro. 2013. Mar;27(2):864–873.

[39] Srivastava RK, Rahman Q, Kashyap MP, Singh AK, Jain G, Jahan S, Lohani M, Lantow M, Pant AB. Nano-titanium dioxide induces genotoxicity and apoptosis in human lung cancer cell line, A549. Hum Exp Toxicol. 2013. Feb;32(2):153–166.

[40] Shukla RK, Kumar A, Gurbani D, Pandey AK, Singh S, Dhawan A. TiO(2) nanoparticles induce oxidative DNA damage and apoptosis in human liver cells, Nanotoxicol. 2013. Feb;7(1):48–60.

[41] Stefaan J Soenen 1, Marcel De Cuyper, Stefaan C De Smedt, Kevin Braeckmans. Investigating the toxic effects of iron oxide nanoparticles. Methods Enzymol. 2012. 509: 195–224.

[42] Kaygisiz ŞY, Ciğerci İH. Genotoxic evaluation of different sizes of iron oxide nanoparticles and ionic form by SMART, Allium and comet assay. Toxicol Ind Health. 2017. Oct;33(10):802–809.

[43] Paolini A, Guarch CP, Ramos-López D, de Lapuente J, Lascialfari A, Guari Y, Larionova J, Long J, Nano R. Rhamnose-coated superparamagnetic iron-oxide nanoparticles: an evaluation of their in vitro cytotoxicity, genotoxicity and carcinogenicity. J Appl Toxicol. 2016.Apr;36(4):510–520.

[44] Cowie H, Magdolenova Z, Saunders M, Drlickova M, Correia Carreira S, Halamoda Kenzaoi B, Gombau L, Guadagnini R, Lorenzo Y, Walker L, Fjellsbø LM, Huk A, Rinna A, Tran L, Volkovova K, Boland S, Juillerat-Jeanneret L, Marano F, Collins AR, Dusinska M. Suitability of human and mammalian cells of different origin for the assessment of genotoxicity of metal and polymeric engineered nanoparticles Nanotoxicol. 2015. May;9 Suppl 1:57–65.

[45] Valdiglesias V, Kiliç G, Costa C, Fernández-Bertólez N, Pásaro E, Teixeira JP, Laffon B. Effects of iron oxide nanoparticles: cytotoxicity, genotoxicity, developmental toxicity, and neurotoxicity. Environ Mol Mutagen. 2015 Mar56(2):125–48.

[46] Alarifi S, Ali D, Alkahtani S, Alhader MS. Iron oxide nanoparticles induce oxidative stress, DNA damage, and caspase activation in the human breast cancer cell line. Biol Trace Elem Res. 2014 Jun; 159 (1–3): 416–424.

[47] Chen T, Yan J, Li Y. Genotoxicity of titanium dioxide nanoparticles. J Food Drug Anal. 2014 Mar;22(1):95–104.[48]   Jiménez-Villarreal J, Rivas-Armendáriz DI, Arellano Pérez-Vertti RD, Olivas Calderón E, García-Garza R, Betancourt-Martínez ND, Serrano-Gallardo LB, Morán-Martínez Relationship between lymphocyte DNA fragmentation and dose of iron oxide (Fe2O3) and silicon oxide (SiO$_2$) nanoparticles. J.Genet Mol Res. 2017 Feb 8;16(1).

[48] Magdolenova Z, Drlickova M, Henjum K, Rundén-Pran E, Tulinska J, Bilanicova D, Pojana G, Kazimirova A, Barancokova M, Kuricova M, Liskova A, Staruchova M, Ciampor F, Vavra I, Lorenzo Y, Collins A, Rinna A, Fjellsbø L, Volkovova K, Marcomini A, Amiry-Moghaddam

M, Dusinska M Coating-dependent induction of cytotoxicity and genotoxicity of iron oxide nanoparticles. Nanotoxicol. 2015 May;9 Suppl 1:44–56.

[49] Ahamed M, Alhadlaq HA, Alam J, Khan MA, Ali D, Alarafi S. Iron oxide nanoparticle-induced oxidative stress and genotoxicity in human skin epithelial and lung epithelial cell lines. Curr Pharm Des, 2013;19(37):6681–6690.

[50] Nallanthighal S, Chan C, Bharali DJ, Mousa SA, Vásquez E, Reliene R.Particle coatings but not silver ions mediate genotoxicity of ingested silver nanoparticles in a mouse model. NanoImpact. 2017. Jan;5:92–100.

[51] Chávez-Andrade GM, Tanomaru-Filho M, Rodrigues EM, Gomes-Cornélio AL, Faria G, Bernardi MIB, Guerreiro-Tanomaru JM. Cytotoxicity, genotoxicity and antibacterial activity of poly(vinyl alcohol)-coated silver nanoparticles and farnesol as irrigating solutions. Arch Oral Biol. 2017. Sep 25;84:89–93.

[52] Roszak J, Domeradzka-Gajda K, Smok-Pieniążek A, Kozajda A, Spryszyńska S, Grobelny J, Tomaszewska E, Ranoszek-Soliwoda K, Cieślak M, Puchowicz D, Stępnik M.Genotoxic effects in transformed and non-transformed human breast cell lines after exposure to silver nanoparticles in combination with aluminium chloride, butylparaben or di-n-butylphthalate. Toxicol In Vitro. 2017. Dec;45(Pt 1):181–193.

[53] Brkić Ahmed L, Milić M, Pongrac IM, Marjanović AM, Mlinarić H, Pavičić I, Gajović S, Vinković Vrček I.Impact of surface functionalization on the uptake mechanism and toxicity effects of silver nanoparticles in HepG2 cells. Food Chem Toxicol. 2017. Sep;107(Pt A) :349–361.

[54] Yang L, Kuang H, Zhang W, Aguilar ZP, Wei H, Xu H. Comparisons of the biodistribution and toxicological examinations after repeated intravenous administration of silver and gold nanoparticles in mice. Sci Rep. 2017 Jun 12;7(1):3303.

[55] Luo Q, Zhai B, Fan G, Liu Z, Cheng K, Xin L. Cytotoxicity and genotoxicity of nanosilver in stable GADD45α promoter-driven luciferase reporter HepG2 and A549 cells. Che B Environ Toxicol. 2017. Sep;32(9):2203–2211.

[56] Bastos V, Duarte IF, Santos C, Oliveira H. Genotoxicity of citrate-coated silver nanoparticles to human keratinocytes assessed by the comet assay and cytokinesis blocked micronucleus assay. Environ Sci Pollut Res Int. 2017. Feb;24(5):5039–5048.

[57] Juarez-Moreno K, Gonzalez EB, Girón-Vazquez N, Chávez-Santoscoy RA, Mota-Morales JD, Perez-Mozqueda LL, Garcia-Garcia MR, Pestryakov A, Bogdanchikova N. Comparison of cytotoxicity and genotoxicity effects of silver nanoparticles on human cervix and breast cancer cell lines. Hum Exp Toxicol. 2017. Sep;36(9) :931–948.

[58] Wang J, Che B, Zhang LW, Dong G, Luo Q, Xin L. Comparative genotoxicity of silver nanoparticles in human liver HepG2 and lung epithelial A549 cells. WJ Appl Toxicol. 2017 Apr;37(4):495–501.

[59] Guo X, Li Y, Yan J, Ingle T, Jones MY, Mei N, Boudreau MD, Cunningham CK, Abbas M, Paredes AM, Zhou T, Moore MM, Howard PC, Chen T.Size- and coating-dependent cytotoxicity and genotoxicity of silver nanoparticles evaluated using in vitro standard assays. Nanotoxicol. 2016. Nov;10(9):1373–1384.

[60] Li Y, Qin T, Ingle T, Yan J, He W, Yin JJ, Chen T. Differential genotoxicity mechanisms of silver nanoparticles and silver ions. Arch Toxicol. 2017. Jan;91(1):509–519.

[61] Asare N, Duale N, Slagsvold HH, Lindeman B, Olsen AK, Gromadzka-Ostrowska J, Meczynska-Wielgosz S, Kruszewski M, Brunborg G, Instanes C. Genotoxicity and gene expression modulation of silver and titanium dioxide nanoparticles in mice. Nanotoxicol. 2016 ;10(3) :312–321.

[62] Souza TA, Franchi LP, Rosa LR, da Veiga MA, Takahashi CS. Cytotoxicity and genotoxicity of silver nanoparticles of different sizes in CHO-K1 and CHO-XRS5 cell lines. Mutat Res Genet Toxicol Environ Mutagen. 2016. Jan 1;795:70–83.

[63] Butler KS, Peeler DJ, Casey BJ, Dair BJ, Elespuru RK. Silver nanoparticles: correlating nanoparticle size and cellular uptake with genotoxicity. Mutagenesis. 2015. Jul;30(4):577–591.

[64] Ivask A, Voelcker NH, Seabrook SA, Hor M, Kirby JK, Fenech M, Davis TP, Ke PC.DNA melting and genotoxicity induced by silver nanoparticles and graphene. Chem Res Toxicol. 2015. May 18;28(5):1023–1035.

[65] El Mahdy MM, Eldin TA, Aly HS, Mohammed FF, Shaalan MI. Evaluation of hepatotoxic and genotoxic potential of silver nanoparticles in albino rats. Exp Toxicol Pathol. 2015. Jan;67(1):21–29.

[66] Suliman Y AO, Ali D, Alarifi S, Harrath AH, Mansour L, Alwasel SH. Environ toxicol. evaluation of cytotoxic, oxidative stress, proinflammatory and genotoxic effect of silver nanoparticles in human lung epithelial cells. 2015 .Feb;30(2):149–160.

[67] Li X, Xu L, Shao A, Wu G, Hanagata N. Cytotoxic and genotoxic effects of silver nanoparticles on primary Syrian hamster embryo (SHE) cells. J Nanosci Nanotechnol. 2013. Jan;13(1):161–170.

[68] Kim HR, Park YJ, Shin DY, Oh SM, Chung KH.Appropriate in vitro methods for genotoxicity testing of silver nanoparticles. Environ Health Toxicol. 2013;28:e2013003.

[69] Ghosh M, J M, Sinha S, Chakraborty A, Mallick SK, Bandyopadhyay M,Mukherjee A.In vitro and in vivo genotoxicity of silver nanoparticles. Mutat Res. 2012 ,Dec 12; 749 (1–2): 60–69.

[70] Kim HR, Kim MJ, Lee SY, Oh SM, Chung KH. Genotoxic effects of silver nanoparticles stimulated by oxidative stress in human normal bronchial epithelial (BEAS-2B) cells. Mutat Res. 2011 Dec 24;726(2):129–135.

[71] Park MV, Neigh AM, Vermeulen JP, de la Fonteyne LJ, Verharen HW, Briedé JJ, van Loveren H, de Jong WH. The effect of particle size on the cytotoxicity, inflammation, developmental toxicity and genotoxicity of silver nanoparticles. Biomaterials. 2011. Dec;32(36):9810–9817.

[72] Srivastav AK, Kumar A, Prakash J, Singh D, Jagdale P, Shankar J, Kumar M. Genotoxicity evaluation of zinc oxide nanoparticles in Swiss mice after oral administration using chromosomal aberration, micronuclei, semen analysis, and RAPD profile. Toxicol Ind Health. 2017. 1; Jan:748233717717842.

[73] Naeem A, Khan T, Alam T, Husain Q. An insight into biophysical characterization and genotoxicity assessment of Concanavalin A immobilized on zinc oxide nanoparticles. Protein Pept Lett. 2017. Sep; :19.

[74] Shalini D, Senthilkumar S, Rajaguru P. Effect of size and shape on toxicity of zinc oxide (ZnO) nanomaterials in human peripheral blood lymphocytes. Toxicol Mech Methods. 2017. Sep; 3:1–8.

[75] Kononenko V, Repar N, Marušič N, Drašler B, Romih T, Hočevar S, Drobne D. Comparative in vitro genotoxicity study of ZnO nanoparticles, ZnO macroparticles and ZnCl2 to MDCK kidney cells: Size matters. Toxicol In Vitro. 2017. Apr;40:256–263.

[76] Ghosh M, Sinha S, Jothiramajayam M, Jana A, Nag A, Mukherjee A. Cyto-genotoxicity and oxidative stress induced by zinc oxide nanoparticle in human lymphocyte cells in vitro and Swiss albino male mice in vivo. Food Chem Toxicol. 2016. Nov;97:286–296.

[77] Condello M, De Berardis B, Ammendolia MG, Barone F, Condello G, Degan P, Meschini S. ZnO nanoparticle tracking from uptake to genotoxic damage in human colon carcinoma cells. Toxicol In Vitro. 2016. Sep;35:169–179.

[78] Yin H, Casey PS, McCall MJ, Fenech M. Size-dependent cytotoxicity and genotoxicity of ZnO particles to human lymphoblastoid (WIL2-NS) cells. Environ Mol Mutagen. 2015. Dec;56(9):767–76.

[79] Demir E, Creus A, Marcos R. Genotoxicity and DNA repair processes of zinc oxide nanoparticles. J Toxicol Environ Health A. 2014;.77(21):1292–1303.

[80] Zijno A, De Angelis I, De Berardis B, Andreoli C, Russo MT, Pietraforte D, Scorza G, Degan P, Ponti J, Rossi F, Barone F. Toxicol in vitro different mechanisms are involved in oxidative DNA damage and genotoxicity induction by ZnO and TiO2 nanoparticles in human colon carcinoma cells. 2015. Oct;29(7):1503–1512.

[81] Franchi LP, Manshian BB, de Souza TA, Soenen SJ, Matsubara EY, Rosolen JM, Takahashi CS. Cyto- and genotoxic effects of metallic nanoparticles in untransformed human fibroblast. Toxicol In Vitro. 2015 Oct;29(7):1319–1331.

[82] Uzar NK, Abudayyak M, Akcay N, Algun G, Özhan G. Zinc oxide nanoparticles induced cyto- and genotoxicity in kidney epithelial cells. Toxicol Mech Methods. 2015.;25(4):334–339.

[83] Heim J, Felder E, Tahir MN, Kaltbeitzel A, Heinrich UR, Brochhausen C, Mailänder V, Tremel W, Brieger J.Genotoxic effects of zinc oxide nanoparticles,. Nanoscale,. 2015. May 21;7(19):8931–8938.

[84] Khan M, Naqvi AH, Ahmad M. Comparative study of the cytotoxic and genotoxic potentials of zinc oxide and titanium dioxide nanoparticles. Toxicol Rep. 2015. Feb 19;2:765–774.

[85] Sliwinska A, Kwiatkowski D, Czarny P, Milczarek J, Toma M, Korycinska A, Szemraj J, Sliwinski T. Genotoxicity and cytotoxicity of ZnO and Al2O3 nanoparticles. Toxicol Mech Methods. 2015. Mar;25(3):176–183.

[86] Kumbıçak U, Cavaş T, Cinkılıç N, Kumbıçak Z, Vatan O, Yılmaz D. Evaluation of in vitro cytotoxicity and genotoxicity of copper-zinc alloy nanoparticles in human lung epithelial cells. Food Chem Toxicol. 2014. Nov;73:105–112.

[87] Valdiglesias V, Costa C, Kiliç G, Costa S, Pásaro E, Laffon B, Teixeira JP. Neuronal cytotoxicity and genotoxicity induced by zinc oxide nanoparticles.

[88] Galeone A, Vecchio G, Malvindi MA, Brunetti V, Cingolani R, Pompa PP. In vivo assessment of CdSe-ZnS quantum dots: coating dependent bioaccumulation and genotoxicity. Nanoscale. 2012. Oct 21;4(20):6401–6407.

[89] Matsuda S, Matsui S, Shimizu Y, .Genotoxicity of colloidal fullerene C$_{60}$ Matsuda T. Environ Sci Technol. 2011. May 1;45(9):4133–4138.

[90] Sharma V, Anderson D, Dhawan A. Zinc oxide nanoparticles induce oxidative stress and genotoxicity in human liver cells (HepG2). J Biomed Nanotechnol. 2011 .Feb7(1):98–9.

[91] Prylutska S, Politenkova S, Afanasieva K, Korolovych V, Bogutska K, Sivolob A, Skivka L, Evstigneev M, Kostjukov V, Prylutskyy Y,. Ritter U.A nanocomplex of C$_{60}$ fullerene with cisplatin: design, characterization and toxicity. Beilstein J Nanotechnol. 2017. Jul 20;8:1494–1501.

[92] Gerencsér G, Varjas T, Szendi K, Varga C. In Vivo Induction of Primary DNA Lesions upon Subchronic Oral Exposure to Multi-walled Carbon Nanotubes In Vivo. 2016. 11–12;30(6):863–867.

[93] León-Mejía G, Silva LF, Civeira MS, Oliveira ML, Machado M, Villela IV, Hartmann A, Premoli S, Corrêa DS, Da Silva J, Henriques JA. Cytotoxicity and genotoxicity induced by coal and coal fly ash particles samples in V79 cells. Environ Sci Pollut Res Int. 2016. Dec;23(23):24019–24031.

[94] Karpeta-Kaczmarek J, Dziewięcka M, Augustyniak M, Rost-Roszkowska M, Pawlyta M. Oxidative stress and genotoxic effects of diamond nanoparticles. Environ Res. 2016. Jul;148:264–272.

[95] Manshian BB, Soenen SJ, Brown A, Hondow N, Wills J, Jenkins GJ, Doak SH. Genotoxic capacity of Cd/Se semiconductor quantum dots with differing surface chemistries. Mutagenesis. 2016. Jan;31(1):97–106.

[96] Hinzmann M, Jaworski S, Kutwin M, Jagiełło J, Koziński R, Wierzbicki M, Grodzik M, Lipińska L, Sawosz E, Chwalibog AHinzmann M, Jaworski S, Kutwin M, Jagiełło J, Koziński R, Wierzbicki M, Grodzik M, Lipińska L, Sawosz E, Chwalibog .A Nanoparticles containing allotropes of carbon

have genotoxic effects on glioblastoma multiforme cells. Int J Nanomedicine. 2014. May; 159:2409–2417.

[97]   Aye M, Di Giorgio C, Mekaouche M, Steinberg JG, Brerro-Saby C, Barthélémy P, De Méo M, Jammes Y. Genotoxicity of intraperitoneal injection of lipoamphiphile CdSe/ZnS quantum dots in rats. Mutat Res. 2013. Dec 12; 758 (1–2): 48–55.

[98]   Aye M, Di Giorgio C, Berque-Bestel I, Aime A, Pichon BP, Jammes Y, Barthélémy P, De Méo M.Genotoxic and mutagenic effects of lipid-coated CdSe/ZnS quantum dots. Mutat Res. 2013. Jan 20; 750 (1–2): 129–138.

[99]   Liu Y, Yang X, Wang W, Wu X, Zhu H, Liu F. Melatonin counteracts cobalt nanoparticle-induced cytotoxicity and genotoxicity by deactivating reactive oxygen species-dependent mechanisms in the NRK cell line. Mol Med Rep. 2017. Oct;16(4):4413–4420.

[100]   Abudayyak M, Gurkaynak TA, Özhan G. In vitro evaluation of cobalt oxide nanoparticle-induced toxicity. Toxicol Ind Health. 2017. Aug;33(8):646–654.

[101]   Uboldi C, Orsière T, Darolles C, Aloin V, Tassistro V, George I, Malard V.Poorly soluble cobalt oxide particles trigger genotoxicity via multiple pathways. Part Fibre Toxicol. 2016. Feb 3;13:5.

[102]   Cavallo D, Ciervo A, Fresegna AM, Maiello R, Tassone P, Buresti G, Casciardi S, Iavicoli S, Ursini CL. Investigation on cobalt-oxide nanoparticles cyto-genotoxicity and inflammatory response in two types of respiratory cells. J Appl Toxicol. 2015. Oct;35(10):1102–1113.

[103]   Magaye R, Zhao J, Bowman L, Ding M. Genotoxicity and carcinogenicity of cobalt-, nickel- and copper-based nanoparticles,. Exp Ther Med. 2012. Oct;4(4):551–561.

[104]   Carpentier R, Platel A, Maiz-Gregores H, Nesslany F, Betbeder D. Vectorization by nanoparticles decreases the overall toxicity of airborne pollutants. PLoS One. 2017 Aug 15;12(8):e0183243.

[105]   Graça D, Louro H, Santos J, Dias K, Almeida AJ, Gonçalves L, Silva MJ, Bettencourt A.Toxicity screening of a novel poly(methylmethacrylate)-Eudragit nanocarrier on L929 fibroblasts. Toxicol Lett. 2017. Jul 5;276:129–137.

[106]   Iglesias T, Dusinska M, El Yamani N, Irache JM, Azqueta A, López de Cerain A Int J Pharm,. In Vitro Evaluation Of The Genotoxicity Of Poly(anhydride) NanoparticlesDesigned for Oral Drug Delivery. Int J Pharm. 2017.May15;523(1):418–426. 2017 May 15;523(1):418–426.

[107]   Iglesias T, López de Cerain A, Irache JM, Martín-Arbella N, Wilcox M, Pearson J, Azqueta A. Evaluation of the cytotoxicity, genotoxicity and mucus permeation capacity of several surface modified poly(anhydride) nanoparticlesdesigned for oral drug delivery. int J pharm. 2017. Jan 30; 517 (1–2): 67–79.

[108]   Manuja A, Kumar B, Chopra M, Bajaj A, Kumar R, Dilbaghi N, Kumar S, Singh S, Riyesh T, Yadav SC. Cytotoxicity and genotoxicity of a trypanocidal drug quinapyramine sulfate loaded-sodium alginate nanoparticles in mammalian cells. Int J Biol Macromol. 2016 .Jul;88:146–155.

[109]   Zheng Y, Chen Y, Jin LW, Ye HY, Liu G. Cytotoxicity and Genotoxicity in Human Embryonic Kidney Cells Exposed to Surface Modify Chitosan Nanoparticles Loaded with Curcumin. AAPS PharmSciTech. 2016 .Dec;17(6):1347–1352.

[110]   Platel A, Carpentier R, Becart E, Mordacq G, Betbeder D, Nesslany F. Influence of the surface charge of PLGA nanoparticles on their in vitro genotoxicity, cytotoxicity, ROS production and endocytosis. J Appl Toxicol. 2016. Mar;36(3):434–444.

[111]   Kazimirova A, Magdolenova Z, Barancokova M, Staruchova M, Volkovova K, Dusinska M. Genotoxicity testing of PLGA-PEO nanoparticles in TK6 cells by the comet assay and the cytokinesis-block micronucleus assay.,. Mutat Res. 2012 .Oct 9; 748 (1–2): 42–47.

[112]   Shah V, Taratula O, Garbuzenko OB, Patil ML, Savla R, Zhang M, Minko T. Genotoxicity of different nanocarriers: Possible modifications for the delivery of nucleic acids. Curr Drug Discov Technol. 2013 Mar;10(1):8–15.

[113] Alarifi S, Ali D, Alkahtani S, Almeer RS. ROS-Mediated Apoptosis and Genotoxicity Induced by Palladium Nanoparticles in Human Skin Malignant Melanoma Cells,. Oxid Med Cell Longev, 2017.;2017:8439098.

[114] Woźniak A, Walawender M, Tempka D, Coy E, Załęski K, Grześkowiak BF, Mrówczyński R. In vitro genotoxicity and cytotoxicity of polydopamine-coated magnetic nanostructures. Toxicol In Vitro. 2017. Oct;44:256–265.

[115] Liu Y, Xia Q, Liu Y, Zhang S, Cheng F, Zhong Z, Wang L, Li H, Xiao K. Genotoxicity assessment of magnetic iron oxide nanoparticles with different particle sizes and surface coatings. Nanotechnol. 2014. Oct 2425(42):425101.

[116] Zhang Q, Wang H, Ge C, Duncan J, He K, Adeosun SO, Xi H, Peng H, Niu Q. J Appl Toxicol Alumina at 50 and 13 nm nanoparticle sizes have potential genotoxicity. 2017. Sep;37(9):1053–1064.

[117] Hashimoto M, Imazato S. Cytotoxic and genotoxic characterization of aluminum and silicon oxide nanoparticles in macrophages. Dent Mater. 2015. May;31(5):556–564.

[118] Wills JW, Hondow N, Thomas AD, Chapman KE, Fish D, Maffeis TG, Penny MW, Brown RA, Jenkins GJ, Brown AP,. Genetic toxicity assessment of engineered nanoparticles using a 3D in vitro skin model (EpiDerm™). Part Fibre Toxicol. 2016; 13: 50.

[119] Niu M, Zhong H, Shao H, Hong D, Ma T, Xu K, Chen X, Han J, Sun J.Shape-dependent genotoxicity of mesoporous silica nanoparticlesand cellular mechanisms. J Nanosci Nanotechnol. 2016. Mar;16(3):2313–2318.

[120] Maser E, Schulz M, Sauer UG, Wiemann M, Ma-Hock L, Wohlleben W, Hartwig A, Landsiedel R. In vitro and in vivo genotoxicity investigations of differently sized amorphous SiO_2 nanomaterials. Mutat Res Genet Toxicol Environ Mutagen. 2015. Dec;794:57–74.

[121] Zhang Q, Xu H, Zheng S, Su M, Wang J. Genotoxicity of mesoporous silica nanoparticles in human embryonic kidney 293 cells. Drug Test Anal. 2015. Sep;7(9):787–796.

[122] Tarantini A, Huet S, Jarry G, Lanceleur R, Poul M, Tavares A, Vital N, Louro H, João Silva M, Fessard V.Genotoxicity of synthetic amorphous silica nanoparticles in rats following short-term exposure Part 1: oral route. Environ Mol Mutagen. 2015. Mar;56(2):218–227.

[123] Guichard Y, Maire MA, Sébillaud S, Fontana C, Langlais C, Micillino JC, Darne C, Roszak J, Stępnik M, Fessard V, Binet S, Gaté L. Genotoxicity of synthetic amorphous silica nanoparticles in rats following short-term exposure Part 2: intratracheal instillation and intravenous injection. Environ Mol Mutagen. 2015. Mar;56(2):228–244.

[124] Tarantini A, Lanceleur R, Mourot A, Lavault MT, Casterou G, Jarry G, Hogeveen K, Fessard V. Toxicity, genotoxicity and proinflammatory effects of amorphous nanosilica in the human intestinal Caco-2 cell line. Toxicol In Vitro. 2015. Mar;29(2):398–407.

[125] Sahu SC, Roy S, Zheng J, Yourick JJ,. Sprando RL Comparative genotoxicity of nanosilver in human liver HepG2 and colon Caco2 cells evaluated by fluorescent microscopy of cytochalasin B-blocked micronucleus formation. J Appl Toxicol. 2014. Nov;34(11):1200–1208.

[126] Sergent JA, Paget V, Chevillard S. Toxicity and genotoxicity of nano-SiO_2 on human epithelial intestinal HT-29 cell line. Ann Occup Hyg. 2012. Jul;56(5):622–630.

[127] Libralato G, Galdiero E, Falanga A, Carotenuto R, de Alteriis E, Guida M.Toxicity Effects of Functionalized Quantum Dots, Gold and Polystyrene Nanoparticles on Target Aquatic Biological. Models: A Review. Molecules. 2017. Aug 31;22(9).pii: E1439.

[128] Deng J, Yao M, Gao C. Cytotoxicity of gold nanoparticles with different structures and surface-anchored chiral polymers. Acta Biomater. 2017. Apr 15;53:610–618.

[129] George JM, Magogotya M, Vetten MA, Buys AV, Gulumian M. From the Cover: An Investigation of the Genotoxicity and Interference of Gold Nanoparticles in Commonly Used In Vitro Mutagenicity and Genotoxicity Assays. Toxicol Sci. 2017.Mar 1;156(1):149–166.

[130] The effect of particle size on the genotoxicity of gold nanoparticles. Xia Q, Li H, Liu Y, Zhang S, Feng Q, Xiao K. J Biomed Mater Res A. 2017. Mar;105(3):710–719.

[131] Paino IM, Marangoni VS, de Oliveira Rde C, Antunes LM, Zucolotto V.Cyto and genotoxicity of gold nanoparticles in human hepatocellular carcinoma and peripheral blood mononuclear cells. Toxicol Lett. 2012. Nov 30;215(2):119–125.

[132] Di Guglielmo C, De Lapuente J, Porredon C, Ramos-López D, Sendra J, Borràs M. In vitro safety toxicology data for evaluation of gold nanoparticles-chronic cytotoxicity, genotoxicity and uptake. J Nanosci Nanotechnol. 2012. Aug;12(8):6185–6191.

[133] Abudayyak M, Guzel E, Özhan G. Nickel Oxide. Nanoparticles Induce Oxidative DNA Damage and Apoptosis in Kidney Cell Line (NRK-52E). Biol Trace Elem Res. 2017. Jul;178(1):98–104.

[134] Latvala S, Hedberg J, Di Bucchianico S, Möller L, Odnevall Wallinder I, Elihn K Karlsson HL. Nickel Release, ROS Generation and Toxicity of Ni and NiO Micro- and Nanoparticles PLoS One. 2016. Jul 19;11(7):e0159684.

[135] Magaye R, Zhao J. Recent progress in studies of metallic nickel and nickel-based nanoparticles' genotoxicity and carcinogenicity. Environ Toxicol Pharmacol. 2012. Nov34(3):644–650.

[136] Liu YK, Deng XX, Yang HL. Cytotoxicity and genotoxicity in liver cells induced by cobalt nanoparticles and ions. Bone Joint Res. 2016. Oct;5(10):461–469.

[137] Sundaramoorthy R, Velusamy Y, Balaji AP, Mukherjee A, Chandrasekaran N. Comparative cytotoxic and genotoxic effects of permethrin and its nanometric form on human erythrocytes and lymphocytes in vitro. Chem Biol Interact. 2016. Sep 25;257:119–124.

[138] Zhang W, Jiang P, Chen W, Zheng B, Mao Z, Antipov A, Correia M, Larsen EH, Gao C.J. Genotoxicity of Copper Oxide Nanoparticles with Different Surface Chemistry on Rat Bone Marrow Mesenchymal Stem Cells. 2016. Jun;16(6):5489–5497.

[139] Semisch A, Ohle J, Witt B, Hartwig A. Cytotoxicity and genotoxicity of nano - and microparticulate copper oxide: role of solubility and intracellular bioavailability. Part Fibre Toxicol. 2014. Feb 13;11:10.

[140] Akhtar MJ, Kumar S, Alhadlaq HA, Alrokayan SA, Abu-Salah KM, Ahamed M. Dose-dependent genotoxicity of copper oxide nanoparticles stimulated by reactive oxygen species in human lung epithelial cells. Toxicol Ind Health. 2016. May;32(5):809–821.

[141] Alarifi S, Ali D, Verma A, Alakhtani S, Ali BA.Cytotoxicity and genotoxicity of copper oxide nanoparticles in human skin keratinocytes cells. Int J Toxicol. 2013. Jul;32(4):296–307.

[142] Di Bucchianico S, Fabbrizi MR, Misra SK, Valsami-Jones E, Berhanu D, Reip P, Bergamaschi E, Migliore L. Multiple cytotoxic and genotoxic effects induced in vitro by differently shaped copper oxide nanomaterials. Mutagenesis. 2013. May;28(3):287–299.

[143] Perreault F, Pedroso Melegari S, .Henning da Costa C, de Oliveira Franco Rossetto AL. Popovic R, Gerson Matias W.Genotoxic effects of copper oxide nanoparticles in Neuro 2A cell cultures. Sci Total Environ. 2012. Dec 15;441:117–124.

[144] Wang Z, Li N, Zhao J, White JC, Qu P, Xing B.CuO nanoparticle interaction with human epithelial cells: cellular uptake, location, export, and genotoxicity. Chem Res Toxicol. 2012. Jul 16;25(7):1512–1521.

[145] Singh SP, Chinde S, Kamal SS, Rahman MF, Mahboob M, Grover P.Genotoxic effects of chromium oxide nanoparticles and microparticles in Wistar rats after 28 days of repeated oral exposure. Environ Sci Pollut Res Int. 2016. Feb;23(4):3914–3924.

[146] Akyıl D, Eren Y, Konuk M, Tepekozcan A, Sağlam E. Determination of mutagenicity and genotoxicity of indium tin oxide nanoparticles using the Ames test and micronucleus assay. Toxicol Ind Health. 2016 .Sep;32(9):1720–1728.

[147] Eskandani M, Barar J, Dolatabadi JE, Hamishehkar H, Nazemiyeh H. Formulation, characterization, and geno/cytotoxicity studies of galbanic acid-loaded solid lipid nanoparticles. Pharm Biol. 2015;53(10):1525–1538.

[148] Moche H, Chevalier D, Vezin H, Claude N, Lorge E, Nesslany F. Genotoxicity of tungsten carbide-cobalt (WC-Co) nanoparticles in vitro: mechanisms-of-action studies. Mutat Res Genet Toxicol Environ Mutagen. 2015. Feb;779:15–22.

[149] Paget V, Moche H, Kortulewski T, Grall R, Irbah L, Nesslany F, Chevillard S. Human cell line--dependent WC-Co nanoparticle cytotoxicity and genotoxicity: a key role of ROS production. Toxicol Sci. 2015. Feb;143(2):385–397.

[150] Akbaba GB, Turkez H, Sönmez E, Tatar A, Yilmaz M. Genotoxicity in primary human peripheral lymphocytes after exposure to lithium titanate nanoparticles in vitro. Toxicol Ind Health. 2016. Aug;32(8):1423–1429.

[151] Siddiqui MA, Saquib Q, Ahamed M, Farshori NN, Ahmad J, Wahab R, Khan ST, Alhadlaq HA, Musarrat J, Al-Khedhairy AA, Pant AB. Molybdenum nanoparticles-induced cytotoxicity, oxidative stress, G2/M arrest, and DNA damage in mouse skin fibroblast cells (L929). Colloids Surf B Biointerfaces. 2015. Jan 1;125:73–81.

[152] Kumari M, Kumari SI, Kamal SS, Grover P. Genotoxicity assessment of cerium oxide nanoparticles in female Wistar rats after acute oral exposure. Mutat Res Genet Toxicol Environ Mutagen. 2014. Dec;775–776:7–19.

[153] Benameur L, Auffan M, Cassien M, Liu W, Culcasi M, Rahmouni H, Stocker P, Tassistro V, Bottero JY, Rose J, Botta A, Pietri S. DNA damage and oxidative stress induced by CeO_2 nanoparticles in human dermal fibroblasts: Evidence of a clastogenic effect as a mechanism of genotoxicity. Nanotoxicology. 2015;9(6):696–705.

[154] De Marzi L, Monaco A, De Lapuente J, Ramos D, Borras M, Di Gioacchino M, Santucci S, Poma A. Cytotoxicity and genotoxicity of ceria nanoparticles on different cell lines in vitro. Int J Mol Sci. 2013. Feb 1;14(2):3065–3077.

[155] Catalán J, Ilves M, Järventaus H, Hannukainen KS, Kontturi E, Vanhala E, Alenius H, Savolainen KM, Norppa H. Genotoxic and immunotoxic effects of cellulose nanocrystals in vitro. Environ Mol Mutagen. 2015. Mar;56(2):171–182.

[156] Selvaraj V, Bodapati S, Murray E, Rice KM, Winston N, Shokuhfar T, Zhao Y, Blough E.Cytotoxicity and genotoxicity caused by yttrium oxide nanoparticles in HEK293 cells. Int J Nanomedicine. 2014 Mar 12;9:1379.

[157] Ahamed M, Alhadlaq HA. Nickel nanoparticle-induced dose-dependent cyto-genotoxicity in human breast carcinoma MCF-7 cells. Onco Targets Ther. 2014. Feb 14;7:269–280.

[158] Mesárošová M, Kozics K, Bábelová A, Regendová E, Pastorek M, Vnuková D, Buliaková B, Rázga F, Gábelová A.The role of reactive oxygen species in the genotoxicity of surface-modified magnetite nanoparticles. Toxicol Lett. 2014. May2;226(3):303–313.

[159] Gomulak P, Klajnert B, Bryszewska M, Majoral JP, Caminade AM, Blasiak J. Cytotoxicity and genotoxicity of cationic phosphorus-containing dendrimers. Curr Med Chem. 2012;19(36):6233–6240.

[160] Patil G, Khan MI, Patel DK, Sultana S, Prasad R, Ahmad I.Evaluation of cytotoxic, oxidative stress, proinflammatory and genotoxic responses of micro- and nano-particles of dolomite on human lung epithelial cells A(549). Environ Toxicol Pharmacol. 2012. Sep;34(2):436–445.

[161] Hong SC, Lee JH, Lee J, Kim HY, Park JY, Cho J, Lee J, Han DW.Subtle cytotoxicity and genotoxicity differences in superparamagnetic iron oxide nanoparticles coated with various functional groups. Int J Nanomedicine. 2011;6:3219–3231.

[162] Singh N, Jenkins GJ, Nelson BC, Marquis BJ, Maffeis TG, Brown AP, Williams PM, Wright CJ, Doak SH.The role of iron redox state in the genotoxicity of ultrafine superparamagnetic iron oxide nanoparticles. Biomaterials. 2012. Jan;33(1):163–170.

[163] Roszak J, Catalán J, Järventaus H, Lindberg HK, Suhonen S, Vippola M, Stępnik M, Norppa H. Effect of particle size and dispersion status on cytotoxicity and genotoxicity of zinc oxide in human bronchial epithelial cells. Mutat Res Genet Toxicol Environ Mutagen. 2016 Jul;805:7–18.

[164] Battal D, Çelik A, Güler G, Aktaş A, Yildirimcan S, Ocakoglu K, Çömelekoğlu Ü. SiO$_2$. Nanoparticule-induced size-dependent genotoxicity - an in vitro study using sister chromatid exchange, micronucleus and comet assay. Drug Chem Toxicol. 2015 Apr;38(2):196–204.

[165] Shyamasree Ghosh Green synthesis of nanoparticle and fungal infection: Book Green Synthesis. Characterization and Applications of Nanoparticles. Springer in press: Edited by Shukla and Iravani.

8 Nanoparticles and occupational health hazards

Abstract: While nanotechnology holds immense promises in the almost all sectors of the human life from medications, agriculture, industry, usage in personal and healthcare products and so on, they pose health hazards to the exposed workers associated with the production and handling of nanoparticles (NPs) and are a major cause of global concern. Both unintentional and intentional exposures to NPs at work pose hazards to the health of workers. Although they are not completely characterized and their detailed effects on health remain largely unknown, quite a number of studies have shown that they pose hazards to the lungs, cardiovascular system, brain and systemic functions when exposed from the environment. Thus, medical surveillance in the workers exposed to NPs finds utmost importance. Safety measures when exposed to NPs need to be taken into consideration to safeguard the health of workers in industries and sectors involving production and handling of NPs.

Keywords: xenobiotics, carbon nanotubes, fly ash, ultrafine nanoparticles, epidemiology, workers, occupation, silica nanoparticles, genotoxicity, Clara cell protein 16, occupational health hazard, occupational medical surveillance

8.1 Introduction

Nanotechnology is known to have promising solutions to the existing problems and is a boon to the agricultural, forestry and healthcare industry. It has played a significant role in solving issues related to urbanization, energy constraints and sustainable use of resources. Researchers have developed new nanoparticles (NPs) of less than 100 nm of different sizes, shapes, surface charges, chemical characteristics, solubility and degree of agglomeration with desired and improved properties with diverse applications. Nanofertilizers, nanosized nutrients, engineered nanomaterials, nanopesticides or nanoformulations of active ingredients seem to hold promises in applications to our day-to-day lives, but exposure to NPs by inhalation at the work place can cause acute or chronic toxicity and can affect the respiratory system, causing respiratory disorders including granuloma formation, peribronchial inflammation, progressive interstitial fibrosis, chronic inflammatory responses, collagen deposition, oxidative stress, cancer and death. This exposure can also affect the workers in the field of agriculture who are exposed to xenobiotics during their work. Although nanotechnology holds promising solutions to agricultural activities, it nevertheless faces challenges of toxicity and health effects adversely; therefore, occupational risk assessment and management is an emerging field. Workers in the mining sector are also prone to exposed to NPs. Coal and fly ash samples from a power plant in Meghalaya, India, had components of clay minerals, glass fragments, spinel, quartz, fly ash

https://doi.org/10.1515/9783110579093-008

carbons as chars, ultrafine particles, multiwalled carbon nanotubes (CNTs) and other minerals [1]. In this chapter, we discuss different hazards in occupations associated with NP exposure.

8.2 NP exposure and occupational Hazard

Epidemiological studies on air pollution have revealed that ultrafine particles pose potential health hazards. However, studies defining the exposure level to ultrafine NPs at work [2] are the need of the hour to determine safety levels Figure 8.1.

Hazards associated with nano particles exposure at work place

- Information about health hazards of nanoparticles remains largely unknown
- Currently there is no universal safety limits of exposure for different types of nanoparticles
- Effects of nanoparticles by different routes of exposure by ingestion, skin and inhalation not known
- Effect of diverse engineered nanoparticles remains unknown
- Different agencies like NIOSH, OSHA, EPA are working to frame regulations to protect health of workers and the environment

Figure 8.1: Concerns with NPs as occupational hazard.

In a study on workers exposed to NPs in a manufacturing plant in Taiwan revealed expressions of pulmonary, cardiovascular disease markers including vascular cell adhesion molecules, paraoxonase enzymes alteration, inflammation and oxidative stress markers, antioxidant enzymes and genotoxicity markers, antioxidant enzymes including superoxide dismutase, glutathione peroxidase and cardiovascular markers, expression of small airway damage marker including Clara cell protein 16 and lung function test parameters to be associated in individuals handling such nanomaterials when compared to the workers who did not handle such NPs. The study also proposed the importance of markers and lung function tests as possible markers that could be useful for surveillance of health of workers involved in nanomaterial handling and manufacturing [3].

Silica NPs and nanosilicates have been evidenced to be toxic with fatal consequences to occupational workers exposed to them without any effective personal protective equipment [4]. Unintentional or intentional entry of engineered nanomaterials pose health hazards and have led to the understanding of importance of health surveillance and medical monitoring [5] in workers.

Engineered nanoparticles (ENPs) entering through skin, injection, olfactory, respiratory and intestinal routes can lead to various health problems; however, the downstream effects remain largely untested. They result in health threats to workers by posing both acute and chronic health hazards due to their uncharacterized nature and form a major occupational health hazard and risk to the workers who are exposed. Both airborne and ENPs influence their immunotoxicity and interact with the cells, and their biological interactions govern the toxic effects. Thus to understand or estimate the toxic effects of NPs as occupational hazard, their interactions with the host body need to be understood [6].

The National Institute for Occupational Safety and Health Nanotechnology Field Studies Team in 2011 reported the use of Nanomaterial Exposure Assessment Technique (2.0) to assess the exposure and effects of alumina and amorphous silica NPs from air on workers [7]. ENPs including titanium dioxide (TiO_2 NPs), zinc oxide (ZnO), silver (Ag) and other metals or their oxides present in commercial healthcare, personal and fuel products need to be handled with care and caution [8]. ENPs and CNTs entering through inhalation routes have been found to deposit in the alveolar area, and they interact with epithelial and alveolar macrophages [9]. CNTs comprising of single-walled and multiwalled CNTs administered in rodents through intratracheal or intrapharyngeal routes reveal pulmonary toxicity with inflammation, formation of epithelioid granulomas, symptoms of pulmonary fibrosis and toxicity in lungs [10]. Fullerenes, CNTs, metal oxides of iron and titanium and natural inorganic compounds, including asbestos and quartz, can affect the health and environment. Assessment of their risk to the human body requires evaluation of their release into the environment, reactivity, environmental toxicity and stability [11].

Newer studies are being designed to detect and understand the effects of NPs exposed to individuals by occupation. A combined standardized mortality ratio and Cox proportional hazards studies from the US, UK and German carbon black (CB) production workers have revealed traits of mortality, heart disease, ischemic heart disease and acute myocardial infarction when exposed to CB, indicating that CB exposure is a major professional hazard [12].

Industry workers associated with manufacturing of AgNPs and manufacturing and handling of $AgNO_3$NPs and silver dusts, when exposed to NPs, pose health hazards, resulting in damages in liver, skin and lungs. AgNPs, shows an occupational exposure limit (OEL) of 0.19 µg/m [3] from subchronic rat inhalation toxicity studies and exposure below this limit might protect workers from potential health hazards, including lung, liver, and skin damage [13].

Nano-TiO_2 exposed workers show the presence of affected nucleic acid oxidation including 8-hydroxy-2-deoxyguanosine, 8-hydroxyguanosine, 5-hydroxymethyl uracil and proteins including o-tyrosine, 3-chlorotyrosine and 3-nitrotyrosine in exhaled air, and overall oxidative stress is proposed as biomarkers for workers exposed to TiO_2 [14].

Exposures to TiO_2NPs through inhalation lead to adverse effects in lungs in rats. However, epidemiology studies could not correlate the exposure of TiO_2 to workers and its adverse health effects including lung cancer and chronic respiratory disorders. In some studies, it is revealed that cosmetics and sunscreens that lead to exposure of TiO_2 to skin could not penetrate skin and therefore do not have adverse effects [15]. The exposure to metals including TiO_2NPs can affect the upper and lower respiratory tract, causing asthma, rhinosinusitis, acute bronchitis, chronic bronchitis, chronic beryllium disease, acute pneumonitis, bronchogenic carcinoma and interstitial lung disease and is proposed to be a major occupational health hazard. Exposure to indium at workplace has been proposed to be a novel occupational hazard [16]. TiO_2NPs have wide-scale applications in a diverse range of products and can enter the body by inhalation, orally or by skin. Epidemiological studies indicate that it could because major exposure to and methodologies for setting OELs need to be set, and accurate effects in humans exposed to these at work need to be studied in greater detail [17].

Workers associated with works involving welding devices of structural steel are exposed to metallic NPs, iron fume hazards and inhalable welding fume concentration exceeding the American Conference of Governmental Industrial Hygienists threshold limit values [18].

The need of the hour is the awareness of workplace exposure to NPs and new technologies to identify effects of NPs in the workplace and environment. Owing to the exposure to asbestos, workers involved in sandblasting and coalmines suffer from lung diseases, accelerated silicosis and coal worker's pneumoconiosis. Exposures to heavy metals from hydraulic fracturing can affect the overall health of workers. Photochemical smog with identified NPs can lead to respiratory sickness. Therefore, the need of the hour is awareness of mass on the ill effects of NP contamination and risks associated to remain vigilant [19]. Scientific Committee of National Council on Radiation Protection and Measurements are drafting the available information and guidance and measures for radiation safety by an informatics-based method to access the risks of radiation exposure and safety measures [20]. Cellulose nanomaterials in composites and biomedicine has been known to affect the lungs. Cellulose nanocrystal, two cellulose nanofibril and cellulose microcrystal on artificial lung airway lining fluid using serum ultrafiltrate or SUF at pH 7.3 and alveolar macrophage phagolysosomal fluid (PSF, pH 4.5), increased the free radicals [21]. Usage of chemical fume hood in work places and laboratories has been instrumental in insignificantly reducing the exposure of the body of the workers to NPs from chemicals [22].

Using nanotechnology, electronics associated with printed materials lead to exposure of NPs to the workers associated with them and can be an occupational health hazard [23].

Exposure to ZnO metal fumes is known to have ill human health effects, leading to cardiopulmonary injury in Sprague–Dawley rats exposed by IT instillation and inhalation. Bronchoalveolar lavage fluid revealed intrapulmonary zinc levels and pulmonary oxidative damage after 72 h of ZnONP instillation, altering the overall zinc balance, and increased neutrophils, lactate dehydrogenase and total protein with increased 8-hydroxy-2′-deoxyguanosine in blood, with predominant accumulation of ZnONPs in lungs, heart, liver, kidneys and blood. ZnONPs intake by inhalation route result in lung inflammation. Cardiac inflammation, fibrosis, degeneration and necrosis of the myocardium occur on prolonged exposure of 30 days and are health hazards to the workers working with ZnONP [24].

In the New Energy and Industrial Technology Development Organization Project of Japan, the properties of manufactured nanomaterials are being detected and tested for their potential hazards [25]. Metal oxide NPs including cerium oxide (CeO_2NP), TiO_2NP, carbon black, silicon dioxide (SiO_2NP), nickel oxide (NiONP), ZnONP, copper oxide (CuONP) and amine-modified polystyrene beads with wide scale of medical and industrial applications when instilled into rat lungs lead to inflammatory responses, with inflammation at higher doses with CeO_2NP, NiONP, ZnONP and CuONP, and prove to be a potential health hazard [26]. ENPs in diesel exhaust particulate were indicative of adverse effects to people working in the automobile sector exposed to diesel exhaust, leading to inflammation in lung, systemic inflammation, cardiac activity impairment, thrombogenesis, vascular function and affected brain activity [27]. Experiments with the help of a model system to assess risk of exposure to NPs to individuals at work reveal that there are two types: individuals at high and intermediate risks [28]. Studies from pulmonary toxicity with lungs exposed to ultrafine or NPs lead to adverse inflammatory responses with free radical generation [29] and thus pose to be an important health hazard.

The use of synthetic amorphous silica NPs from industries effected the Chinese hamster lung fibroblasts (V79 cells), causing cytotoxic and genotoxic effects, decreased cell viability after 24 h and resulted in induction of apoptosis and DNA damage [30]. NPs have been tested in experimental animals by pharyngeal instillation, injection, inhalation, cell culture lines and gavage exposures routes [31].

8.3 Detection of NPs as hazard in workplace

NPs have been identified as potential hazards in workplace. A new safety management programme on, Management, Information, Control and Emergency, study has enabled preparation of exhaustive hazard inventory and is a promising step toward workplace health protection [32]. On the other hand, nanosensors and nanoremediation methods have been employed to detect and remove environmental contaminants. However, little is known about their biosafety, their adverse effects, fate in the environment and acquired biological reactivity when released into the environment and so requires further scientific researches to assess possible nanoagricultural risks.

Therefore, there is a need for toxicological research to define nanomaterial hazards, levels of exposure along the life-cycle and assessment of physicochemical properties governing its toxicity [33].

Studies in animals have revealed that NPs in air could lead to lung injury and extrapulmonary toxicity, leading to bilateral chest fluid, pulmonary fibrosis, pleural granuloma and multiorgan damage. This indicates maximum risk for the workers in industries or agricultural sectors exposed to NPs in air.

Today, studies on characterization, detection of surface properties of NPs, temporal factors associated with their exposure, their interaction with the biological system, data from epidemiological sources, data on workers exposed to ENPs and occupational medical surveillance [33] and the development of epidemiological studies on workers exposed to NPs with more focus on NP exposure effects at workplace are important [35–36].

8.4 Occupational hazards: insights from management and regulations – present status

Researchers and authorities are active across the globe to search for effective management and formulate legislations on insights of exposure to NPs as an occupational hazard by chronic inhalation of NPs and how it can be regulated. Recently, a chance-constrained programming integrated with nonlinear programming, thus leading to chance-constrained nonlinear programming that can handle uncertainties involved in the production of nanomaterial and workplace exposure to nanomaterials and controls, has been employed to minimize risks of workers handling them, while maximizing the production of NPs/materials. Manufacturing single-walled CNT, studied using this method, revealed the optimal production strategies meeting the standards of environmental health and safety standard regulations, and the chance-constrained nonlinear programming has proved to be an important tool in optimizing such unwanted exposures and effective production of nanomaterials in industries [37]. Time and again it has been emphasized that risk management associated with workplace exposure and its effective management should be the topmost priority of manufacturing units, and different tools useful for the health and safety management for the workers have been suggested [38].

Aircraft engine exhaust releases NPs as exhaust to the surrounding environment, thus raising the concerns regarding health of workers at the airport. A study to detect NPs, characterized by size, particle number concentration and distribution in the exhaled breath condensate (EBC) in a noninvasive way in France detected by dynamic light scattering and scanning electron microscopy coupled to X-ray spectroscopy, revealed the presence of aluminum, cadmium and chromium predominantly in the EBC contents, emphasizing the role of EBC as a potential useful tool for the noninvasive monitoring of health of workers exposed to NP and metals [39]. Risks to diseases have been reported from chronic exposures to aerosols of combustion particles and asbestos, but knowledge gaps exit in methods for determining such toxicity by diverse types of NPs largely uncharacterized with yet-to-be studied effects on long-term exposure to human health, posing difficulties for the regulators in framing appropriate regulations [40]. Information about effects of nanomaterials on health across countries is limited and scattered. From a Swiss industry a study on health and safety of workers in nonmaterial handling/manufacturing unit has enabled assessment of risks [41].

A study of CNT-handling workers revealed compliance to limits as in current American Conference of Governmental Industrial Hygienists threshold limit values and OELs set by the Korean Ministry of Labor for carbon black (3.5 mg/m^3), particles not otherwise specified (PNOS) (3 mg/m^3) and asbestos (0.1 fiber/cc) and effectively controlled operations with appropriate engineering controls [42] revealed minimum risks to health of workers [42]. However, a study in the Australian workplace reveals the threats on workplace exposure posed by asbestos manufacturing unit and it has been suggested that current Australian regulation needs to take into consideration workplace dangers associated with size of manufactured NPs [43]. China being the major producer of nano-based manufactured products revealed compliance maintained by manufacturing units to general safety regulations, but needs for occupational safety legislation, safety guideline for handling and manufacturing of NPs and materials and training on nanosafety while at work were highlighted [44].

Fullerenes, known for their antioxidant and radical scavenging properties, have been extensively applied in targeted drug delivery, energy applications, designing with polymers and cosmetic products of which pristine fullerenes with low toxicity and negligible risks to humans even in workplace [45]. The European Commission has drafted an "incremental approach," focusing on adapting existing laws to regulate nanotechnology production, analyzing fullerenes (C_{60}) and CNTs to understand status of current applicable regulations, applicability, identify their gaps and to suggest solutions, which is known as the Safety at Workplace Directives [46]. After studying the toxicological effects, OELs for manufactured nanomaterials are being drafted [47]. Scientific findings and ethical values and occupational health practice have found importance in the Italian law 81/2008 [48]. Studies reported from California on safety, knowledge gaps, regulations, future prospects and challenges in NP research have been initiated [49].

8.5 Discussion

Scientists, regulators, industrialists and academicians need to work in an integrated manner to achieve the four major goals: (i) to characterize the diverse NPs; (ii) to understand their effects in human tissues; (iii) to time-to-time monitor the health of the workers exposed to NPs and understand the impact of workplace exposure to NPs and (iv) to understand the necessity of regulations to control workplace exposure to hazardous NPs. It has been aptly studied that while regulations are important, the nature of diverse nature of NPs and their effects need to be considered to arrive at a conclusion. These measures and regulations should be followed across the globe since the environmental contaminants of NPs can affect people across the world and thus global measures need to be taken; this is the need of the hour as NPs are finding roles in almost every aspect of our day-to-day lives.

References

[1] Oliveira ML, Marostega F, Taffarel SR, Saikia BK, Waanders FB, DaBoit K, Baruah BP, Silva LF. Nano-mineralogical investigation of coal and fly ashes from coal-based captive power plant (India): an introduction of occupational health hazards. Sci Total Environ. 2014 Jan;15:468–469:1128–1137.

[2] Peters A, Rückerl R, Cyrys J. Lessons from air pollution epidemiology for studies of engineered nanomaterials. J Occup Environ Med. 2011 Jun;53(6 Suppl):S8–S13.

[3] Liao HY, Chung YT, Lai CH, Wang SL, Chiang HC, Li LA, Tsou TC, Li WF, Lee HL, Wu WT, Lin MH, Hsu JH, Ho JJ, Chen CJ, Shih TS, Lin CC, Liou SH. Six-month follow-up study of health markers of nanomaterials among workers handling engineered nanomaterials. Nanotoxicol.2014. Aug;8 Suppl 1:100–110.

[4] Song Y, Tang S. Nanoexposure, unusual diseases, and new health and safety concerns. Scientific World Journal. 2011;11:1821–1828.

[5] Schulte P, Geraci C, Zumwalde R, Hoover M, Kuempel E. Occupational risk management of engineered nanoparticles. J Occup Environ Hyg. 2008 Apr;5(4):239–249.

[6] Pedata P, Petrarca C, Garzillo EM, Di Gioacchino M. Immunotoxicological impact of occupational and environmental nanoparticles exposure: The influence of physical, chemical, and combined characteristics of the particles. Int J Immunopathol Pharmacol. 2016 Sep;29(3):343–353.

[7] Brenner SA, Neu-Baker NM, Eastlake AC, Beaucham CC, Geraci CL NIOSH field studies team assessment: Worker exposure to aerosolized metal oxide nanoparticles in a semiconductor fabrication facility. J Occup Environ Hyg. 2016 Nov;13(11):871–880.

[8] Teow Y, Asharani PV, Hande MP, Valiyaveettil S. Health impact and safety of engineered nanomaterials. Chem Commun (Camb). 2011 Jul 7;47(25):7025–7038.

[9] Bergamaschi E, Bussolati O, Magrini A, Bottini M, Migliore L, Bellucci S, Iavicoli I, Bergamaschi A. Nanomaterials and lung toxicity: interactions with airways cells and relevance for occupational health risk assessment. Int J Immunopathol Pharmacol. 2006 Oct-Dec;19 (4 Suppl):3–10.

[10] Lam CW, James JT, McCluskey R, Arepalli S, Hunter RL. A review of carbon nanotube toxicity and assessment of potential occupational and environmental health risks. Crit Rev Toxicol. 2006 Mar;36(3):189–217.

[11] Taghavi SM, Momenpour M, Azarian M, Ahmadian M, Souri F, Taghavi SA, Sadeghain M, Karchani M. Effects of nanoparticles on the environment and outdoor workplaces. Electron Physician. 2013 Nov 1;5(4):706–712. eCollection 2013 Oct-Dec.

[12]   Morfeld P, Mundt KA, Dell LD, Sorahan T, McCunney RJMeta-analysis of cardiac mortality in three cohorts of carbon black production workers. Int J Environ Res Public Health. 2016 Mar 9; 13(3).pii:E302.

[13]   Weldon BA, M Faustman E, Oberdörster G, Workman T, Griffith WC, Kneuer C, Yu IJ. Occupational exposure limit for silver nanoparticles: considerations on the derivation of a general health-based value. Nanotoxicol. 2016 Sep;10(7):945–956.

[14]   Pelclova D, Zdimal V, Fenclova Z, Vlckova S, Turci F, Corazzari I, Kacer P, Schwarz J, Zikova N, Makes O, Syslova K, Komarc M, Belacek J, Navratil T, Machajova M, Zakharov S. Markers of oxidative damage of nucleic acids and proteins among workers exposed to TiO_2 (nano) particles.Occup Environ Med. 2016 Feb;73(2):110–8. -2015–103161.

[15]   Warheit DB, Donner EM. Risk assessment strategies for nanoscale and fine-sized titanium dioxide particles: Recognizing hazard and exposure issues. Food Chem Toxicol. 2015 Nov;85:138–147.

[16]   Mayer A, Hamzeh N. Beryllium and other metal-induced lung disease. Curr Opin Pulm Med. 2015 Mar;21(2):178–184.

[17]   Warheit DB How to measure hazards/risks following exposures to nanoscale or pigment-grade titanium dioxide particles. Toxicol Lett. 2013 Jul 4;220(2):193–204.

[18]   Fethke NB, Peters TM, Leonard S, Metwali M, Mudunkotuwa IA. Reduction of biomechanical and welding ume exposures in stud welding. Ann Occup Hyg. 2016 Apr;60(3):387–401.

[19]   Moitra S, Puri R, Paul D, Huang YC. Global perspectives of emerging occupational and environmental lung diseases. Health Phys.2015 Feb;108(2):179–194.

[20]   Hoover MD, Myers DS, Cash LJ, Guilmette RA, Kreyling WG, Oberdörster G, Smith R, Cassata JR, Boecker BB, Grissom MP. Application of an informatics-based decision-making framework and process to the assessment of radiation safety in nanotechnology.

[21]   Stefaniak AB, Seehra MS, Fix NR, Leonard SS. Inhal Toxicol. Lung biodurability and free radical production of cellulose nanomaterials. 2014 Oct;26(12):733–749.

[22]   Dunn KH, Tsai CS, Woskie SR, Bennett JS, Garcia A, Ellenbecker MJ Evaluation of leakage from fume hoods using tracer gas, tracer nanoparticles and nanopowder handling test methodologies. J Occup Environ Hyg. 2014.11(10):D164–D173.

[23]   Lee JH, Sohn EK, Ahn JS, Ahn K, Kim KS, Lee JH, Lee TM, Yu IJ. Exposure assessment of workers in printed electronics workplace. Inhal Toxicol. 2013 Jul;25(8):426–434.

[24]   Chuang HC, Juan HT, Chang CN, Yan YH, Yuan TH, Wang JS, Chen HC, Hwang YH, Lee CH, Cheng TJ. Cardiopulmonary toxicity of pulmonary exposure to occupationally relevant zinc oxide nanoparticles. Nanotoxicol. 2014 Sep;8(6):593–604.

[25]   Morimoto Y, Kobayashi N, Shinohara N, Myojo T, Tanaka I, Nakanishi J. Hazard assessments of manufactured nanomaterials. J Occup Health. 2010;52(6):325–334.

[26]   Cho WS, Duffin R, Poland CA, Howie SE, MacNee W, Bradley M, Megson IL, Donaldson K. Metal oxide nanoparticles induce unique inflammatory footprints in the lung: important implications for nanoparticle testing. Environ Health Perspect. 2010 Dec;118(12):1699–1706.

[27]   Hesterberg TW, Long CM, Lapin CA, Hamade AK, Valberg PA. Diesel exhaust particulate (DEP) and nanoparticle exposures: what do DEP human clinical studies tell us about potential human healthhazards of nanoparticles? .Inhal Toxicol.2010 Jul;22(8):679–694.

[28]   Giacobbe F, Monica L, Geraci D. Risk assessment model of occupational exposure to nanomaterials. Hum Exp Toxicol. 2009 Jun; 28 (6–7):401–406.

[29]   Warheit DB, Sayes CM, Reed KL, Swain KA. Health effects related to nanoparticle exposures: environmental, health and safety considerations for assessing hazards and risks. Pharmacol Ther. 2008 Oct;120(1):35–42.

[30]   Guichard Y, Fontana C, Chavinier E, Terzetti F, Gaté L, Binet S, Darne C. Cytotoxic and genotoxic evaluation of different synthetic amorphous silica nanomaterials in the V79 cell line. Toxicol Ind Health. 2016 Sep;32(9):1639–1650.

[31] Yah CS, Simate GS, Iyuke SE Nanoparticles toxicity and their routes of exposures. Pak J Pharm Sci. 2012 Apr;25(2):477–491.

[32] Marendaz JL, Friedrich K, Meyer T. Safety management and risk assessment in chemical laboratories.Chimia (Aarau). 2011;65(9):734–737.

[33] Marie-Desvergne C, Dubosson M, Touri L, Zimmermann E, Gaude-Môme M, Leclerc L, Durand C, Klerlein M, Molinari N, Vachier I, Chanez P, Mossuz VC.Assessment of nanoparticles and metal exposure of airport workers using exhaled breath condensate. J Breath Res. 2016 Jul 13;10(3):036006.

[34] Iavicoli I, Leso V, Beezhold DH, Shvedova AA. Nanotechnology in agriculture: Opportunities, toxicological implications, and occupational risks. Toxicol Appl Pharmacol. 2017 Aug;15329:96–111.

[35] Nasterlack M. Role of medical surveillance in risk management. J Occup Environ Med. 2011 Jun;53(6 Suppl):S18–S21.

[36] Schulte PA, Schubauer-Berigan MK, Mayweather C, Geraci CL, Zumwalde R, McKernan JL. Issues in the development of epidemiologic studies of workers exposed to engineered nanoparticles. J Occup Environ Med. 2009 Mar;51(3):323–335.

[37] Schulte P, Geraci C, Zumwalde R, Hoover M, Castranova V, Kuempel E, Murashov V, Vainio H, Savolainen K. Sharpening the focus on occupational safety and health in nanotechnology. Scand J Work Environ Health. 2008 Dec;34(6):471–478.

[38] Chen Z, Yuan Y, Zhang SS, Chen Y, Yang FL Management of occupational exposure to engineered nanoparticles through a chance-constrained nonlinear programming approach. Int J Environ Res Public Health. 2013 Mar 26;10(4):1231–1249.

[39] van Broekhuizen P. Dealing with uncertainties in the nanotech workplace practice: making the precautionary approach operational. J Biomed Nanotechnol. 2011 Feb;7(1):15–17.

[40] Seaton A, Tran L, Aitken R, Donaldson K. Nanoparticles, human health hazard and regulation. J R Soc Interface. 2010 Feb 6;7 Suppl 1:S119–S129.

[41] Schmid K, Danuser B, Riediker M Nanoparticle usage and protection measures in the manufacturing industry–a representative survey. J Occup Environ Hyg. 2010 Apr;7(4):224–232.

[42] Lee JH, Lee SB, Bae GN, Jeon KS, Yoon JU, Ji JH, Sung JH, Lee BG, Lee JH, Yang JS, Kim HY, Kang CS, Yu IJ. Exposure assessment of carbon nanotube manufacturing workplaces. Inhal Toxicol. 2010 Apr;22(5):369–381.

[43] Ludlow K One size fits all? Australian regulation of nanoparticle exposure in the workplace. J Law Med. 2007 Aug;15(1):136–152.

[44] Zhang C, Zhang J, Wang G Current safety practices in nano-research laboratories in China. J Nanosci Nanotechnol. 2014 Jun;14(6):4700–4705.

[45] Aschberger K, Johnston HJ, Stone V, Aitken RJ, Tran CL, Hankin SM, Peters SA, Christensen FM. Review of fullerene toxicity and exposure–appraisal of a human health risk assessment, based on open literature. Regul Toxicol Pharmacol. 2010 Dec;58(3):455–473.

[46] Franco A, Hansen SF, Olsen SI, Butti L Limits and prospects of the "incremental approach" and the European legislation on the management of risks related to nanomaterials. Regul Toxicol Pharmacol. 2007 Jul;48(2):171–183.

[47] Mihalache R, Verbeek J, Graczyk H, Murashov V, van Broekhuizen P. Occupational exposure limits for manufactured nanomaterials, a systematic review. Nanotoxicol. 2017 Feb;11(1):7–19.

[48] Franco G. Occupational health practice and exposure to nanoparticles: reconciling scientific evidence, ethical aspects, and legal requirements. Arch Environ Occup Health. 2011;66(4):236–240.

[49] Fan AM, Alexeeff G. Nanotechnology and nanomaterials: toxicology, risk assessment, and regulations. J Nanosci Nanotechnol. 2010 Dec;10(12):8646–8657.

9 Toxicity and effect of nanoparticles on liver and kidneys, and kidney disorders

Abstract: The renal system is associated with the excretion of toxic wastes from the body and the liver is associated with the metabolism of drugs, xenobiotics and their removal, and thus they are susceptible to the adverse effects of these harmful chemicals. While reports from human cell lines and animal models have confirmed the nephrotoxic and hepatotoxic potential of nanoparticles (NPs), the threshold safety limits of their exposure are yet not known. Some studies have confirmed the toxic effects but biomarkers in detection of early toxicity effects are yet to be known. No guidelines exist on the threshold limit of any NP exposure causing nephrotoxic effect. We discuss in this chapter the different effects of NPs in imposing toxic effects in liver and renal system and the gap in our knowledge in this domain of science. The further scope remains in the estimation of threshold limits and understanding of nephrotoxic potential associated with NP exposure.

Keywords: Bowman's space, nephrotoxicity, mesangial cells, glomeruli, inorganic nanoparticles, kidneys, CNTs, CD1 mice, mesoporous silica nanoparticles, dimercaptosuccinic acid, oxidative stress, inflammation, alanine aminotransferase, aspartate aminotransferase, triglyceride, total bilirubin, total bile acid

9.1 Introduction

Nephrotoxicity or renal toxicity involves the toxicity in the kidneys due to the effects of toxic chemicals, such as lead, mercury, cadmium and drugs and medicines, on renal function in different ways, including loss of excretion and removal of wastes and detoxification of toxins. These chemicals capable of inducing nephrotoxic effects are known as nephrotoxins. The most commonly used parameters for the detection of nephrotoxicity are (i) blood–urea nitrogen (BUN), which reflects the total amount of nitrogen present in the form of urea that rises due to the inability of the kidneys to filter them out of the body, and (ii) creatinine where it is unable to filter the large amounts of creatinine and the body creatinine level rises in case of kidney failure.

Liver plays a pivotal role in the metabolism of xenobiotics, chemicals and drugs and is susceptible to their adverse effects. Hepatotoxicity or liver damage induced by drugs and medicines can cause acute and chronic liver diseases affecting liver functions, and functions of liver enzymes are detected by liver function tests [1]. These harmful toxins or chemicals are also known as hepatotoxins.

Although studies on toxicity of nanoparticles (NPs) exist, few studies have been carried out on NPs and their effect on the liver and excretory system. While it is known that kidneys are affected by xenobiotics, studies reveal their hepatotoxic and

https://doi.org/10.1515/9783110579093-009

nephrotoxic effects in human cell lines and animal models on exposure to most NPs by different routes in the host body. In this chapter, we discuss the different effects of NPs that induce the nephrotoxic effects.

9.2 Nephrotoxicity and NP exposure

NP exposure is reflected by accumulation in the different organs and is transmitted through blood circulation, affecting the liver, kidneys and lungs, together with the generation of oxidative stress, inflammation and DNA damage [2].

Mesoporous silica NPs have been reported to cause acute toxicity and biodistribution leading to systemic absorption and excretion, decreased liver distribution and urinal excretion, renal system damage including hemorrhage, vascular congestion and renal tubular necrosis [3].

Elimination of NPs by renal excretion has led to adverse effects on kidneys, and also induced nephrotoxicity. Carbon, metal and/or silica NPs have been reported to exert cytotoxic effects such as decreased cell viability, oxidative stress, mitochondrial and cytoskeleton dysfunction and DNA damage, and caused nephrotoxicity, fibrosis in the renal tubules of the kidneys with tubular epithelial cell degeneration, formation of cellular fragments and proteinaceous liquid in tubule lumen, swollen glomeruli, affected Bowman's space and proliferation of mesangial cells [4].

Inorganic NPs have been reported to affect kidneys. Degradable luminescent glutathione (GSH)-coated copper NPs (CuNPs) have been reported to be easily cleared through the renal system with less accumulation in the liver [5].

Superparamagnetic iron oxide nanoparticles (SPIONs) are also known for their nanotoxicity causing drug-induced liver injury. Hepatotoxicity of polyethylene glycol (PEG)-8000-coated ultrasmall SPIONs (USPIONs) was revealed on experimental Wistar rats [6] of species *Rattus norvegicus* (brown rat), which is bred and kept for scientific research and also finds importance as an animal model in psychology and biomedical science research.

Silica NPs have been reported for toxic effects exerted on various tissues, including liver, kidneys, lungs and testis [7].

Carbon nanotubes (CNTs) are known to rapidly disseminate with initial accumulation in the lungs and brain and transmitted to liver and kidneys in CD1 mice, an outbred mouse line originated from the colony of Swiss mice in 1926 and commonly used in toxicology and in chemical carcinogenicity studies, through the blood circulation. The MITO-Luc bioluminescence reporter mice model is a transgenic mice model designed with MITO-Luc, engineered in the actively proliferating cells to express the luciferase reporter gene, thus enabling visualization of body events noninvasively by bioluminescence imaging methods in vivo [8]. These mice could induce systemic cell proliferation, bone marrow and the immune system when exposed to single-walled CNTs, with transient accumulation in the lungs, spleen, kidneys and liver associated with increased predominant liver enzymes such as aspartate aminotransferase, also known as SGOT, alanine aminotransferase and bilirubinemia or presence of bilirubin in the blood, indicative of liver [9]. Biodistribution, biocompatibility and nanotoxicity of maghemite NPs stabilized with dimercaptosuccinic acid revealed their biocompatibility without affecting hepatic and renal normal activities with preferentially distributed to the lung, liver and kidneys [10].

CuNPs caused hepatotoxicity and nephrotoxicity; increased citrate, succinate, trimethylamine-N-oxide, glucose and amino acids; decreased creatinine levels in urine, elevated lactate, 3-hydroxybutyrate, acetate, creatinine, triglycerides and phosphatide; and reduced serum glucose levels. The hepatotoxicity and nephrotoxicity were manifested by increased triglycerides, mitochondrial failure, enhanced ketogenesis, beta-oxidation of fatty acids and glycolysis [11]. Toxicity of CuNPs, but not microparticles, is reported to affect the kidney, liver and spleen [12]. CuNPs cause acute toxicity in mice, with the liver and kidney affected. Male Wistar rats exposed to subacute toxic effects of CuNPs revealed hepatotoxicity with increase in alanine aminotransferase, aspartate aminotransferase, triglyceride, total bilirubin, total bile acid, and with decrease in body weight with scattered, dotted hepatocytic necrotic histopathology, oxidative stress and affected signal transduction [13].

Cytotoxicity of surface-biofunctionalized colloidal cadmium sulfide quantum dots (QDs) with chitosan on human osteosarcoma cell line or SAOS cell line, human non-Hodgkin's B-cell lymphoma or Toledo cell line, and human embryonic kidney cell line or HEK293T cell line have been reported to be manifested by a dose- and time-dependent inhibition of cell viability of cells [14].

Orally administered PVP-coated silver nanoparticles (AgNPs), silver nitrate (AgNO$_3$), ionic silver in male and female rats revealed no significant toxic effects of AgNPs and AgNO$_3$ up to 1 mg/kg of body weight; however, Ag is observed to deposit and affect the liver, kidneys, testis and spleen [15]. AgNPs, administered by a single intratracheal instillation, in a rat model revealed overall toxic response but neither AgNPs nor AgNO$_3$ could induce oxidative stress effects in kidneys and plasma [16].

9.3 Applications of NPs to diseases of kidney

Both organic and inorganic NPs are finding application in drug delivery and imaging as theranostic agents in different kidney-related disorders [17, 18] and in treatment of nephropathologies [18].

Studies on biodistribution of PEG-coated QDs revealed circulation and renal accumulation in mice blood and deposited in the kidney intraglomerular mesangial cells when injected intravenously, which find importance in site-specific treatment strategies in kidney diseases including diabetic nephropathy (DN) [19]. Titanium dioxide nanoparticles (TiO$_2$NPs) when administered orally to the kidneys of rats have been revealed to affect serum kidney function indicated by biomarkers including urea, creatinine and uric acid, increased serum glucose and serum immunoinflammatory biomarkers such as tumor necrosis factor-α, interleukin-6, C-reactive protein, immunoglobulin G, vascular endothelial growth factor (angiogenic factor) and nitric oxide, decreased renal GSH content. However, administration of quercetin or idebenone has been revealed to provide protective and prophylactic role [20]. Mineralo-organic NPs from blood and urine may cause the formation of kidney stone by an alternative mechanism [21]. NPs are finding applications in noninvasive detection and monitoring of chronic kidney disease by the marker protein, ferritin, through magnetic resonance imaging-based approaches [22]. Selenium NPs have been reported to confer protective effect in the progression of DN induced in male Sprague–Dawley rats by lowering the levels of BUN, creatinine, fibronectin and collagen and elevated the levels of albumin in diabetic rats, elevated heat shock protein-70, longevity protein SIRT 1 and also modulated apoptotic proteins Bax and Bcl-2 in diabetic kidney [23]. Chitosan-based delivery system of small interfering RNA (siRNA) in the kidney revealed that chitosan/siRNA particle uptake was mediated by a megalin-based endocytic pathway with a promising potential therapeutic application in treatment of different kidney diseases [24].

Polymeric NPs including poly(lactic-co-glycolic) acid (PLGA) NPs and chitosan (Chi)-covered PLGA NPs in Madin-Darby bovine kidney and human colorectal adenocarcinoma (Colo 205) cells were reported to be nontoxic up to 2,500 µg/mL [25].

TiO$_2$NPs have been reported to cause oxidative stress in the liver and kidney but simultaneous treatment with quercetin on male Wistar rats revealed the renal protective effect of quercetin [26]. Diverse nanomedicine formulations have been reported to be effective for kidney disorders [27], particularly glomerulonephritic diseases [28] and NP-based thrombin inhibitor, phenylalanine-proline-arginine-chloromethylketone, would improve kidney reperfusion and protect renal function after injury in rodent models [29]. Cyclosporine (CsA) entrapped in biodegradable NPs revealed to reduce nephrotoxicity of CsA alone [30]. Nonnephrotoxic perfluorocarbon NPs have found applications in noninvasive detection and therapy of acute and chronic kidney diseases [31].

Sorafenib-loaded PLGA and hydrophobically modified chitosan (HMC)-coated 1,2-dipalmitoyl-sn-glycero-3-phosphocholine liposomes or HMC-coated liposomes were reported to be promising nanocarriers of the drug Sorafenib in renal carcinoma [32]. Coordination complexes derived from catechol-derived chitosan (HCA-Chi), metal ions and active drug molecules fabricated with doxorubicin (DOX) forming HCA-Chi-Cu-DOX proved as promising drug delivery systems for treating renal fibrosis [33]. In a study of rats, arsenic-induced damage to the kidney and brain could be prevented by the coadministration of nanoformulation of curcumin (CUR-NP) [34]. Nanocontrast agents are finding applications in detection of inflammation markers in kidney diseases [35]. MicroRNA (miR), miR-146a, delivery by polyethylenimine NPs could inhibit renal fibrosis emphasizing their therapeutic importance in preventing renal fibrosis [36].

Although nanomedicine finds its importance in improved targeting of renal cells, there remains scope of research for challenges and applications of diverse range of nanomedicines in targeting renal diseases [37]. Cerium oxide (CeO_2) NPs have been reported for its importance in the treatment of intra-abdominal infection or peritonitis-induced acute kidney injury [38]. Podocytes of kidney have been successfully targeted by cyclo(RGDfC)-modified QDs and reported for selective binding to the $\alpha v \beta 3$ integrin receptor [39]. NPs ranging from 75 ± 25 nm diameters have been reported to be successfully targeting the mesangium of the kidney [40]. Macromolecular contrast agents including gadolinium (Gd)-bound albumin, Gd-bound dendrimer and ultrasmall particles of iron oxide have been reported to find importance in monitoring renal parenchymal diseases in animal models [41]. SPIONs have enabled studies by imaging in nephritic kidneys in animal model [42]. USPIONs including ferumoxytol offers promise in enabling the study of nephrogenic systemic fibrosis disease by imaging approaches [43].

9.4 Discussion

Nephrotoxicity is associated with the loss of excretion and detoxification due to the adverse effects of toxins, xenobiotics and NPs. Drug-induced nephrotoxicity has been identified by the specific biomarkers including (i) affected glomerular hemodynamics, (ii) toxicity of the tubules, (iii) renal inflammation physiology, (iv) intratubular precipitation of crystals, resulting in obstruction called as crystal nephropathy, (v) rhabdomyolysis caused due to direct or indirect muscle injury leading to death of muscle fibers thereby releasing their contents into the bloodstream leading to kidney failure and (vii) thrombotic microangiopathy, a pathological condition that leads to capillary and arteriole thrombosis due to endothelial injury. But biomarkers to detect an early renal damage are still the domain of active basic and applied research [44].

Although some studies have indicated the pronounced effects of NPs in causing hepatotoxicity and nanotoxicity and biomarkers also exist for general drug-induced nephrotoxicity, the specific biomarkers for NP-induced nephrotoxicity and/or hepatotoxicity remains the another major unexplored area. The currently unknown domain of study revolves around the detection and estimation of threshold limits of NP exposure either through intentional purposes as in the form of drugs, healthcare products or exposures by unintentional means through airborne or exposure at work needs to be explored and remains the road ahead in this domain of biology.

References

[1]   Thapa BR, Walia A. Liver function tests and their interpretation. Ind J Pediatr. 2007 July;74:663–671.
[2]   Wu T, Tang M. Review of the effects of manufactured nanoparticles on mammalian target organs. J Appl Toxicol. 2018 Jan;38(1):25–40.
[3]   Li L, Liu T, Fu C, Tan L, Meng X, Liu H. Biodistribution, excretion, and toxicity of mesoporous silica nanoparticles after oral administration depend on their shape. Nanomedicine. 2015 Nov; 11(8):1915–1924.
[4]   Iavicoli I, Fontana L, Nordberg G. The effects of nanoparticles on the renal system. Crit Rev Toxicol. 2016 Jul;46(6):490–560.
[5]   Yang S, Sun S, Zhou C, Hao G, Liu J, Ramezani S, Yu M, Sun X, Zheng J. Renal clearance and degradation of glutathione-coated copper nanoparticles. Bioconjug Chem. 2015 Mar 18; 26(3):511–519.
[6]   Rajan B, Sathish S, Balakumar S, Devaki T. Synthesis and dose interval dependent hepatotoxicity evaluation of intravenously administered polyethylene glycol-8000 coated ultra-small superparamagnetic iron oxide nanoparticle on Wistar rats. Environ Toxicol Pharmacol. 2015 Mar;39(2):727–735.
[7]   Hassankhani R, Esmaeillou M, Tehrani AA, Nasirzadeh K, Khadir F, Maadi H. In vivo toxicity of orally administered silicon dioxide nanoparticles in healthy adult mice. Environ Sci Pollut Res Int. 2015 Jan;22(2):1127–1132.
[8]   de Latouliere L, Manni I, Iacobini C, Pugliese G, Grazi GL, Perri P, Cappello P, Novelli F, Menini S, Piaggio G. A bioluminescent mouse model of proliferation to highlight early stages of pancreatic cancer: a suitable tool for preclinical studies. Ann Anat. 2016 Sep; 207:2–8.
[9]   Principi E, Girardello R, Bruno A, Manni I, Gini E, Pagani A, Grimaldi A, Ivaldi F, Congiu T, De Stefano D, Piaggio G, de Eguileor M, Noonan DM, Albini A. Systemic distribution of single-walled carbon nanotubes in a novel model: alteration of biochemical parameters, metabolic functions, liver accumulation, and inflammation in vivo. Int J Nanomedicine. 2016 Sep 1; 11:4299–4316.
[10]   Monge-Fuentes V, Garcia MP, Tavares MC, Valois CR, Lima EC, Teixeira DS, Morais PC, Tomaz C, Azevedo RB. Biodistribution and biocompatibility of DMSA-stabilized maghemite magnetic nanoparticles in nonhuman primates (Cebus spp.). Nanomedicine (Lond). 2011 Nov. 6(9):1529–1544.
[11]   Lei R, Wu C, Yang B, Ma H, Shi C, Wang Q, Wang Q, Yuan Y, Liao M. Integrated metabolomic analysis of the nano-sized copper particle-induced hepatotoxicity and nephrotoxicity in rats: a rapid in vivo screening method for nanotoxicity. Toxicol Appl Pharmacol. 2008 Oct 15;232(2):292–301.
[12]   Chen Z, Meng H, Xing G, Chen C, Zhao Y, Jia G, Wang T, Yuan H, Ye C, Zhao F, Chai Z, Zhu C, Fang X, Ma B, Wan L. Acute toxicological effects of copper nanoparticles in vivo. Toxicol Lett. 2006 May 25; 163(2):109–120.
[13]   Yang B, Wang Q, Lei R, Wu C, Shi C, Wang Q, Yuan Y, Wang Y, Luo Y, Hu Z, Ma H, Liao M. Systems toxicology used in nanotoxicology: mechanistic insights into the hepatotoxicity of nano-copper particles from toxicogenomics. J Nanosci Nanotechnol. 2010 Dec; 10(12):8527–8537.
[14]   de Carvalho SM, Mansur AAP, Mansur HS, Guedes MIMC, Lobato ZIP, Leite MF. In vitro and in vivo assessment of nanotoxicity of CdS quantum dot/aminopolysaccharide bionanoconjugates. Mater Sci Eng C Mater Biol Appl. 2017 Feb 1; 71:412–424.
[15]   Qin G, Tang S, Li S, Lu H, Wang Y, Zhao P, Li B, Zhang J, Peng L. Toxicological evaluation of silver nanoparticles and silver nitrate in rats following 28 days of repeated oral exposure. Environ Toxicol. 2017 Feb; 32(2):609–618.

[16] Roda E, Barni S, Milzani A, Dalle-Donne I, Colombo G, Coccini T. Single silver nanoparticle instillation induced early and persisting moderate cortical damage in rat kidneys. Int J Mol Sci. 2017 Oct 10;18(10):pii: E2115.

[17] Kim SY and Moon A. Drug-induced nephrotoxicity and its biomarkers. Biomol Ther (Seoul). 2012 May; 20(3): 268–272.

[18] Lee SH, Lee JB, Bae MS, Balikov DA, Hwang A, Boire TC, Kwon IK, Sung HJ, Yang JW. Current progress in nanotechnology applications for diagnosis and treatment of kidney diseases. Adv Healthcare Mater. 2015 Sep 16; 4(13):2037–2045.

[19] Pollinger K, Hennig R, Bauer S, Breunig M, Tessmar J, Buschauer A, Witzgall R, Goepferich A. Biodistribution of quantum dots in the kidney after intravenous injection. J Nanosci Nanotechnol. 2014 May; 14(5):3313–3319.

[20] Al-Rasheed NM, Faddah LM, Mohamed AM, Abdel Baky NA, Al-Rasheed NM, Mohammad RA. Potential impact of quercetin and idebenone against immuno- inflammatory and oxidative renal damage induced in rats by titanium dioxide nanoparticles toxicity. J Oleo Sci. 2013; 62(11):961–971.

[21] Martel J, Wu CY, Young JD. Translocation of mineralo-organic nanoparticles from blood to urine: a new mechanism for the formation of kidney stones?. Nanomedicine (Lond). 2016 Sep;11(18):2399–2404.

[22] Charlton JR, Beeman SC, Bennett KM. MRI-detectable nanoparticles: the potential role in the diagnosis of and therapy for chronic kidney disease. Adv Chronic Kidney Dis. 2013 Nov; 20(6):479–487.

[23] Kumar GS, Kulkarni A, Khurana A, Kaur J, Tikoo K. Selenium nanoparticles involve HSP-70 and SIRT1 in preventing the progression of type 1 diabetic nephropathy. Chem Biol Interact. 2014 Nov 5; 223:125–133.

[24] Gao S, Hein S, Dagnæs-Hansen F, Weyer K, Yang C, Nielsen R, Christensen EI, Fenton RA, Kjems J. Megalin-mediated specific uptake of chitosan/siRNA nanoparticles in mouse kidney proximal tubule epithelial cells enables AQP1 gene silencing. Theranostics. 2014 Aug 13; 4(10):1039–1051.

[25] Trif M, Florian PE, Roseanu A, Moisei M, Craciunescu O, Astete CE, Sabliov CM. Cytotoxicity and intracellular fate of PLGA and chitosan-coated PLGA nanoparticles in Madin-Darby bovine kidney (MDBK) and human colorectal adenocarcinoma (Colo 205) cells. J Biomed Mater Res A. 2015 Nov; 103(11):3599–3611.

[26] González-Esquivel AE, Charles-Niño CL, Pacheco-Moisés FP, Ortiz GG, Jaramillo-Juárez F, Rincón-Sánchez AR. Beneficial effects of quercetin on oxidative stress in liver and kidney induced by titanium dioxide (TiO2) nanoparticles in rats. Toxicol Mech Methods. 2015 Mar; 25(3):166–175.

[27] Williams RM, Jaimes EA, Heller DA. Nanomedicines for kidney diseases. Kidney Int. 2016 Oct; 90(4):740–745.

[28] Zuckerman JE, Davis ME. Targeting therapeutics to the glomerulus with nanoparticles. Adv Chronic Kidney Dis. 2013 Nov; 20(6):500–507.

[29] Chen J, Vemuri C, Palekar RU, Gaut JP, Goette M, Hu L, Cui G, Zhang H, Wickline SA. Antithrombin nanoparticles improve kidney reperfusion and protect kidney function after ischemia-reperfusion injury. Am J Physiol Renal Physiol. 2015 Apr 1; 308(7):F765–F773.

[30] Ankola DD, Wadsworth RM, Ravi Kumar MN. Nanoparticulate delivery can improve peroral bioavailability of cyclosporine and match Neoral Cmax sparing the kidney from damage. J Biomed Nanotechnol. 2011 Apr; 7(2):300–307.

[31] Chen J, Pan H, Lanza GM, Wickline SA. Perfluorocarbon nanoparticles for physiological and molecular imaging and therapy. Adv Chronic Kidney Dis. 2013 Nov; 20(6):466–478.

[32]   Liu J, Boonkaew B, Arora J, Mandava SH, Maddox MM, Chava S, Callaghan C, He J, Dash S, John VT, Lee BR. Comparison of sorafenib-loaded poly (lactic/glycolic) acid and DPPC liposome nanoparticles in the in vitro treatment of renal cell carcinoma. J Pharm Sci. 2015 Mar; 104(3):1187–1196.

[33]   Qiao H, Sun M, Su Z, Xie Y, Chen M, Zong L, Gao Y, Li H, Qi J, Zhao Q, Gu X, Ping Q. Kidney-specific drug delivery system for renal fibrosis based on coordination-driven assembly of catechol-derived chitosan. Biomaterials. 2014 Aug; 35(25):7157–71.

[34]   Sankar P, Telang AG, Kalaivanan R, Karunakaran V, Suresh S, Kesavan M. Oral nanoparticulate curcumin combating arsenic-induced oxidative damage in kidney and brain of rats. Toxicol Ind Health. 2016 Mar;32(3):410–421.

[35]   Thurman JM, Serkova NJ. Nanosized contrast agents to noninvasively detect kidney inflammation by magnetic resonance imaging. Adv Chronic Kidney Dis. 2013 Nov;20(6):488–499.

[36]   Morishita Y, Imai T, Yoshizawa H, Watanabe M, Ishibashi K, Muto S, Nagata D. Delivery of microRNA-146a with polyethylenimine nanoparticles inhibits renal fibrosis in vivo. Int J Nanomedicine. 2015 May 11; 10:3475–3488.

[37]   Kamaly N, He JC, Ausiello DA, Farokhzad OC. Nanomedicines for renal disease: current status and future applications. Nat Rev Nephrol. 2016 Dec; 12(12):738–753.

[38]   Manne ND, Arvapalli R, Nepal N, Shokuhfar T, Rice KM, Asano S, Blough ER. Cerium oxide nanoparticles attenuate acute kidney injury induced by intra-abdominal infection in Sprague-Dawley rats. J Nanobiotechnol. 2015 Oct 24; 13:75.

[39]   Pollinger K, Hennig R, Breunig M, Tessmar J, Ohlmann A, Tamm ER, Witzgall R, Goepferich A. Kidney podocytes as specific targets for cyclo(RGDfC)-modified nanoparticles. Small. 2012 Nov 5; 8(21):3368–3375.

[40]   Choi CH, Zuckerman JE, Webster P, Davis ME. Targeting kidney mesangium by nanoparticles of defined size. Proc Natl Acad Sci U S A. 2011 Apr 19; 108(16):6656–6661.

[41]   Choyke PL, Kobayashi H. Abdom Imaging. 2006 Mar-Apr;31(2):224–31. Functional magnetic resonance imaging of the kidney using macromolecular contrast agents.

[42]   Serkova NJ, Renner B, Larsen BA, Stoldt CR, Hasebroock KM, Bradshaw-Pierce EL, Holers VM, Thurman JM. Renal inflammation: targeted iron oxide nanoparticles for molecular MR imaging in mice. Radiology. 2010 May;255(2):517–526.

[43]   Neuwelt EA, Hamilton BE, Varallyay CG, Rooney WR, Edelman RD, Jacobs PM, Watnick SG Ultrasmall superparamagnetic iron oxides (USPIOs): a future alternative magnetic resonance (MR) contrast agent for patients at risk for nephrogenic systemic fibrosis (NSF)?. Kidney Int. 2009 Mar;75(5):465–474.

[44]   Brede C, Labhasetwar V. Applications of nanoparticles in the detection and treatment of kidney diseases. Adv Chronic Kidney Dis. 2013 Nov; 20(6):454–465.

10 Nanomaterials and aquatic toxicity

Abstract: Toxic effects of nanoparticles (NPs) spread to the environment, affecting the aquatic plants and animals. Environmental safety is thus very important prior to the safe use of nanoparticles. Engineered NPs and their unknown effects on plants and animals in the aqueous environment are alarming and limit their safe use. Although numerous studies have shown their toxic effects, knowledge is far from being complete. This chapter highlights the effects of NPs on aquatic animals and how they affect the food chain, leading to bioaccumulation affecting the health, growth, reproduction and physiology of aquatic animals.

Keywords: food chain, *Chlorella, Daphnia*, aminoclay nanoparticles, medaka fish, *Mytilus edulis, Ceriodaphnia* sp., *ROS, Danio* sp., CNT, PAH, *Eurytemora affinis, Oryzias latipes, Raphidocelis subcapitata, Lymnaea stagnalis, Platynereis dumerilii*, silver nanoparticles, *Potamogeton diversifolius, Desmodesmus subspicatus*, nTiO₂, silver nanowires (AgNWs), *Chlamydomonas reinhardtii*, polyvinylpyrrolidone, *Lymnaea luteola, Tetrahymena pyriformis, Pimephales promelas, Strongylocentrotus droebachiensis, Fundulus heteroclitus, Euphorbia heterophylla, Oryzias melastigma, Artemia salina nauplii*, SOD, *O. mykiss, Caenorhabditis elegans, Cyprinus carpio, Oreochromis mossambicus, Cophixalus riparius, Chironomus dilutes, Lumbriculus variegatus*, graphene nanomaterials, MWCNT, SWCNT, OH-MWCNTs, *Ceriodaphnia dubia*, fullerene, graphene oxide, oxidative stress.

10.1 Introduction

Toxicity of nanoparticles (NPs) is known to affect plants, animals and aquatic lives tremendously. Both in vitro and in vivo studies reveal that engineered nanomaterials (ENMs) and their toxic effects [1] as well as contamination by carbon nanotubes (CNTs) have been proved to be harmful to aquatic animals, bacteria and higher plants [2]. Guidelines drafted by the Organisation for Economic Co-operation and Development (OECD) have shown hazards of manufactured nanomaterials (NMs) in aquatic environments [3]. NPs can affect the environment as well as the growth, development, physiology and reproduction of terrestrial and aquatic organisms, with the toxicity accumulating through trophic levels of organisms, thereby disrupting the entire food chain Figure 10.1. Few studies on benthic aquatic lifeforms including daphnids and crabs reported effects of nanotoxicology due to their benthic lifestyle and tendency to uptake contaminants, which show bioaccumulation that is indicative of toxicity in aquatic food chain [4] (Figure 10.2). In this chapter we discuss the different toxic NPs and their effects on the aquatic species.

NP size, shape, chemical architecture, agglomeration, crystal structure, surface area, chemical and physical surface properties all play a major role in NP toxicity

https://doi.org/10.1515/9783110579093-010

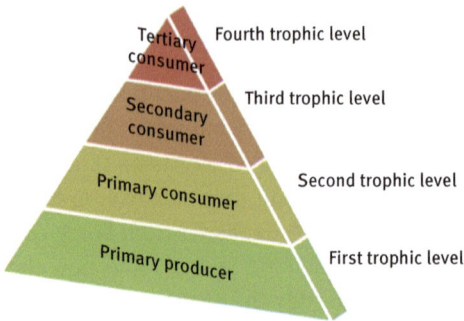

Figure 10.1: Food chain: trophic levels indicate the position of organisms in the food chain.

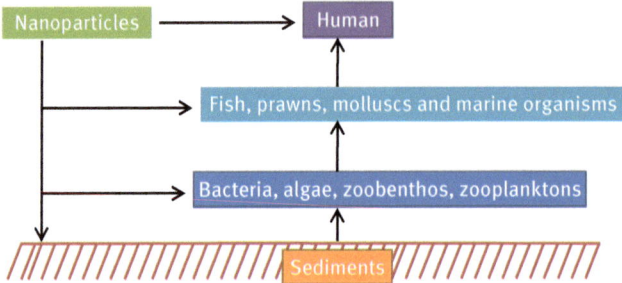

Figure 10.2: Nanoparticle, trophic level interactions.

(Figure 10.2) [5]. NPs have been reported to affect reproduction, fertility and embryogenesis of animals and cause genetic alteration of the progeny [1]. Studies based on proteomics and OMICS approach on nanotoxic effects reveal protein corona formation in different organisms, including bacteria, plants and mammals [6]. Reverse transcription-polymerase chain reaction studies have revealed that enabled nanotoxicity could lead to altered transcription [7]. However, detailed reports of toxicity of different types of NMs in the environment in the vast diversity of animals and ecosystem are largely incomplete. Overall, NP-mediated toxicity has been reported to activate the oxidative stress pathways by reactive oxygen species (ROS) generation and overall loss of cellular and system homeostasis [8]. There is a need of a standard protocol for evaluation of each type of nanotoxicity on each type of organisms to understand its complete nature of toxicity and its effects on the ecosystem.

Among the most studied vertebrate as model in aquatic toxicity is the zebrafish, *Danio rerio (D. rerio)* (Table 10.1), measuring 2.5–4 cm in length with a transparent larval stage and body with stripes in adults is a member of the family under order Cypriniformes and is a native of Southeast Asia.

The body of the adult male is slender, torpedo-shaped while females are pinkish and fatter due to carrying eggs. The whole genome consisting of 1,505,581,940 base

Table 10.1: Taxonomic position of Danio rerio.

Kingdom	Animalia
Phylum	Chordata
Class	Actinopterygii
Order	Cypriniformes
Family	Cyprinidae
Subfamily	Danioninae
Genus	*Danio*
Species	*D. rerio*

pairs (bp) with 26,247 genes coding protein was published in 2013. With the advantage of being small, robust, low-cost maintenance, mating triggered within a day with faster embryo and progeny formation, fast growth of embryos and transparent embryos enabling easy visualization of system, they are largely being used in studies of developmental biology. The development of fertilized embryos outside the mother's body enables the study of early development. About 70% genetic similarities with the human and in genes with human disease, they enable the understanding of the human system. Their unique repair system of the heart enables the study of restoration events post human heart failures. Mutation studies in zebrafish have enabled better understanding of life processes. Initiating in 1960, they have proved efficient animal models in different diseases and find importance as a vertebrate in biological research for understanding diseases such as muscular dystrophy and cancer.

In the recent years, they have found importance in detecting effects of drugs. Different NPs have revealed diverse effects on zebrafish and their embryos. Graphene oxide (GO) is reported to affect embryogenesis, and treated *D. rerio* embryos reveal blocked pore canals of chorionic membrane generating hypoxia, delayed hatching, mitochondrial damage, malformed eyes, edema of the cardiac/yolk sac, tail flexure and reduced heart rate, ROS generation, increased oxidative stress, DNA damage and apoptosis [9]. In zebrafish embryos (Figure 10.3), in vivo study, SiNPs caused toxicity of the membrane enclosing the heart or pericardia as well as bradycardia or slower heartbeat and posed a threat to the cardiovascular system [10]. Toxicity of CNTs, dendrimers, emulsions, liposomes, metal NPs and solid lipid NPs when evaluated on zebrafish revealed their accumulation after ingestion, inhalation or absorption [11].

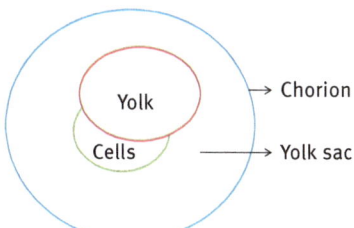

Figure 10.3: Schematic representation of *D. rerio* embryo.

10.2 NPs and their effects on aquatic animals

10.2.1 Lipid NPs

Calanoid copepods, zooplankton in marine and brackish aquatic environments are exposed to high levels of pollution. Lipid nanocapsules used to deliver PAHs (pyrene, phenanthrene and fluoranthene) and PCB 153 when inserted into the digestive track of males and females of *Eurytemora affinis* (*E. affinis*) led to nanocapsules adhesion on the exoskeleton, accumulation in digestive tracts, contact and systemic toxicity, increased swimming activity and its velocity and escaping stress response. Females were observed to be less sensitive than males [12].

10.2.2 Silver nanoparticles

Silver nanoparticles (AgNPs) are most toxic to aquatic life forms [13,14]. AgNPs including organic-coated silver are reported to cause genotoxicity and ecotoxicity across the organisms in the food chain including plants, bacteria and aquatic and terrestrial organisms due to damaging of cell membranes, disruption of ATP production, affected DNA replication, alternate gene expressions, release of toxic Ag^+ ion and ROS production to oxidize biological components of the cell [15]. AgNPs and silver nanowires (AgNWs) released into freshwater environments have led to toxicities detected by the OECD test guidelines, including an acute immobilization test for *Daphnia* sp., acute toxicity test for fish and growth inhibition test for freshwater alga and cyanobacteria. AgNPs have been designated as category acute 1 for *Daphnia magna* (*D. magna*), category acute 2 for *Oryzias latipes* (*O. latipes*) and category acute 1 for *Raphidocelis subcapitata* (*R. subcapitata*), while the AgNWs are classified as category acute 1 for *D. magna*, category acute 2 for *O. latipes* and category acute 2 for *R. subcapitata*, according to the Globally Harmonized System of Classification and Labeling of Chemicals [16]. Nanosilver (nAg) in the form of polyvinylpyrrolidone (PVP)-coated nAg and monodispersed nAg affect the algal species *R. subcapitata*, Crustacean, *Chydorus sphaericus* and a freshwater fish larva of *D. rerio* [17]. Humic substances (HS) have been reported to increase the aquatic toxicity of AgNPs [18]. *R. subcapitata*, algae, *Chydorus sphaericus* and *D. rerio*, when exposed to colloids of the PVP-coated AgNPs in the presence of HS, revealed increased aquatic toxicity of the AgNP colloids induced by HS to all the organisms in different trophic levels in a dose-dependent manner [18]. Suwannee River humic acid (SRHA) adsorbed to nano-Ag increased dispersion and agglomeration when exposed to *Daphnia* and revealed to decrease toxicity with increasing organic matter [19].

Aquatic organisms can be exposed to AgNPs through diet, which can adversely affect their growth. Water hardness and humic acids led to the bioaccumulation and toxicity of AgNPs coated with PVP when exposed to the freshwater snail *Lymnaea stagnalis (L. stagnalis)*. They have been observed to assimilate Ag from the PVP-AgNPs mixed with diatoms [20]. Aggregated silver released into the environment when deposited on suspended particles has been reported to affect the deposit feeder annelid *Platynereis dumerilii (P. dumerilii)*. Eggs, larvae, juveniles and adults of *P. dumerilii* when exposed to citrate (cit-AgNPs) or humic acid (HA-AgNPs) capped AgNPs and Ag showed that young ones were more affected by mortality and abnormal development rate, indicating their role in affecting the life cycle of marine coastal organisms [21]. AgNP toxicity affected two species of aquatic plants *Potamogeton diversifolius (P. diversifolius)* and *Egeria densa (E. densa)* [22]. AgNPs have been reported to adversely affect aquatic crustaceans, *D. magna* and *Thamnocephalus platyurus (T. platyurus)* [23]. Acute toxicity of capped AgNPs and capped and uncapped titanium dioxide ($nTiO_2$) to *D. magna* was revealed by affected growth and reduced reproduction [24]. Chronic toxicity of AgNPs varies from species to species in common cladocerans including *D. magna* [25], *D. galeata* and *Bosmina longirostris (B. longirostris)* [26]. Dispersed nAg has been observed for toxic effects on *Ceriodaphnia dubia (C. dubia)* and *Pseudokirchneriella subcapitata (P. subcapitata)* [27]. AgNPs coated with citrate could reduce toxic effects of AgNPs on *D. magna*, although both caused increased protein thiol content vitellogenin (vtg) indicating their role as a general stress sensor [28]. AgNPs tested on bacterium *Vibrio fischeri (V. fischeri)*, alga *Desmodesmus subspicatus (D. subspicatus)* and *D. magna* reveal the uptake and toxic effects of AgNPs [29].

P25-$nTiO_2$ affected the toxicity and uptake of Ag, As and Cu in developmental stages of *D. magna* and that $nTiO_2$ increased toxicity of Ag in younger forms but reduced the toxicities of As and Cu [30]. AgNPs were found to be more toxic on mussel hemocytes and gill cells, as compared to ionic and bulk Ag and overall toxicity of AgNPs are conferred through dissolved Ag [31]. Toxicity of polyvinylpyrrolidone-coated AgNWs to freshwater organisms including alga *Chlamydomonas reinhardtii (C. reinhardtii)*, the water flea *D. magna*, and the zebrafish revealed that toxicity is transferred through the food chain, individually inhibiting growth of algae and damaged the water fleas digestive system and may affect the health of humans through the food web [32]. Photo-induced toxicity of TiO_2 [33] and AgNPs [34] could cause chronic toxicity affecting survival, growth and reproduction of *D. magna* [33,34]. Freshwater snail *Lymnaea luteola* showed signs of acute toxicity and genotoxicity on exposure to AgNPs with reduced glutathione, glutathione S-transferase (GST) and glutathione peroxidase activity, increased malondialdehyde (MDA) and catalase (CAT) in its digestive gland together with DNA damage and oxidative stress [35]. AgNPs have less toxic effect on *D. magna* as compared to free Ag^+ ions [36]. Citrate-coated AgNPs when treated on *D. magna* Straus clone MBP996, revealed leass toxicity of AgNP as compared to free Ag^+ ions [36]. AgNPs led to bioaccumulation in *D. magna* but no mortality after 48 h of exposure up to 500 µg Ag/L as AgNP cause delay in reproduction, inhibition in growth and reproduction in *D. magna* fed with $AgNO_3$ [37]. *Tetrahymena pyriformis*, the ciliated protozoa exposed to TiO_2NP, indicated illumination affecting its toxic effects in AgNPs with no effects in dark, antagonism in natural light and synergism in continuous light [38]. Exposure to nAg and ionic silver (Ag^+) on *D. magna* and

Pimephales promelas revealed an association between specific particle surface area and acute toxic effects [39]. Ag^+, AgNPs and functionalized single-walled CNT (f-SWCNT) have been reported to affect embryos of sea urchin, *Strongylocentrotus droebachiensis*, by extrusion of mesenchymal cells on early embryos. Combined effect of f-SWCNTs–Ag^+ affect the gut development and hamper developmental processes at higher concentration with increased mortality [40]. Ag^+ and AgNPs could induce toxicity in *D. magna* affected by sunlight [41]. AgNP and Ag^+ affect photosynthesis in *C. reinhardtii* [42].

AgNPs when exposed to *Catla catla* (*C. catla*) and *Labeo rohita* (*L. rohita*) including *C. catla* heart cell line, Indian *C. catla* gill cell line and *L. rohita* gill cell line revealed DNA damage and nuclear fragmentation and oxidative stress [43]. AgNPs affect the development of zebrafish *D. rerio*, which led to death, delayed hatching of embryos, depleted glutathione levels and induction of oxidative stress [44]. AgNPs in the form of gum arabic-coated AgNPs, PVP-coated AgNPs or $AgNO_3$ in mesocosms when exposed to embryos and larvae of Atlantic killifish *Fundulus heteroclitus* (*F. heteroclitus*) and the nematode *Caenorhabditis elegans* (*C. elegans*) affect the juveniles [45].

However, green synthesis of AgNP from active latex of *Euphorbia heterophylla* (Poinsettia) when tested on *D. magna* showed that latex-synthesized AgNPs were comparatively very less toxic than chemically synthesized AgNPs, indicating that green synthesis is a better method in generating less toxic molecules as compared to chemical methods [46]. Trophic transfer of AgNPs from brine shrimp *Artemia salina nauplii* to marine medaka *Oryzias melastigma* (*O. melastigma*) revealed both aggregation, dispersion and accumulation by *Artemia* [47]. AgNPs could inhibit the whole-body Na^+/ K^+-ATPase and superoxide dismutase (SOD) activity in the fish [48]. Citrate-coated AgNP (cit-AgNP) when exposed to rainbow trout *Oncorhynchus mykiss* (*O. mykiss*) gill, RTgill-W1 revealed agglomeration of cit-AgNP, affecting the lysosomal membrane integrity [49]. Silver, gold and copper nanocolloids are found to be absorbed and accumulated by rainbow trout leukocyte and probably affect the immunocompetent cells, decrease the respiratory burst activity of splenocyte with adverse effects on their proliferation; however, low concentrations of silver nanocolloid could stimulate lymphocyte proliferation [50]. AgNPs revealed agglomeration and could induce acute toxicity with reduced brain enzymatic activity, altered antioxidant enzymes for CAT in liver and GST in fish gills and oxidative stress in juvenile common carp *Cyprinus carpio* (*C. carpio*), however, with no lethal effect even after 4 days of exposure [51].

Adult tilapia *Oreochromis mossambicus* (*O. mossambicus*) exposed to AgNPs lowered antioxidant enzymes in fish liver and gills, leading to increased free radical accumulation and imbalance of oxidative and antioxidant system as toxic effects observed [52]. Both coated and bare AgNPs caused acute and chronic toxicity when exposed to *Cophixalus riparius* (*C. riparius*) revealed by DNA damage and oxidative stress, with effects on development and reproduction [53]. Fathead minnow exposed to silver (Ag) revealed the affected circulatory system, increased total mucous goblet cells and disrupted ion regulation [54]. Radiolabeled AgNPs with silver isotopes in Japanese medaka revealed bioaccumulation [55].

10.2.3 Carbon NPs

CNTs could induce toxicity to cells, organs, tissues and whole organisms, including mammals and other species such as aquatic species, plants and bacteria [56]. Hydrophobic CNTs in sediments accumulate in gut and outer surface of sediment-dwelling invertebrates including *Hyalella azteca* (*H. azteca*), amphipod, *Chironomus dilutus* (*C. dilutus*), midge, *Lumbriculus variegatus*, oligochaete and *Villosa iris* (*V. iris*), a rainbow mussel indicative of both contact and systemic toxicities that affect their health [57]. Nanotoxicity may pass down through generations. Acute and chronic toxicity of carbon nanomaterials (CNMs) including fullerenes (C_{60}), SWCNTs and multiwalled CNTs (MWCNTs) with neutral, positive and negative functional groups have been reported to pass down from F1 and F2 generations in *D. magna* [58]. Graphene NMs including pristine graphene, reduced GO and GO in aquatic environments adversely affected the health of aquatic organisms including bacteria, algae, plants, invertebrates and fish [59]. CNT toxicity in fruit fly *Drosophila* embryo was revealed when injected in early *Drosophila* embryos, with uptake of CNTs on cytoplasm but not nucleus cell communication, tissue and organ formation in living embryos. Ectodermal and neural stem cells in MWCNT-injected embryos revealed normal division patterns and differentiation capacity with an increase in cell death of ectodermal but not of neural stem cells, indicating stem cell-specific vulnerability to MWCNT exposure [60] of NP-affected fish invertebrates by nano–bio interactions [61]. The CNM fullerene (C_{60}) forming aggregates caused stress in gills of *C. carpio* (Cyprinidae) by causing oxidative stress and $^1O_2$ formation [62]. SWCNT and fullerenol adversely affected freshwater zebrafish by oxidative stress responses on fish brain, decreased total antioxidant capacity and expression of the transcription factor Nrf2 [63]. Aqueous suspensions of CNMs on *P. subcapitata* and *C. dubia* revealed toxic effects [64]. Hydroxylated multiwalled carbon nanotubes (OH-MWCNTs) with hydrophilic groups caused acute toxicity to *D. magna*, accumulation in tissues and increased Ni toxicity [65]. OH-MWCNTs can cause synergistic effects of toxicity together with other pollutants in aquatic environments. OH-MWCNTs could enhance the toxicity of arsenic to *D. magna* [66].

NMs of carbon, silicon and metals such as gold, cadmium and selenium cause toxicity by releasing reactive radicals [67]. SWCNTs are found to be toxic to freshwater microalgae, including *R. subcapitata* and *Chlorella vulgaris*, *D. magna* and fish *O. latipes* detected by OECD test guidelines (201, 202 and 203) by inhibiting growth of the algae *R. subcapitata* and *C. vulgaris* [61]. CNTs affected the biokinetics and toxicity of cadmium (Cd) in *D. magna*, decreasing its tolerance to Cd [62]. CNMs leach to the aquatic environment through pesticides, delivery systems and nanosensors. Functionalized fullerene ([1,2-methanofullerene C_{60}]-61-carboxylic acid) (fC_{60}) and the hydrophobic pesticides bifenthrin and tribufos when tested on *D. magna*, reduced its survival and reproduction with fullerenes significantly increasing bifenthrin acute toxicity [68]. CNTs exposed to *D. magna* could lead to ecotoxicity and that SWCNTs were more toxic as compared to MWCNTs [69].

Functionalized SWCNTs led to increased mortality and developmental defects in minnow embryos and larvae, affected antioxidant enzyme activities including SOD, CAT and GST, upregulation of caspases, increased apoptosis, release of ROS and DNA damage [70]. CNT and fullerenes in sediments affect benthic life forms. MWCNT-contaminated sediments affected the benthic community level [71]. Exposures to raw MWNTs decreased the viability of *C. dubia*, increased mortality of amphipods *Leptocheirus plumulosus* and *Hyalella azteca* when exposed to sediment contaminated with carbon particles [72]. CNT can affect the transport of other pollutants including triclocarban (TCC). TCC has ben reported to affect *D. magna* by reducing the production of juveniles and adults and size-dependent mortality. MWCNT could lower the toxic effects of TCC [73]. GO could effect the zebrafish embryogenesis by causing delay in hatching, cardiac edema, damages in the chorion, oxidative stress and damage of mitochondria. HA, however, could reduce the adverse effects [74]. CNTs when exposed to rainbow trout *O. mykiss* could induce oxidative stress, disturbances in ion balance and homeostasis; cause toxicity in respiration, altered ventilation, mucus secretion; and affect gill pathology and systemic toxicity, thereby affecting the liver, spleen, blood circulatory system, kidney, gut and brain [75]. Fullerenes (C_{60}) and metallofullerene waste solids have been proved to be potentially toxic for aquatic environment determined through protocols of U.S. Environmental Protection Agency on two organisms *Pimephales promelas* and *C. dubia* [76].

10.2.4 Copper and Zinc NPs

CuNPs and ZnONPs caused toxic effects on *D. magna* [77,78] by affecting mortality, reproduction and reduced glycogen, lipid and protein concentration [79], which formed NP aggregates and exerted toxicity by generation of copper ions [80] in a host of organisms including crustaceans, algae, fish and bacteria [81]. Engineered NPs affect the food webs including different organisms like bacteria, plants and multicellular aquatic/terrestrial organisms [82]. Engineered NPs like n-TiO_2 suspensions exposed to *D. magna* led to adverse effects [83]. Metal oxide NPs including ZnO, TiO_2, Fe_2O_3 and CuO NMs (20–100 nm) treated on amphibians using the Frog Embryo Teratogenesis Assay Xenopus protocol caused developmental abnormalities including gastrointestinal, spinal and other abnormalities and affected negatively in amphibian development [84]. Acute toxicity of ZnO and TiO_2NPs has been known in earthworms *Eisenia fetida* [85]. Toxicity of ZnONP on Pacific oysters *Crassostrea gigas* is reported to be mediated by zinc assimilation with loss of electron-dense vesicles and loss of mitochondrial cristae, swollen mitochondria in the gill ultrastructure and prolonged exposure leading to disrupted mitochondrial membranes and increased cytosolic vesicles, damaged digestive glands affecting thiol homeostasis or immunological parameters such as phagocytosis, hemocyte viability and activation and total hemocyte count, decreased glutathione reductase (GR) activity in gills, low protein thiols, increased lipid peroxidation and GPx activity and oxidative stress [86]. Adverse effects of aquatic CuO nanosuspensions are reported from macrophytic algae cells of *Nitellopsis obtusa*, *Chlorella*, shrimp *Thamnocephalus platyurus* and rotifer *Brachionus calyciflorus* [87]. Uptake and toxicity of ZnONPs, ZnO bulk and $ZnCl_2$ salt in earthworms, *D. magna* and *C. vulgaris* was revealed by increased Zn accumulation, affected reproduction and inhibited cocoon production in earthworms mostly by $ZnCl_2$ [88]. The ingestion of engineered CuONPs by *Lymnaea stagnalis* led to bioaccumulation [89]. Dietary and waterborne exposure to CuO and ZnONPs in zooplankton *Artemia salina* and goldfish *Carassius auratus* revealed bioaccumulation in digestive tract including intestine, gills and liver and toxic effects transmitted through trophic levels, increased MDA in liver and gills, increased oxidative stress in liver due to chronic exposure to Cu ions [90]. Inhaled ZnOENMs lead to cardiorespiratory dysfunctioning in mammalian models and energy metabolism in the white sucker fish *Catostomus commersonii* together with oxidative and cellular stress in gill tissue, increased MDA levels, heat shock protein (HSP) expression, caspase 3/7 activity with increased gill Na^+/K^+-ATPase activity, increased tissue damage or cellular stress on gill neuroepithelial cells, systemic hypoxic response and increased parasympathetic nervous signaling will decrease heart rate [91]. Perfluorooctane sulfonate and ZnONPs caused synergistic acute and chronic toxic effects on *D. rerio* revealing developmental toxicity and DNA damage with toxicity correlating in a dose–response manner with histological tissue damage in adult zebrafish [92]. Environmental health and safety of NPs including water-stable CuO nanospheres with properties of slow-released ions and higher surface area were more toxic, which after combustion may lower toxicity potential to pelagic organisms due to deposition from water to sediment and reduced bioavailability after complexation with sedimentary organic matter [93]. CuONPs with CuO core–shells generate acute and chronic toxicity in *D. magna* and bioluminescent bacteria *V. fischeri* exhibiting both systemic and toxicity [94]. WAG cell line generated from gill tissue of *Wallago attu* exposed to TiO_2 and ZnONPs revealed cytotoxic, genotoxic and oxidative stress, increased DNA damage, lipid peroxidation and protein carbonylation, decrease in activity of SOD, CAT, total glutathione levels and total antioxidants [95]. Nanoscale zinc oxide (nano-ZnO) could cause developmental toxicity, oxidative stress and DNA damage in embryos and larvae of zebrafish, causing oxidative stress, generation of ROS and DNA damage [96]. Copper nanoparticles (CuNPs) exposed to juvenile fish *Epinephelus coioides* caused Cu accumulation and toxicity in liver and gills [97]. Copper oxide (CuO) is known to affect *D. magna* as both micro- and nano-sized particles affecting their growth and reproduction [98]. ZnONPs when exposed to earthworm *Eisenia veneta* caused Zn accumulation and affected the immune system [99]. CuONPs affected the juvenile carp *C. carpio* by systemic distribution, with increased biomass of liver, inhibition of cholinesterase activity with potential neurotoxic effects [100]. CuNPs (50 nm) on juvenile rainbow trout *O. mykiss*, fathead minnow *Pimephales promelas* and *D. rerio*, following standardized OECD 203 guideline tests have revealed that toxic effects are manifested by affected gill filaments and gill pavement cells [101]. CuNPs, AgNPs and TiO$_2$NPs affected the lateral line system of zebrafish embryos reducing functional lateral line neuromasts (LLN) but unaffected by TiO$_2$NPs or AgNPs. LLN affected the fish behavior or rheotaxis that is important for the survival of the fish [102]. *Mytilus galloprovincialis* when exposed to 10 μg Cu/L as CuONPs and Cu^{2+} for 15 days has been reported to cause oxidative stress, lipid peroxidation, Cu accumulation in the digestive glands, lowered SOD activity and CAT activation after 7 days [103]. Brine shrimp *Artemia salina* larvae when exposed to zinc and ZnONPs revealed ingested NPs, increased mortality with size of particles affecting nanotoxicity together with increased lipid peroxidation levels indicative of oxidative stress [104]. Synergistic toxic effects are reported from combined exposure of Cadmium telluride (CdTe) QDs and copper ion Cu^{2+} in zebrafish [105]. CuONP and dissolved copper, as copper chloride (CuCl$_2$), on the sea anemone *Exaiptasia pallida* revealed Cu accumulation and affected activities of CAT, glutathione peroxidase, GR and reduced carbonic anhydrase [105]. Chronic and long-term exposure to CuONPs on *Xenopus laevis* affected its growth, development, metamorphosis and even leading to death [106]. Solar UV radiation increased ZnONP toxicity through photocatalytic ROS generation and photo-induced dissolution when studied in *D. magna* under simulated solar UV radiation [107]. CuONP are toxic, and induce oxidative stress and other pathophysiological conditions on *Mozambique tilapia* (*O. mossambicus*) [108]. Two differently shaped Cu$_2$O micro/nanocrystals (cubes and octahedrons with side lengths of 900 nm), when exposed to *D. magna* for 72 h, revealed symptoms of oxidative stress including generation of ROS, CAT, total antioxidant capacity and MDA. The octahedron with larger surface area was found to be more toxic [109].

10.2.5 TiO$_2$NPs

TiO$_2$NPs can lead to toxic effects [110] in a wide range of organisms including bacteria, algae, invertebrates such as nematodes and vertebrates such as rainbow trout [111]. In the presence of HA and NaCl concentrations, they have been reported to alter SOD, CAT, MDA and glutathione in gills, liver and intestine of *D. rerio*. However, adding HA or increasing the ionic strength could suppress the oxidative stress [112]. They were proved to be toxic to two aquatic species *D. magna* and Japanese medaka [113]. The toxicity of dietary exposure to TiO$_2$ NM commercial product T-Lite™, a sunscreen component contaminating alga *P. subcapitata* cultures, could cause toxicity in *D. magna* due to chronic exposure of these contaminated algae in diet revealing decreased growth and reproduction [114]. Toxic effects of Ti and its accumulation causing embryo hypoxia in juvenile fish exposed to contaminated food [115] have been studied.

SRHA affected TiO$_2$NPs toxicity on developing *D. rerio* in dark and under simulated sunlight illumination, wherein HA adsorption decreased exposure to TiO$_2$NP and added to toxic effects. Absence of simulated sunlight increased the lethality in TiO$_2$NPs-HA-exposed *D. rerioI*, thus revealing that NPs together with environmental factors can affect the overall toxicity [116]. TiO_2NPs could affect the bioavailability, metabolism, oxidative damage and development, and could cause toxicity of organic pollutant pentachlorophenol in *D. rerio* embryos or larvae alone or in combination [117].

TiO_2NPs have been reported for their bioaccumulation and transfer to different trophic levels in the food chain. *C. dubia*, feeding on green algae *Scenedesmus obliquus* contaminated with TiO_2NPs, showed intake, contamination and toxic effects [118]. Anatase crystal of TiO_2NP could induce toxic effects on *D. magna* to produce ROS in the presence of UV light [119]. *C. elegans* when exposed to TiO_2NPs revealed the increase in SOD3, oxidative stress, generation of ROS and increase in toxicity of the NM [120]. UV light could increase toxicity of nano-TiO_2NPs and observed in *Hyalella azteca*, *Lumbriculus variegatus* and *Chironomus* dilutes when exposed in the presence of sunlight [121].

Zebrafish embryos in combination with Pb and nano-TiO_2 (0.1 mg/L) for 6 days post-fertilization revealed synergistic effects in imposing toxic effects with decreased thyroid hormone levels (T4 and T3), downregulation of thyroid-related factor TG. $TSH\beta$ upregulation and TTR downregulation in the hypothalamic-pituitary-thyroid, downregulation of genes involved in CNS development including α-tubulin, mbp, gfap and shha and affected locomotion were reported in Pb contamination [122]. Nano-TiO_2 enhanced the toxicity of copper to *D. magna* [123].

TiO_2NPs caused genotoxicity in zebrafish after 2 weeks of adaptation [124]. Zebrafish embryos microinjected in the otic vesicle with a sublethal dose of engineered TiO_2NPs and hydroxylated fullerenes/$C_{60}OH_{24}$ affected gene regulation of circadian rhythm, kinase activity, vesicular transport and immune response [125]. Nanotoxicity of TiO_2 has been reported on erythrocytes of *O. mykiss* trout [126]. Acute (96 h) and chronic (21 days) toxicities by TiO_2NPs were observed on *D. magna* [127]. Suspensions of nanosized TiO_2 led to toxicity in *D. magna* [119].

10.2.6 SeNPs

Elemental SeNPs on waterborne and dietary exposures to larvae of *Chironomus dilutus*, a common benthic invertebrate, led to Se accumulation in larvae, inhibiting larval growth in both waterborne and dietary exposures, indicative of risk of Se toxicity to egg-laying vertebrates (fish and piscivorous birds) in Se-contaminated aquatic systems [128].

10.2.7 Ceria

Engineered NPs of ceria on an aquatic system model consisting of sediments, water, hornworts, fish and snails using a radiotracer technique revealed that both snail and fish could fast absorb and clear with maximum accumulation recorded in Hornwort and that ceria NPs were obtained from sediments [129]. Bivalves have been studied for the fate and transport and effects of ENMs in aquatic environments due to their ability to filter and concentrate particles from water. CeO_2 ENMs when exposed either directly or through feed of phytoplankton to a marine mussel, *Mytilus galloprovincialis*, phytoplankton sorbed ENMs in less than 1 h, captured and excreted in pseudofeces and average pseudofeces mass doubled in response to CeO_2 exposure. [130]. Silver and CeO_2 particles have been observed to show toxicity in the pattern from high to low nano-Ag > micro-Ag > nano-CeO_2 = micro-CeO_2 in different organisms [131].

10.2.8 Polymers

Embryos of anurans and *Rhinella arenarum* larvae, when exposed to polyaniline NPs (PANI-Np), have been revealed to cause embryo toxicity and teratogenicity [132]. Food coated with the biopolymers zein and chitosan NPs containing rhodamine B in amphipod *Hyalella azteca*, however, has not been observed to affect the survival, growth or feeding behavior [133]. The triazine class of herbicides including compounds such as ametryn, atrazine and simazine used as weed control in crops like maize, sorghum and sugarcane act as environmental hazard affecting the animals. Poly(epsilon-caprolactone) NPs containing ametryn and atrazine affected the aquatic organisms and showed toxicity on the alga *P. subcapitata* and the microcrustacean *D. similis* [134].

10.2.9 Other NPs

Nano-sized zero valent iron (nZVI) had temporary adverse effects on collembola *Folsomia candida* and ostracods *Heterocypris incongruens* [135]. ENMs by sulfidation have been reported to decrease particle toxicity when tested on aquatic *D. rerio*, *F. heteroclitus* (killifish), *C. elegans* and plant *Lemna minuta* [136].

CdTe QDs showed developmental and behavioral toxicities to zebrafish [137]. ENMs on bivalve molluscs including *Mytilus* reveal accumulation in the digestive system. Different bivalves have been observed to affect the endosomal–lysosomal system and mitochondria, immunotoxicity, oxidative stress and cellular injury to proteins, membrane and DNA damage [138]. *Daphnia pulex* when exposed for 48 h to sublethal doses of QDs (25% and 50% of LC_{50}) with differing spectral properties (CdTe and CdSe/ZnS QDs) and Cd and Zn salts, $CdSO_4$ leads to Cd uptake and accumulation with increased expression of metallothionein-1 gene for stress defense and DNA repair [139]. Acute toxicity of lanthanum oxide NPs (La_2O_3 NP) on microalgae *Chlorella* sp. and *D. magna* revealed overall toxicity and adsorption to surface at higher concentrations [140]. Contaminated sediment containing 14 nm AuNPs exposed to *D. rerio* leads to oxidative stress, mitochondrial metabolism, DNA damage and altered neurotransmission with increased brain acetylcholinesterase activity [141].

SiO$_2$NPs have been proved to be toxic to human and mammalian cells that affect the aquatic biota, fish. Twelve adherent fish cell lines derived from rainbow trout, fathead minnow, zebrafish, goldfish, haddock and American eel revealed that SiO$_2$NPs show size-dependent, time-dependent, temperature-dependent, dose-dependent and tissue-specific toxicities [142].

Aminoclay nanoparticles (ANPs) of ~51 ± 31 nm on eukaryotic microalga *P. subcapitata*, *D. magna* and *V. fischeri* revealed entrapped algal cells in ANP aggregates that affected the growth of *P. subcapitata* [143]. Gold nanoparticles have been tested on blue mussel, *Mytilus edulis*, and caused oxidative stress [144]. AuNPs affect *D. magna* causing ROS production and oxidative stress and general cellular stress: GST, CAT, heat shock protein 70 and metallothionein 1 [145].

Medaka fish (*O. latipes*) larvae when treated with carboxymethyl cellulose-stabilized nZVI, aged nanoscale iron oxides (nFe-oxides) or ferrous ion (Fe[II]) for 12–14 days revealed toxic effects with increased oxidative stress, increased ROS generation, decreased dissolved oxygen, hypoxia in tissues and oxidative damage [146]. Iron oxide NPs exposed to *D. rerio* affected the embryonic development causing mortality, delayed hatching and developmental defects [147]. Nano-sized, fluorescent, latex particles when exposed to *O. latipes* revealed uptake, excretion, bioaccumulation and affected survival rate [148]. *C. dubia* when exposed to aluminum oxide NPs induced acute toxicity and generation of oxidative stress [149]. Plastic NPs including 24 and 27 nm polystyrene NPs administered to fish through an aquatic food chain, from algae through *Daphnia*, affected the behavior and metabolism [150].

10.3 Discussion

Currently although many NPs are being studied for their toxic effects on the aquatic species including invertebrates and fish, there are many gaps to our understanding of the harmful effects of NPs. Titanium dioxide in aquatic system has revealed sublethal effects on the immune system. However, the immunotoxicity generated by NPs in aquatic species is not well studied [148].

References

[1]   He X, Aker WG, Leszczynski J, Hwang HM. Using a holistic approach to assess the impact of engineered nanomaterials inducing toxicity in aquatic systems. J Food Drug Anal. 2014 Mar;22(1):128–146.

[2]   Du J, Wang S, You H, Zhao X. Understanding the toxicity of carbon nanotubes in the environment is crucial to the control of nanomaterials in producing and processing and the assessment of health risk for human: A review. Environ Toxicol Pharmacol. 2013 Sep;36(2):451–462.

[3]   Petersen EJ, Diamond SA, Kennedy AJ, Goss GG, Ho K, Lead J, Hanna SK, Hartmann NB, Hund-Rinke K, Mader B, Manier N, Pandard P, Salinas ER, Sayre P. Adapting OECD aquatic toxicity tests for use with manufactured nanomaterials: Key issues and consensus recommendations. Environ Sci Technol. 2015 Aug 18;49(16):9532–9547.

[4]   Walters CR, Pool EJ, Somerset VS. Ecotoxicity of silver nanomaterials in the aquatic environment: a review of literature and gaps in nano-toxicological research. J Environ Sci Health A Tox Hazard Subst Environ Eng. 2014;49(13):1588–1601.

[5]   Gatoo MA, Naseem S, Arfat MY, Dar AM, Qasim K, and Swaleha Zubair S. Physicochemical properties of nanomaterials: Implication in associated toxic manifestations. Biomed Res Int. 2014; 498420:8.

[6]   Matysiak M, Kapka-Skrzypczak L, Brzóska K, Gutleb AC, Kruszewski M. Proteomic approach to nanotoxicity. J Proteomics. 2016 Mar 30;137:35–44.

[7]   Mo Y, Wan R, Zhang Q. Application of reverse transcription-PCR and real-time PCR in nanotoxicity research. Methods Mol Biol. 2012;926:99–112.

[8]   Peter P. Fu, Qingsu Xia, Huey-MinHwang, Paresh C. Ray, Hongtao Yu. Mechanisms of nanotoxicity: Generation of reactive oxygen species. J Food Drug Anal. 2014; Vol 22(1): 64–75.

[9]   Chen Y, Hu X, Sun J, Zhou Q. Specific nanotoxicity of graphene oxide during zebrafish embryogenesis. Nanotoxicol. 2016;10(1):42–52.

[10]   Duan J, Yu Y, Li Y, Yu Y, Sun Z. Cardiovascular toxicity evaluation of silica nanoparticles in endothelial cells and zebrafish model. Biomater. 2013 Jul;34(23):5853–5862.

[11]   Jackson P, Jacobsen NR, Baun A, Birkedal R, Kühnel D, Jensen KA, Vogel U, Håkan Wallin. Bioaccumulation and ecotoxicity of carbon nanotubes. Chem Cent J. 2013; 7: 154.

[12]   Michalec FG, Holzner M, Souissi A, Stancheva S, Barras A, Boukherroub R, Souissi S. Lipid nanocapsules for behavioural testing in aquatic toxicology: Time-response of Eurytemora affinis to environmental concentrations of PAHs and PCB. Aquat Toxicol. 2016 Jan;170:310–322.

[13]   Völker C, Oetken M, Oehlmann J. The biological effects and possible modes of action of nanosilver. J. Rev Environ Contam Toxicol. 2013;223:81–106.

[14]   Fabrega J, Luoma SN, Tyler CR, Galloway TS, Lead JR. Silver nanoparticles: Behaviour and effects in the aquatic environment. Environ Int. 2011 Feb;37(2):517–531.

[15]   Sharma VK, Siskova KM, Zboril R, Gardea-Torresdey JL. Organic-coated silver nanoparticles in biological and environmental conditions: Fate, stability and toxicity. Adv Colloid Interface Sci. 2014 Feb;204:15–34.

[16]   Sohn EK, Johari SA, Kim TG, Kim JK, Kim E, Lee JH, Chung YS, Yu IJ. Aquatic toxicity comparison of silver nanoparticles and silver nanowires. Biomed Res Int. 2015;2015:893049.

[17]   Wang Z, Chen J, Li X, Shao J, Peijnenburg WJ. Aquatic toxicity of nanosilver colloids to different trophic organisms: Contributions of particles and free silver ion. Environ Toxicol Chem. 2012 Oct;31(10):2408–2413.

[18]   Wang Z, Quik JT, Song L, Van Den Brandhof EJ, Wouterse M, Peijnenburg WJ. Humic substances alleviate the aquatic toxicity of polyvinylpyrrolidone-coated silver nanoparticles to organisms of different trophic levels. Environ Toxicol Chem. 2015 Jun;34(6):1239–1245.

[19]   Gao J, Powers K, Wang Y, Zhou H, Roberts SM, Moudgil BM, Koopman B, Barber DS. Influence of Suwannee River humic acid on particle properties and toxicity of silver nanoparticles. Chemosphere. 2012 Sep;89(1):96–101.

[20]   Oliver AL, Croteau MN, Stoiber TL, Tejamaya M, Römer I, Lead JR, Luoma SN. Does water chemistry affect the dietary uptake and toxicity of silver nanoparticles by the freshwater snail Lymnaea stagnalis?. Environ Pollut. 2014 Jun;189:87–91.

[21]   García-Alonso J, Rodriguez-Sanchez N, Misra SK, Valsami-Jones E, Croteau MN, Luoma SN, Rainbow PS. Toxicity and accumulation of silver nanoparticles during development of the marine polychaete Platynereis dumerilii. Sci Total Environ. 2014 Apr 1;476–477:688–695.

[22]   Bone AJ, Colman BP, Gondikas AP, Newton KM, Harrold KH, Cory RM, Unrine JM, Klaine SJ, Matson CW, Di Giulio RT. Biotic and abiotic interactions in aquatic microcosms determine fate and toxicity of Ag nanoparticles: Part 2-toxicity and Ag speciation. Environ Sci Technol. 2012 Jul 3;46(13):6925–6933.

[23]   Blinova I, Niskanen J, Kajankari P, Kanarbik L, Käkinen A, Tenhu H, Penttinen OP, Kahru A. Toxicity of two types of silver nanoparticles to aquatic crustaceans Daphnia magna and Thamnocephalus platyurus. Environ Sci Pollut Res Int. 2013 May;20(5):3456–3463.

[24]   Das P, Xenopoulos MA, Metcalfe CD. Toxicity of silver and titanium dioxide nanoparticle suspensions to the aquatic invertebrate, Daphnia magna. Bull Environ Contam Toxicol. 2013 Jul;91(1):76–82.

[25]   Seitz F, Rosenfeldt RR, Storm K, Metreveli G, Schaumann GE, Schulz R, Bundschuh M. Effects of silver nanoparticle properties, media pH and dissolved organic matter on toxicity to Daphnia magna. Ecotoxicol Environ Saf. 2015 Jan;111:263–270.

[26]   Sakamoto M, Ha JY, Yoneshima S, Kataoka C, Tatsuta H, Kashiwada S. Free silver ion as the main cause of acute and chronic toxicity of silver nanoparticles to cladocerans. Arch Environ Contam Toxicol. 2015 Apr;68(3):500–509.

[27]   McLaughlin J, Bonzongo JC. Effects of natural water chemistry on nanosilver behavior and toxicity to Ceriodaphnia dubia and Pseudokirchneriella subcapitata. Environ Toxicol Chem. 2012 Jan;31(1):168–175.

[28]   Rainville LC, Carolan D, Varela AC, Doyle H, Sheehan D. Proteomic evaluation of citrate-coated silver nanoparticles toxicity in Daphnia magna. Analyst. 2014 Apr 7;139(7):1678–1686.

[29]   Georgantzopoulou A, Balachandran YL, Rosenkranz P, Dusinska M, Lankoff A, Wojewodzka M, Kruszewski M, Guignard C, Audinot JN, Girija S, Hoffmann L, Gutleb AC. Ag nanoparticles: size- and surface-dependent effects on model aquatic organisms and uptake evaluation with NanoSIMS. Nanotoxicol. 2013 Nov;7(7):1168–1178.

[30]   Rosenfeldt RR, Seitz F, Schulz R, Bundschuh M. Heavy metal uptake and toxicity in the presence of titanium dioxide nanoparticles: A factorial approach using Daphnia magna. Environ Sci Technol. 2014 Jun 17;48(12):6965–6972.

[31]   Katsumiti A, Gilliland D, Arostegui I, Cajaraville MP. Mechanisms of toxicity of Ag nanoparticles in comparison to bulk and ionic Ag on mussel hemocytes and gill cells. PLoS One. 2015 Jun 10;10(6):e0129039.

[32]   Chae Y, An YJ. Toxicity and transfer of polyvinylpyrrolidone-coated silver nanowires in an aquatic food chain consisting of algae, water fleas, and zebrafish. Aquat Toxicol. 2016 Apr;173:94–104.

[33]   Mansfield CM, Alloy MM, Hamilton J, Verbeck GF, Newton K, Klaine SJ, Roberts AP. Photo-induced toxicity of titanium dioxide nanoparticles to Daphnia magna under natural sunlight. Chemosphere. 2015 Feb;120:206–210.

[34]   Sakka Y, Skjolding LM, Mackevica A, Filser J, Baun A. Behavior and chronic toxicity of two differently stabilized silver nanoparticles to Daphnia magna. Aquat Toxicol. 2016 Aug;177:526–535.

[35]   Ali D, Yadav PG, Kumar S, Ali H, Alarifi S, Harrath AH. Sensitivity of freshwater pulmonate snail Lymnaea luteola L., to silver nanoparticles. Chemosphere. 2014 Jun;104:134–140.

[36]   Ulm L, Krivohlavek A, Jurašin D, Ljubojević M, Šinko G, Crnković T, Žuntar I, Šikić S, Vinković Vrček I. Response of biochemical biomarkers in the aquatic crustacean Daphnia magna exposed to silver nanoparticles. Environ Sci Pollut Res Int. 2015 Dec;22(24):19990–19999.

[37]   Zhao CM, Wang WX. Comparison of acute and chronic toxicity of silver nanoparticles and silver nitrate to Daphnia magna. Environ Toxicol Chem. 2011 Apr;30(4):885–892.

[38]   Zou X, Shi J, Zhang H. Coexistence of silver and titanium dioxide nanoparticles: enhancing or reducing environmental risks? Aquat Toxicol. 2014 Sep;154:168–175.

[39] Hoheisel SM, Diamond S, Mount D. Comparison of nanosilver and ionic silver toxicity in Daphnia magna and Pimephales promelas. Environ Toxicol Chem. 2012 Nov;31(11):2557–2563.

[40] Magesky A, Pelletier É. Toxicity mechanisms of ionic silver and polymer-coated silver nanoparticles with interactions of functionalized carbon nanotubes on early development stages of sea urchin. Aquat Toxicol. 2015 Oct;167:106–123.

[41] Zhang Z, Yang X, Shen M, Yin Y, Liu J. Sunlight-driven reduction of silver ion to silver nanoparticle by organic matter mitigates the acute toxicity of silver to Daphnia magna. J Environ Sci (China). 2015 Sep 1;35:62–68.

[42] Navarro E, Piccapietra F, Wagner B, Marconi F, Kaegi R, Odzak N, Sigg L, Behra R. Toxicity of silver nanoparticles to Chlamydomonas reinhardtii. Environ Sci Technol. 2008 Dec 1;42(23):8959–8964.

[43] Taju G, Abdul Majeed S, Nambi KS, Sahul Hameed AS. In vitro assay for the toxicity of silver nanoparticles using heart and gill cell lines of Catla catla and gill cell line of Labeo rohita. Comp Biochem Physiol C Toxicol Pharmacol. 2014 Apr;161:41–52.

[44] Massarsky A, Dupuis L, Taylor J, Eisa-Beygi S, Strek L, Trudeau VL, Moon TW. Assessment of nanosilver toxicity during zebrafish (Danio rerio) development. Chemosphere. 2013 Jun;92(1):59–66.

[45] Bone AJ, Matson CW, Colman BP, Yang X, Meyer JN, Di Giulio RT. Silver nanoparticle toxicity to Atlantic killifish (Fundulus heteroclitus) and Caenorhabditis elegans: A comparison of mesocosm, microcosm, and conventional laboratory studies. Environ Toxicol Chem. 2015 Feb;34(2):275–282.

[46] Borase HP, Patil CD, Salunkhe RB, Suryawanshi RK, Salunke BK, Patil SV. Mercury sensing and toxicity studies of novel latex fabricated silver nanoparticles. Bioprocess Biosyst Eng. 2014 Nov;37(11):2223–2233.

[47] Wang J, Wang WX. Low bioavailability of silver nanoparticles presents trophic toxicity to marine medaka (Oryzias melastigma). Environ Sci Technol. 2014 Jul 15;48(14):8152–8161.

[48] Yue Y, Behra R, Sigg L, Fernández Freire P, Pillai S, Schirmer K. Toxicity of silver nanoparticles to a fish gill cell line: Role of medium composition. Nanotoxicol. 2015 Feb;9(1):54–63.

[49] Małaczewska J, Siwicki AK. The in vitro effect of commercially available noble metal nanocolloids on the rainbow trout (Oncorhynchus mykiss) leukocyte and splenocyte activity. Pol J Vet Sci. 2013;16(1):77–84.

[50] Nations S, Wages M, Cañas JE, Maul J, Theodorakis C, Cobb GP.Acute effects of Fe_2O_3, TiO_2, ZnO and CuO nanomaterials on Xenopus laevis. Chemosphere. 2011 May;83(8):1053–1061.

[51] Ale Analía, Bacchetta, Carla, Rossi Andrea, Galdoporpora Juan, Desimone Martin, de la Torre Fernando, Gervasio Susana, Cazenave Jimena. Nanosilver toxicity in gills of a neotropical fish: Metal accumulation, oxidative stress, histopathology and other physiological effects. Ecotoxicol Environ Safety. 2018;148:976–984. 10.1016/j.ecoenv.2017.11.072.

[52] Govindasamy R, Rahuman AA. Histopathological studies and oxidative stress of synthesized silver nanoparticles in Mozambique tilapia (Oreochromis mossambicus). J Environ Sci (China). 2012;24(6):1091–1098.

[53] Park SY, Chung J, Colman BP, Matson CW, Kim Y, Lee BC, Kim PJ, Choi K, Choi J. Ecotoxicity of bare and coated silver nanoparticles in the aquatic midge, Chironomus riparius. Environ Toxicol Chem. 2015 Sep;34(9):2023–2032.

[54] Natàlia Garcia-Reyero, Cammi Thornton, Adam D. Hawkins, Lynn Escalon, Alan J Kennedy, Jeffery A. Steevens, Kristine L. Willett, Assessing the exposure to nanosilver and silver nitrate on fathead minnow gill gene expression and mucus production. Environ. Nanotechnol. Monit. Manage. 2015;Volume 4, Pages58–66,

[55]   Jung YJ, Kim KT, Kim JY, Yang SY, Lee BG, Kim SD. Bioconcentration and distribution of silver nanoparticles in Japanese medaka (Oryzias latipes). J Hazard Mater. 2014 Feb 28;267:206–213.

[56]   Zhao X, Liu R. Recent progress and perspectives on the toxicity of carbon nanotubes at organism, organ, cell, and biomacromolecule levels. Environ Int. 2012 Apr;40:244–255.

[57]   Mwangi JN, Wang N, Ingersoll CG, Hardesty DK, Brunson EL, Li H, Deng B. Toxicity of carbon nanotubes to freshwater aquatic invertebrates. Environ Toxicol Chem. 2012 Aug;31(8):1823–1830.

[58]   Arndt DA, Chen J, Moua M, Klaper RD. Multigeneration impacts on Daphnia magna of carbon nanomaterials with differing core structures and functionalizations. Environ Toxicol Chem. 2014 Mar;33(3):541–547.

[59]   Zhao J, Wang Z, White JC, Xing B. Graphene in the aquatic environment: Adsorption, dispersion, toxicity and transformation. Environ Sci Technol. 2014 Sep 2;48(17):9995–10009.

[60]   Liu B, Campo EM, Bossing T. Drosophila embryos as model to assess cellular and developmental toxicity of multi-walled carbon nanotubes (MWCNT) in living organisms. PLoS One. 2014 Feb 18;9(2):e88681.

[61]   Ma S, Lin D. The biophysicochemical interactions at the interfaces between nanoparticles and aquatic organisms: adsorption and internalization. Environ Sci Process Impacts. 2013 Jan;15(1):145–160.

[62]   Socoowski Britto R, Garcia ML, Martins da Rocha A, Flores JA, Pinheiro MV, Monserrat JM, Ferreira JL. Effects of carbon nanomaterials fullerene C_{60} and fullerol $C_{60}(OH)_{18-22}$ on gills of fish Cyprinus carpio (Cyprinidae) exposed to ultraviolet radiation. Aquat Toxicol. 2012 Jun 15;114–115:80–87.

[63]   da Rocha AM, Ferreira JR, Barros DM, Pereira TC, Bogo MR, Oliveira S, Geraldo V, Lacerda RG, Ferlauto AS, Ladeira LO, Pinheiro MV, Monserrat JM. Gene expression and biochemical responses in brain of zebrafish Danio rerio exposed to organic nanomaterials: carbon nanotubes (SWCNT) and fullerenol ($C_{60}(OH)18-22(OK4)$). Comp Biochem Physiol A Mol Integr Physiol. 2013 Aug;165(4):460–467.

[64]   Jennifer L Bouldin, Taylor M Ingle, Anindita Sengupta, Regina Alexander, Robyn E. Hannigan, Roger A. Buchanan. Aqueous toxicity and food chain transfer of quantum dots™ in freshwater algae and Ceriodaphnia Dubia. Environ Toxicol Chem. 2008 Sep; 27(9): 1958–1963.

[65]   Wang C, Wei Z, Feng M, Wang L, Wang Z. The effects of hydroxylated multiwalled carbon nanotubes on the toxicity of nickel to Daphnia magna under different pH levels. Environ Toxicol Chem. 2014 Nov;33(11):2522–2528.

[66]   Wang X, Qu R, Allam AA, Ajarem J, Wei Z, Wang Z. Impact of carbon nanotubes on the toxicity of inorganic arsenic [AS(III) and AS(V)] to Daphnia magna: The role of certain arsenic species. Environ Toxicol Chem. 2016 Jul;35(7):1852–1859.

[67]   Farré Marinella, Schrantz Krisztina, Kantiani Lina, Barcelo Damia. Ecotoxicity and analysis of nanomaterials in the aquatic environment. Anal Bioanal Chem. 2008; 393: 81–95.

[68]   Sohn EK, Chung YS, Johari SA, Kim TG, Kim JK, Lee JH, Lee YH, Kang SW, Yu IJ. Acute toxicity comparison of single-walled carbon nanotubes in various freshwater organisms. Biomed Res Int. 2015;2015:323090.

[69]   Jackson P, Jacobsen NR, Baun A, Birkedal R, Kühnel D, Jensen KA, Ulla Vogel U, Håkan Wallin H Bioaccumulation and ecotoxicity of carbon nanotubes. Chem Cent J. 2013; 7: 154.

[70]   Zhu B, Liu GL, Ling F, Song LS, Wang GX. Development toxicity of functionalized single-walled carbon nanotubes on rare minnow embryos and larvae. Nanotoxicol. 2015;9(5):579–590.

[71]   Velzeboer I, Kupryianchyk D, Peeters ET, Koelmans AA. Community effects of carbon nanotubes in aquatic sediments. Environ Int. 2011 Aug;37(6):1126–1130.

[72]   Kennedy AJ, Hull MS, Steevens JA, Dontsova KM, Chappell MA, Gunter JC, Weiss CA Jr. Factors influencing the partitioning and toxicity of nanotubes in the aquatic environment. Environ Toxicol Chem. 2008 Sep;27(9):1932–1941.

[73]   Simon A, Preuss TG, Schäffer A, Hollert H, Maes HM. Population level effects of multiwalled carbon nanotubes in Daphnia magna exposed to pulses of triclocarban. Ecotoxicol. 2015 Aug;24(6):1199–1212.

[74]   Chen Y, Ren C, Ouyang S, Hu X, Zhou Q. Mitigation in multiple effects of graphene oxide toxicity in Zebrafish embryogenesis driven by humic acid. Environ Sci Technol. 2015 Aug 18;49(16):10147–10154.

[75]   Smith CJ, Shaw BJ, Handy RD. Toxicity of single walled carbon nanotubes to rainbow trout, (Oncorhynchus mykiss): Respiratory toxicity, organ pathologies, and other physiological effects. Aquat Toxicol. 2007 May 1;82(2):94–109.

[76]   Hull MS, Kennedy AJ, Steevens JA, Bednar AJ, Weiss CA Jr, Vikesland PJ. Release of metal impurities from carbon nanomaterials influences aquatic toxicity. Environ Sci Technol. 2009 Jun 1;43(11):4169–4174.

[77]   Xiao Y, Vijver MG, Chen G, Peijnenburg WJ. Toxicity and accumulation of Cu and ZnO nanoparticles in Daphnia magna. Environ Sci Technol. 2015 Apr 7;49(7):4657–4664.

[78]   Lopes S, Ribeiro F, Wojnarowicz J, Łojkowski W, Jurkschat K, Crossley A, Soares AM, Loureiro S. Zinc oxide nanoparticles toxicity to Daphnia magna: Size-dependent effects and dissolution. Environ Toxicol Chem. 2014 Jan;33(1):190–198.

[79]   Adam N, Vergauwen L, Blust R, Knapen D. Gene transcription patterns and energy reserves in Daphnia magna show no nanoparticle specific toxicity when exposed to ZnO and CuO nanoparticles. Environ Res. 2015 Apr;138:82–92.

[80]   Adam N, Vakurov A, Knapen D, Blust R. The chronic toxicity of CuO nanoparticles and copper salt to Daphnia magna. J Hazard Mater. 2015;283:416–422.

[81]   Bondarenko O, Juganson K, Ivask A, Kasemets K, Mortimer M, Kahru A. Toxicity of Ag, CuO and ZnO nanoparticles to selected environmentally relevant test organisms and mammalian cells in vitro: A critical review. Arch Toxicol. 2013 Jul;87(7):1181–1200.

[82]   Maurer-Jones MA, Gunsolus IL, Murphy CJ, Haynes CL. Toxicity of engineered nanoparticles in the environment. Anal Chem. 2013 Mar 19;85(6):3036–3049.

[83]   Salieri B, Pasteris A, Baumann J, Righi S, Köser J, D'Amato R, Mazzesi B, Filser J. Does the exposure mode to ENPs influence their toxicity to aquatic species? A case study with TiO2 nanoparticles and Daphnia magna. Environ Sci Pollut Res Int. 2015 Apr;22(7):5050–5058.

[84]   Nations S, Wages M, Cañas JE, Maul J, Theodorakis C, Cobb GP. Acute effects of Fe₂O₃, TiO₂, ZnO and CuO nanomaterials on Xenopus laevis. Chemosphere. 2011 May;83(8):1053–1061.

[85]   Cañas JE, Qi B, Li S, Maul JD, Cox SB, Das S, Green MJ. Acute and reproductive toxicity of nano-sized metal oxides (ZnO and TiO₂) to earthworms (Eisenia fetida). J Environ Monit. 2011 Dec;13(12):3351–3357.

[86]   Trevisan R, Delapedra G, Mello DF, Arl M, Schmidt ÉC, Meder F, Monopoli M, Cargnin-Ferreira E, Bouzon ZL, Fisher AS, Sheehan D, Dafre AL. Gills are an initial target of zinc oxide nanoparticles in oysters Crassostrea gigas, leading to mitochondrial disruption and oxidative stress. Aquat Toxicol. 2014 Aug;153:27–38.

[87]   Manusadžianas L, Caillet C, Fachetti L, Gylytė B, Grigutytė R, Jurkonienė S, Karitonas R, Sadauskas K, Thomas F, Vitkus R, Férard JF. Toxicity of copper oxide nanoparticle suspensions to aquatic biota. Environ Toxicol Chem. 2012 Jan;31(1):108–114.

[88]   Li LZ¹, Zhou DM, Peijnenburg WJ, van Gestel CA, Jin SY, Wang YJ, Wang P. Toxicity of ZnO nanoparticles, ZnO bulk, and ZnCl₂ on earthworms in a spiked natural soil and toxicological effects of leachates on aquatic organisms. Arch Environ Contam Toxicol. 2014 Nov;67(4):465–473.

[89]   Croteau MN, Misra SK, Luoma SN, Valsami-Jones E. Bioaccumulation and toxicity of CuO nanoparticles by a freshwater invertebrate after waterborne and dietborne exposures. Environ Sci Technol. 2014 Sep 16;48(18):10929–109237.

[90]   Ates M, Arslan Z, Demir V, Daniels J, Farah IO. Accumulation and toxicity of CuO and ZnO nanoparticles through waterborne and dietary exposure of goldfish (Carassius auratus). Environ Toxicol. 2015 Jan;30(1):119–128.

[91]   Bessemer RA, Butler KM, Tunnah L, Callaghan NI, Rundle A, Currie S, Dieni CA, MacCormack TJ. Cardiorespiratory toxicity of environmentally relevant zinc oxide nanoparticles in the freshwater fish Catostomus commersonii. Nanotoxicol. 2015;9(7):861–870.

[92]   Du J, Wang S, You H, Jiang R, Zhuang C, Zhang X. Developmental toxicity and DNA damage to zebrafish induced by perfluorooctane sulfonate in the presence of ZnO nanoparticles. Environ Toxicol. 2016 Mar;31(3):360–371.

[93]   Kennedy AJ, Melby NL, Moser RD, Bednar AJ, Son SF, Lounds CD, Laird JG, Nellums RR, Johnson DR, Steevens JA. Fate and toxicity of CuO nanospheres and nanorods used in Al/CuO nanothermites before and after combustion. Environ Sci Technol. 2013 Oct 1;47(19):11258–11267.

[94]   Rossetto AL, Vicentini DS, Costa CH, Melegari SP, Matias WG. Synthesis, characterization and toxicological evaluation of a core-shell copper oxide/polyaniline nanocomposite. Chemosphere. 2014 Aug;108:107–114.

[95]   Dubey A, Goswami M, Yadav K, Chaudhary D. Oxidative stress and nano-toxicity induced by TiO2 and ZnO on WAG cell line. PLoS One. 2015 May 26;10(5):e0127493.

[96]   Zhao X, Wang S, Wu Y, You H, Lv L. Acute ZnO nanoparticles exposure induces developmental toxicity, oxidative stress and DNA damage in embryo-larval zebrafish. Aquat Toxicol. 2013 Jul 15;136–137:49–59.

[97]   Wang T, Long X, Cheng Y, Liu Z, Yan S. The potential toxicity of copper nanoparticles and copper sulphate on juvenile Epinephelus coioides. Aquat Toxicol. 2014 Jul;152:96–104.

[98]   Rossetto AL, Melegari SP, Ouriques LC, Matias WG. Comparative evaluation of acute and chronic toxicities of CuO nanoparticles and bulk using Daphnia magna and Vibrio fischeri. Sci Total Environ. 2014 Aug 15;490:807–814.

[99]   Hooper HL, Jurkschat K, Morgan AJ, Bailey J, Lawlor AJ, Spurgeon DJ, Svendsen C. Comparative chronic toxicity of nanoparticulate and ionic zinc to the earthworm Eisenia veneta in a soil matrix. Environ Int. 2011 Aug;37(6):1111–1117.

[100]   Zhao J, Wang Z, Liu X, Xie X, Zhang K, Xing B. Distribution of CuO nanoparticles in juvenile carp (Cyprinus carpio) and their potential toxicity. J Hazard Mater. 2011 Dec 15;197:304–310.

[101]   Song L, Vijver MG, Peijnenburg WJ, Galloway TS, Tyler CR. A comparative analysis on the in vivo toxicity of copper nanoparticles in three species of freshwater fish. Chemosphere. 2015 Nov;139:181–189.

[102]   McNeil PL, Boyle D, Henry TB, Handy RD, Sloman KA. Effects of metal nanoparticles on the lateral line system and behaviour in early life stages of zebrafish (Danio rerio). Aquat Toxicol. 2014 Jul;152:318–323.

[103]   Gomes T, Pereira CG, Cardoso C, Pinheiro JP, Cancio I, Bebianno MJ. Accumulation and toxicity of copper oxide nanoparticles in the digestive gland of Mytilus galloprovincialis. Aquat Toxicol. 2012 Aug 15; 118–119:72–79.

[104]   Ates M, Daniels J, Arslan Z, Farah IO, Rivera HF. Comparative evaluation of impact of Zn and ZnO nanoparticles on brine shrimp (Artemia salina) larvae: effects of particle size and solubility on toxicity. Environ Sci Process Impacts. 2013 Jan;15(1):225–233.

[105]   Zhang W, Miao Y, Lin K, Chen L, Dong Q, Huang C. Toxic effects of copper ion in zebrafish in the joint presence of CdTe QDs. Environ Pollut. 2013 May;176:158–164.

[106] Nations S, Long M, Wages M, Maul JD, Theodorakis CW, Cobb GP. Subchronic and chronic developmental effects of copper oxide (CuO) nanoparticles on Xenopus laevis. Chemosphere. 2015 Sep;135:166–174.

[107] Ma H, Wallis LK, Diamond S, Li S, Canas-Carrell J, Parra A. Impact of solar UV radiation on toxicity of ZnO nanoparticles through photocatalytic reactive oxygen species (ROS) generation and photo-induced dissolution. Environ Pollut. 2014 Oct;193:165–172.

[108] Villarreal FD, Das GK, Abid A, Kennedy IM, Kültz D. Sublethal effects of CuO nanoparticles on Mozambique tilapia (Oreochromis mossambicus) are modulated by environmental salinity. PLoS One. 2014 Feb 10;9(2):e88723.

[109] Fan W, Wang X, Cui M, Zhang D, Zhang Y, Yu T, Guo L. Differential oxidative stress of octahedral and cubic Cu2O micro/nanocrystals to Daphnia magna. Environ Sci Technol. 2012 Sep 18;46(18):10255–10262.

[110] Minetto D, Libralato G, Volpi Ghirardini A. Ecotoxicity of engineered TiO2 nanoparticles to saltwater organisms: An overview. Environ Int. 2014 May;66:18–27.

[111] Sharma VK. Aggregation and toxicity of titanium dioxide nanoparticles in aquati-cenvironment–a review. J Environ Sci Health A Tox Hazard Subst Environ Eng. 2009 Dec;44(14):1485–1495.

[112] Fang T, Yu LP, Zhang WC, Bao SP. Effects of humic acid and ionic strength on TiO$_2$ nanoparticles sublethal toxicity to zebrafish. Ecotoxicol. 2015 Dec;24(10):2054–2066.

[113] Ma H, Brennan A, Diamond SA. Phototoxicity of TiO2 nanoparticles under solar radiation to two aquatic species: Daphnia magna and Japanese medaka. Environ Toxicol Chem. 2012 Jul;31(7):1621–1629

[114] Fouqueray M, Dufils B, Vollat B, Chaurand P, Botta C, Abacci K, Labille J, Rose J, Garric J. Effects of aged TiO2 nanomaterial from sunscreen on Daphnia magna exposed by dietary route. Environ Pollut. 2012 Apr;163:55–61.

[115] Fouqueray M, Noury P, Dherret L, Chaurand P, Abbaci K, Labille J, Rose J, Garric J. Exposure of juvenile Danio rerio to aged TiO$_2$ nanomaterial from sunscreen. Environ Sci Pollut Res Int. 2013 May;20(5):3340–3350.

[116] Yang SP, Bar-Ilan O, Peterson RE, Heideman W, Hamers RJ, Pedersen JA.Influence of humic acid on titanium dioxide nanoparticle toxicity to developing zebrafish. Environ Sci Technol. 2013 May 7;47(9):4718–4725.

[117] Fang Q, Shi X, Zhang L, Wang Q, Wang X, Guo Y, Zhou B. Effect of titanium dioxide nanoparticles on the bioavailability, metabolism, and toxicity of pentachlorophenol in zebrafish larvae. J Hazard Mater. 2015;283:897–904.

[118] Dalai S, Iswarya V, Bhuvaneshwari M, Pakrashi S, Chandrasekaran N, Mukherjee A. Different modes of TiO2 uptake by Ceriodaphnia dubia: Relevance to toxicity and bioaccumulation. Aquat Toxicol. 2014 Jul;152:139–146.

[119] Shah SNA, Shah Z, Hussain M, Khan M. Hazardous Effects of Titanium Dioxide Nanoparticles in Ecosystem. Bioinorganic Chemistry and Applications. 2017;2017:4101735. doi:10.1155/2017/4101735.

[120] Angelstorf JS, Ahlf W, von der Kammer F, Heise S. Impact of particle size and light exposure on the effects of TiO2 nanoparticles on Caenorhabditis elegans. Environ Toxicol Chem. 2014 Oct;33(10):2288–2296.

[121] Li S, Wallis LK, Diamond SA, Ma H, Hoff DJ. Species sensitivity and dependence on exposure conditions impacting the phototoxicity of TiO$_2$ nanoparticles to benthic organisms. Environ Toxicol Chem. 2014 Jul;33(7):1563–1569.

[122] Miao W, Zhu B, Xiao X, Li Y, Dirbaba NB, Zhou B, Wu H. Effects of titanium dioxide nanoparticles on lead bioconcentration and toxicity on thyroid endocrine system and neuronal development in zebrafish larvae. Aquat Toxicol. 2015 Apr;161:117–126.

[123] Fan W, Cui M, Liu H, Wang C, Shi Z, Tan C, Yang X. Nano-TiO2 enhances the toxicity of copper in natural water to Daphnia magna. Environ Pollut. 2011 Mar;159(3):729–734.

[124] Rocco L, Santonastaso M, Mottola F, Costagliola D, Suero T, Pacifico S, Stingo V. Genotoxicity assessment of TiO2 nanoparticles in the teleost Danio rerio. Ecotoxicol Environ Saf. 2015 Mar;113:223–230.

[125] Jovanović B, Ji T, Palić D. Gene expression of zebrafish embryos exposed to titanium dioxide nanoparticles and hydroxylated fullerenes. Ecotoxicol Environ Saf. 2011 Sep;74(6):1518–1525.

[126] Sekar D, Falcioni ML, Barucca G, Falcioni G. DNA damage and repair following In vitro exposure to two different forms of titanium dioxide nanoparticles on trout erythrocyte. Environ Toxicol. 2014 Jan;29(1):117–127.

[127] Ildikó Fekete-Kertész, Gergő Maros, Katalin Gruiz, Mónika Molnár. The effect of TiO2 nanoparticles on the aquatic ecosystem. Period Polytech Chem Eng. 2016;60 (4):pp. 231–243.

[128] Gallego-Gallegos M, Doig LE, Tse JJ, Pickering IJ, Liber K. Bioavailability, toxicity and biotransformation of selenium in midge (Chironomus dilutus) larvae exposed via water or diet to elemental selenium particles, selenite, or selenized algae. Environ Sci Technol. 2013 Jan 2;47(1):584–592.

[129] Zhang P, He X, Ma Y, Lu K, Zhao Y, Zhang Z Distribution and bioavailability of ceria nanoparticles in an aquatic ecosystem model. Chemosphere. 2012 Oct;89(5):530–535.

[130] Conway JR, Hanna SK, Lenihan HS, Keller AA. Effects and implications of trophic transfer and accumulation of CeO2 nanoparticles in a marine mussel. Environ Sci Technol. 2014;48(3):1517–1524.

[131] Birgit K. Gaiser, Anamika Biswas, Philipp Rosenkranz, Mark A. Jepson, Jamie R. Lead, Vicki Stone, Charles R. Tyler and Teresa F. Fernandes. Effects of silver and cerium dioxide micro- and nano-sized particles on *Daphnia magna*. J. Environ. Monit. 2011; **13**: 1227–1235.

[132] Ibarra LE, Tarres L, Bongiovanni S, Barbero CA, Kogan MJ, Rivarola VA, Bertuzzi ML, Yslas EI. Assessment of polyaniline nanoparticles toxicity and teratogenicity in aquatic environment using Rhinella arenarum model. Ecotoxicol Environ Saf. 2015 Apr;114:84–92.

[133] Gott RC, Luo Y, Wang Q, Lamp WO. Development of a biopolymer nanoparticle-based method of oral toxicity testing in aquatic invertebrates. Ecotoxicol Environ Saf. 2014 Jun;104:226–230.

[134] Clemente Z, Grillo R, Jonsson M, Santos NZ, Feitosa LO, Lima R, Fraceto LF. Ecotoxicological evaluation of poly(epsilon-caprolactone) nanocapsules containing triazine herbicides. J Nanosci Nanotechnol. 2014 Jul;14(7):4911–4917.

[135] Chen PJ, Tan SW, Wu WL. Stabilization or oxidation of nanoscale zerovalent iron at environmentally relevant exposure changes bioavailability and toxicity in medaka fish. Environ Sci Technol. 2012 Aug 7;46(15):8431–8439.

[136] Levard C, Hotze EM, Colman BP, Dale AL, Truong L, Yang XY, Bone AJ, Brown GE Jr, Tanguay RL, Di Giulio RT, Bernhardt ES, Meyer JN, Wiesner MR, Lowry GV. Sulfidation of silver nanoparticles: Natural antidote to their toxicity. Environ Sci Technol. 2013;47(23):13440–13448.

[137] Duan J, Yongbo Yu Y, Li Y, Yu Y, Li Y, Huang P, Zhou X, Peng S, Sun Z. Developmental toxicity of CdTe QDs in zebrafish embryos and larvae. J Nanopart Res. 2013; 15:1700.

[138] Rocha TL, Gomes T, Sousa VS, Mestre NC, Bebianno MJ. Ecotoxicological impact of engineered nanomaterials in bivalve molluscs: An overview. Mar Environ Res. 2015 Oct;111:74–88.

[139] Tang S, Wu Y, Ryan CN, Yu S, Qin G, Edwards DS, Mayer GD. Distinct expression profiles of stress defense and DNA repair genes in Daphnia pulex exposed to cadmium, zinc, and quantum dots. Chemosphere. 2015 Feb;120:92–99.

[140] Balusamy B, Taştan BE, Ergen SF, Uyar T, Tekinay T. Toxicity of lanthanum oxide (La2O3) nanoparticles in aquatic environments. Environ Sci Process Impacts. 2015 Jul;17(7):1265–1270.

[141] Dedeh A, Ciutat A, Treguer-Delapierre M, Bourdineaud JP.Impact of gold nanoparticles on zebrafish exposed to a spiked sediment. Nanotoxicol. 2015 Feb;9(1):71–80.

[142] Vo NT, Bufalino MR, Hartlen KD, Kitaev V, Lee LE. Cytotoxicity evaluation of silica nanoparticles using fish cell lines. In Vitro Cell Dev Biol Anim. 2014;50(5):427–438.

[143] Choi MH, Hwang Y, Lee HU, Kim B, Lee GW, Oh YK, Andersen HR, Lee YC, Huh YS. Aquatic ecotoxicity effect of engineered aminoclay nanoparticles. Ecotoxicol Environ Saf. 2014 Apr;102:34–41.

[144] Tedesco, Sara & Doyle, Hugh & Iacopino, Daniela & O'Donovan, Irene & Keane, Sarah & Sheehan, David. Gold nanoparticles and oxidative stress in the blue mussel, Mytilus edulis. Methods Mol Biol. (Clifton, N.J.). 2013;1028. 197–203.

[145] Dominguez GA, Lohse SE, Torelli MD, Murphy CJ, Hamers RJ, Orr G, Klaper RD. Effects of charge and surface ligand properties of nanoparticles on oxidative stress and gene expression within the gut of Daphnia magna. Aquat Toxicol. 2015 May;162:1–9.

[146] Chen PJ, Tan SW, Wu WL. Stabilization or oxidation of nanoscale zerovalent iron at environmentally relevant exposure changes bioavailability and toxicity in medaka fish. Environ Sci Technol. 2012 Aug 7;46(15):8431–8439.

[147] Zhu X, Tian S, Cai Z. Toxicity assessment of iron oxide nanoparticles in zebrafish (Danio rerio) early life stages. PLoS One. 2012;7(9):e46286.

[148] Manabe M, Tatarazako N, Kinoshita M. Uptake, excretion and toxicity of nano-sized latex particles on medaka (Oryzias latipes) embryos and larvae. Aquat Toxicol. 2011 Oct; 105 (3–4): 576–581.

[149] Pakrashi S, Dalai S, Humayun A, Chakravarty S, Chandrasekaran N, Mukherjee A. Ceriodaphnia dubia as a potential bio-indicator for assessing acute aluminum oxide nanoparticle toxicity in fresh water environment. PLoS One. 2013 Sep 5;8(9):e74003.

[150] Mattsson K, Ekvall MT, Hansson LA, Linse S, Malmendal A, Cedervall T. Altered behavior, physiology, and metabolism in fish exposed to polystyrene nanoparticles. Environ Sci Technol. 2015 Jan 6;49(1):553–561.

11 Nanotoxicity and plants

Abstract: Nanoparticles (NPs) have been known as an environmental hazard affecting the lives of plants, animals and humans. The plant growth, metabolism, biochemistry and physiology are affected in the terrestrial environments. Experimental studies with diverse plant species and different NPs have revealed the effect of NPs on plants. In the aqueous environment, phytoplanktons and alga are the primary producers that affect the entire ecosystem and food chain. Bioaccumulation has also been observed and reported in different organisms. Both the terrestrial and aquatic releases of NPs in the environment that affect the plants are discussed in this chapter.

Keywords: cadmium, rice, *Oryza sativa*, nanosilicon, nano-TiO$_2$, *Rhizobium leguminosarum*, nano-TiO2, zinc oxide nanoparticles, *Brassica napus*, rubber ash nanoparticles, mung bean (*Vigna radiata*), nano-anatase, *Ulmus elongata*, *Triticum aestivum* L., multiwalled carbon nanotubes, titanium silicon oxide nanomaterial, *Medicago arborea*, oxidative stress, phytotoxicity, *Lycopersicon lycopersicum*, *Lactuca sativa*, lanthanum oxide NPs, *Coriandrum sativum*, γ-Fe^{2}O^3 NPs, *Phaseolus vulgaris*, *Lolium perenne*, nanocerium oxide, ascorbate peroxidase, catalase, superoxide dismutase, glutathione reductase, lipid peroxidation, *Lemna minor*, *Scenedesmus obliquus*, *Dunaliella tertiolecta*, *Tetraselmis suecica*, *Phaeodactylum tricornutum*, *Chlamydomonas reinhardtii*, *Pseudokirchneriella subcapitata*, *Dunaliella salina*, (C$_{60}$)-nanoparticles (Buckminster fullerenes), *Pseudokirchneriella subcapitata*, cellulose nanofibers, *Hydrilla verticillata*

11.1 Introduction

The extensive and varied applications of nanoparticles (NPs) in the different aspects of day-to-day life in humans have made their manufacturing, engineering and fabrication units and industries very active. The release of NPs in aquatic environment affects the aquatic plants and animals and in the terrestrial environment affects the growth and survival of land plants and animals. The effect on plants finds importance as plants are exposed to NPs by air and aerial deposits. Researchers have confirmed that diverse types of NPs affect the different plants, which is discussed in this chapter.

11.2 Effect of NPs in plants

Nanoparticulate Yb$_2$O$_3$, bulk Yb$_2$O$_3$ and YbCl$_3$·6H$_2$O have been reported of their phytotoxicity, decreased biomass in cucumber plants, detected in their localization in plant roots and transformation to YbPO$_4$ deposits by transmission electron microscopy,

https://doi.org/10.1515/9783110579093-011

energy-dispersive spectroscopy as well as synchrotron radiation-based methods such as Scanning transmission X-ray microscopy (STXM) and near edge X-ray absorption fine structure (NEXAFS) [1]. Nanofertilizers are used to improve plant nutrition; enhancing nutrition can have potential negative effects on the environment. Cadmium could induce toxicity in rice seedlings and 2.5 mM nanosilicon has been used to circumvent Cd stress in rice seedlings (*Oryza sativa* L. cv Youyou 128) [2]. Nanoscale copper particles having applications as antimicrobial formulations may deposit in the soil, affecting the soil ecosystem microbiota and plants [3]. Bulk and NP Cu and Ag were reported to be highly phytotoxic, affecting growth and transpiration when experimented on plant system [4].

Nano-TiO$_2$ treatment affected the rhizobium legume *Rhizobium leguminosarum* bv. *viciae* 3841 symbiosis studied on garden peas, by delaying root nodule formation and delayed the onset of nitrogen fixation [5]. Engineered nanomaterials (ENMs), including nano-Ag, nano-Au, nano-Si, nano-CeO$_2$, nano-TiO$_2$, nano-CuO, nano-ZnO, and CNTs, have been reported of mixed impact (i.e., both positive and negative) in different plants, depending on other parameters such as plant species, soil dynamics and soil microbial community. ENMs [6], Silver oxide NPs (AgNPs) and their transformation to silver NPs (nano-Ag) and silver ions have been reported to be phytotoxic affecting growth of mung bean (*Vigna radiata*) seedlings and pose threat to the environment and productivity and growth of plants [7]. Nano-silicon dioxide has been reported to cause plant resistance to salt stress by improving the antioxidant system of squash (*Cucurbita pepo* L. cv. white bush marrow) [8]. Iron and copper nanomaterials including core-shell nanoscale materials like (Fe/Fe$_3$O$_4$, Cu/CuO) at 10 and 20 mg/L, and FeSO$_4$·7H$_2$O and CuSO$_4$·5H$_2$O at 10 mg/L affected the physiological parameters and nutritional quality of lettuce (*Lactuca sativa*) negatively, of which copper nano-toxicity effects were the most [9]. Nano-CuO has been reported to be phytotoxic when experimented on rice (*Oryza sativa* cv. Swarna) seedlings inducing biochemical oxidative damages [10].

It has been reported that titanium dioxide NPs (TiO$_2$NPs) do not show phytotoxic effects on germination and root elongation of seed and seedlings in three different plant species [11].

In vitro exposure of citrate-coated silver (citrate-nAg) and zinc oxide (ZnONPs), NPs have been reported to affect developmental responses in maize (*Zea mays* L.) and cabbage (*Brassica oleracea* var. *capitata* L.) [12].

TiO$_2$NPs are reported to be nontoxic to *Brassica napus* L. plants and are suggested to have a growth-promoting effect on oilseed rape plants [13]. Nano-ZnO (nZnO) revealed negative impact on *Rhizobium*-legume symbiosis and is known as a potential hazard to the *Rhizobium*-legume symbiosis system [14]. Rubber ash nanoparticles affected the growth of cucumber and also the concentration of Zn, cadmium (Cd) and lead (Pb) in cucumber, thereby affecting the food chain [15]. Nano-anatase TiO$_2$ solutions have been reported to affect the photosynthetic activity and hence affecting the production of *Ulmus elongata*, a deciduous tree endemic to eastern provinces of China [16].

Engineered NPs including aerosol (TiO$_2$NPs) and colloidal silver (AgNP) cause phytotoxicity in tomatoes (*Lycopersicon esculentum*) [17]. Studies on engineered carbon nanomaterials, including carbon nanotubes, C$_{60}$ and graphene, have confirmed the potential toxic effects on the germination of rice seeds [18]. ENMs have been reported to affect the metabolism of primary and secondary metabolites in plants [19]. CuO or ZnO NPs had phytotoxic effects in wheat (*Triticum aestivum* L.) growth [20].

Arabidopsis plants exposed to TiO$_2$NPs, AgNPs and multiwalled carbon nanotubes (MWCNTs) have been revealed to affect translation with the consequence of increased bacterial colonization during infection, inhibition of root hair development and indications of stress in plants [21]. Nano-CeO$_2$ particles have affected the growth of evergreen shrub *Medicago arborea* and revealed the pathophysiology of stress induced in plants [22].

Zinc oxide NPs including spherical ZnO-30, spherical ZnO-50, columnar ZnO-90 and hexagon rod-like ZnO-15 affected adversely the seed germination, inhibition of the root and shoot elongation of Chinese cabbage (*Brassica pekinensis* L.), production of free hydroxyl groups (·OH) and bioaccumulation of Zn in roots and shoots causing phytotoxicity to Chinese cabbage seedlings [24]. Nano-titanium and cerium oxide (nano-cerium) affected the gene expression and growth in *Arabidopsis thaliana* [24]. However, it has been found from the experimental studies on growth of California red kidney bean (*Phaseolus vulgaris*) and rye grass (*Lolium perenne*) plants that nano-aluminum did not have an adverse effect on California red beans compared to its uptake on rye grass [25].

ENPs including ionic (FeCl$_3$), micro- and nano-sized zerovalent iron (nZVI) have been reported not to affect majorly the development of three macrophytes: *Lepidium sativum*, *Sinapis alba* and *Sorghum saccharatum* detected by parameters of seed germination, seedling elongation, germination index and biomass; however, at treatment with highest toxic concentrations they have been reported to affect the black spots and coatings on plants of all species tested [26].

Titanium silicon oxide nanomaterial (nano-TiSiO$_4$) has been found to be significantly phytotoxic on the growth of dicotyledonous plant species including lettuce, *Lactuca sativa* belonging to the daisy family Asteraceae, and common tomato *Lycopersicon lycopersicum* [27].

Phytotoxicity of lanthanum oxide (La$_2$O$_3$) NPs on cucumber plants has been reported from experiments [28]. Imidacloprid (Admire) pesticide-loaded sodium alginate NPs are safer to use in improving crop yield and is safe to the plants [29]. Cu NPs affect germination, growth and physiology of cilantro (*Coriandrum sativum*), repressing its nutritional value in food [30].

Iron oxide NPs (γ-Fe$_2$O$_3$ NPs) have been reported to cause physiological changes in an experimental study on watermelon seedlings [31]. NPs including MWCNT, aluminum, alumina, zinc and zinc oxide have been reported to affect germination and overall physiology in radish, rape, ryegrass, lettuce, corn and cucumber in an experimental study [32]. nZVI ions caused toxicity in plants including cattail

(*Typha latifolia*) and hybrid poplars (*Populous* sp.) in experimental studies [33]. Nano-TiO$_2$ core in sunscreens, coated with aluminum hydroxide and dimethicone films, experimented on a legume crop as edible bean seeds *Vicia faba* plants has been suggestive of long-term effects [34]. Nano-TiO$_2$ or nano-cerium oxide (nCeO$_2$) revealed different effects on experimented plant species and is suggestive of requirements of new methods in detecting subtle changes mediated by NPs on plant physiology [35]. Physiology of root water transmission has been affected in *Helianthus annuus* L. by the exposure of iron oxide NPs including nanomaghe-mite ([36]). Computational modeling has found importance in the study of toxicity assessment of engineered nanomaterials including nZVI and C(60) from engine oil lubricant [37].

NP mixtures including nano-ZnO + nano-CuO and nano-ZnO/nano-CuO have revealed less phytotoxic effects of mixtures as compared to individual mixtures of NPs as revealed from affected seed germination and inhibition of root growth experimented on *Lepidium sativum*, *Linum utisassimmum*, *Cucumis sativus* and *Triticum aestivum* [38].

ZnONPs have been reported to cause phytotoxic and genotoxic effects on *Allium cepa* or onion detected by parameters revealing affected mitotic (MI), micronuclei (MN), chromosomal aberration indices and lipid peroxidation [39].

Copper oxide NP (CuONP)-mediated toxicity on green pea (*Pisum sativum*) plants could be prevented by the administration of the phytohormone indole acetic acid [41]. Nano-copper particles (<25 nm and 60–80 nm) affected Cu uptake, bioaccumulation (roots, leaves and seeds), activity of ascorbate peroxidase (APX), catalase (CAT), superoxide dismutase (SOD), glutathione reductase (GR) and lipid peroxidation in leaves and roots of cowpea (*Vigna unguiculata*) revealing NP-mediated induced stress and phytotoxicity [42]. Cerium oxide NPs (nCeO$_2$) could modulate boron-induced phytotoxic effects and confer protective effects in sunflower (*Helianthus annuus* L. [43]).

11.3 Effects on marine alga and aquatic plants

NPs released in the aquatic environment have been reported to be toxic to marine and aquatic plants and animals. ZnONPs have been reported to be toxic to *Lemna minor*, commonly termed as duckweed or lesser duckweed, an aquatic freshwater plant of the genus *Lemna* affecting its growth and physiology [44]. Nano-Al$_2$O$_3$ released into the aquatic environment exhibited toxic effects on green algae belonging to the genus *Scenedesmus*, namely *Scenedesmus obliquus* indicative of causing oxidative stress by altered SOD activity and glutathione and malondialdehyde concentration [45]. TiO$_2$ and CuONPs and bulk CuO are toxic to *Lemna minor* L., leading to altered activity of antioxidative enzymes such as guaiacol peroxidase, APX and GR, increased necrosis and bleaching [46]. Polystyrene NPs as nanoplastics exert cytotoxic effects on phytoplanktons, including the green microalga *Dunaliella tertiolecta* [47].

Metal-based NPs like ZnONPs (100 nm) cause toxic effects on green *alga Tetraselmis suecica* and the diatom *Phaeodactylum tricornutum* [49]. CuONPs [49], TiO$_2$NPs [50], nano-TiO$_2$, ZnO [51] nano-Ag (nAg), nano-TiO$_2$, nZnO and CdTe/CdS quantum dots [52] have nanotoxic effects on aquatic microorganisms, microalga *Chlamydomonas reinhardtii* (49–52). ZnO, TiO$_2$ and CuO NPs have been proved to be posing toxic threat to sickle shaped microalga *Pseudokirchneriella subcapitata* [53]. About 5 nm SiO$_2$ NPs posed toxic threats when exposed to green alga *Chlorella kessleri* [54]. ZnONPs are toxic to *Dunaliella tertiolecta* [55] and NPs have been reported to exert toxic effects on saltwater microalga *Dunaliella salina* [56]. C(60)-NPs (Buckminster fullerenes) and atrazine, methyl parathion, pentachlorophenol and phenanthrene have been reported to be toxic to algae *Pseudokirchneriella subcapitata* [57]. Cellulose nanofibers are toxic toward *Klebsormidium flaccidum*, affecting the cell walls and cellular membrane and through the generation of ROS [58]. TiO$_2$NP has been reported to cause toxicity by causing oxidative stress response in *Hydrilla verticillata* revealed by the altered hydrogen peroxide (H$_2$O$_2$), reduced and oxidized glutathione (GSH and GSSG), levels of antioxidative enzymes peroxidase (POD), CAT, and GR [59]. ZnONPs have led to bilateral interactions with single-celled green algae, *Chlorella* sp. [60].

11.4 Discussion

NPs have been reported to have toxic effects on the environment, mostly affecting the green plants in the terrestrial environment and aquatic plants and alga and phytoplanktons in the aquatic environment when they leach into the marine or aquatic environment. These NPs deposited in the soil from industrial release and effluents are taken up by the plants and seedlings and cause bioaccumulation and toxic effects in plants [1–43] and when released in the marine or aquatic environment, they affect the marine alga and phytoplanktons leading to toxicity [44–60]. Green synthesis, that is, production of NPs from natural sources is a newly developing domain by which NPs are biocompatible and have less toxic effects in nature and are being tested on animals and plants to detect their toxic effects, which reveals much reduced or no toxic effects as compared to metal-based forms. ZnONPs are synthesized by treating zinc acetate dihydrate with the flower extract of *Elaeagnus angustifolia* (Russian olive) and its effect has been studied on seeds of tomato *Solanum lycopersicum* [61]. Computational approaches are finding applications in the detection of toxicity of NPs in plants; therefore, their usage in fertilizers can be known prior to their application. The future lies in further research of the effect of diverse range of NPs in plant toxicity.

References

[1]   Zhang P, Ma Y, Zhang Z, He X, Guo Z, Tai R, Ding Y, Zhao Y, Chai Z. Comparative toxicity of nanoparticulate/bulk Yb_2O_3 and $YbCl_3$ to cucumber (Cucumis sativus). Environ Sci Technol. 2012 Feb 7;46(3):1834–1841.

[2]   Wang S, Wang F, Gao S. Foliar application with nano-silicon alleviates Cd toxicity in rice seedlings. Environ Sci Pollut Res Int. 2015 Feb;22(4):2837–2845.

[3]   Anjum NA, Adam V, Kizek R, Duarte AC, Pereira E, Iqbal M, Lukatkin AS, Ahmad I. Nanoscale copper in the soil-plant system – toxicity and underlying potential mechanisms. Environ Res. 2015 Apr;138:306–325.

[4]   Musante C, White JC. Toxicity of silver and copper to Cucurbita pepo: Differential effects of nano and bulk-size particles. Environ Toxicol. 2012 Sep;27(9):510–517.

[5]   Fan R, Huang YC, Grusak MA, Huang CP, Sherrier DJ. Effects of nano-TiO_2 on the agronomically-relevant Rhizobium-legume symbiosis. Sci Total Environ. 2014 Jan 1;466–467:503–512

[6]   Reddy PVL, Hernandez-Viezcas JA, Peralta-Videa JR, Gardea-Torresdey JL. Lessons learned: Are engineered nanomaterials toxic to terrestrial plants? Sci Total Environ. 2016 Oct 15;568:470–479.

[7]   Singh D, Kumar A. Effects of nano silver oxide and silver ions on growth of Vigna radiata. Bull Environ Contam Toxicol. 2015 Sep;95(3):379–384.

[8]   Siddiqui MH, Al-Whaibi MH, Faisal M, Al Sahli AA. Nano-silicon dioxide mitigates the adverse effects of salt stress on Cucurbita pepo L. Environ Toxicol Chem. 2014 Nov;33(11):2429–2437.

[9]   Trujillo-Reyes J, Majumdar S, Botez CE, Peralta-Videa JR, Gardea-Torresdey JL. Exposure studies of core-shell Fe/Fe_3O_4 and Cu/CuO NPs to lettuce (Lactuca sativa) plants: Are they a potential physiological and nutritional hazard? J Hazard Mater. 2014 Feb 28;267:255–263.

[10]   Shaw AK, Hossain Z. Impact of nano-CuO stress on rice (Oryza sativa L.) seedlings. Chemosphere. 2013 Oct;93(6):906–915.

[11]   Song U, Shin M, Lee G, Roh J, Kim Y, Lee EJ. Functional analysis of TiO_2 nanoparticle toxicity in three plant species. Biol Trace Elem Res. 2013 Oct;155(1):93–103.

[12]   Pokhrel LR, Dubey B. Evaluation of developmental responses of two crop plants exposed to silver and zinc oxide nanoparticles. Sci Total Environ. 2013 May 1;452–453:321–332.

[13]   Li J, Naeem MS, Wang X, Liu L, Chen C, Ma N, Zhang C. Nano-TiO2 Is not phytotoxic as revealed by the oilseed rape growth and photosynthetic apparatus ultra-structural response. PLoS One. 2015 Dec 1;10(12):e0143885.

[14]   Huang YC, Fan R, Grusak MA, Sherrier JD, Huang CP. Effects of nano-ZnO on the agronomically relevant Rhizobium-legume symbiosis. Sci Total Environ. 2014 Nov 1;497–498:78–90.

[15]   Moghaddasi S, Hossein Khoshgoftarmanesh A, Karimzadeh F, Chaney R⁴. Fate and effect of tire rubber ash nano-particles (RANPs) in cucumber. Ecotoxicol Environ Saf. 2015 May;115:137–143.

[16]   Gao J, Xu G, Qian H, Liu P, Zhao P, Hu Y. Effects of nano-TiO_2 on photosynthetic characteristics of Ulmus elongata seedlings. Environ Pollut. 2013 May;176:63–70.

[17]   Song U, Jun H, Waldman B, Roh J, Kim Y, Yi J, Lee EJ. Functional analyses of nanoparticle toxicity: a comparative study of the effects of TiO_2 and Ag on tomatoes (Lycopersicon esculentum). Ecotoxicol Environ Saf. 2013 Jul;93:60–67.

[18]   Nair R, Mohamed MS, Gao W, Maekawa T, Yoshida Y, Ajayan PM, Kumar DS. Effect of carbon nanomaterials on the germination and growth of rice plants. J Nanosci Nanotechnol. 2012 Mar;12(3):2212–2220.

[19]   Hatami M, Kariman K, Ghorbanpour M. Engineered nanomaterial-mediated changes in the metabolism of terrestrial plants. Sci Total Environ. 2016 Nov 15;571:275–291.

[20] Dimkpa CO, Latta DE, McLean JE, Britt DW, Boyanov MI, Anderson AJ. Fate of CuO and ZnO nano- and microparticles in the plant environment. Environ Sci Technol. 2013 May 7;47(9):4734–4742.

[21] García-Sánchez S, Bernales I, Cristobal S. Early response to nanoparticles in the Arabidopsis transcriptome compromises plant defence and root-hair development through salicylic acid signalling. BMC Genomics. 2015 Apr 24;16:341.

[22] Gomez-Garay A, Pintos B, Manzanera JA, Lobo C, Villalobos N, Martín L. Uptake of CeO$_2$ nanoparticles and its effect on growth of Medicago arborea In vitro plantlets. Biol Trace Elem Res. 2014 Oct;161(1):143–150.

[23] Xiang L, Zhao HM, Li YW, Huang XP, Wu XL, Zhai T, Yuan Y, Cai QY, Mo CH. Effects of the size and morphology of zinc oxide nanoparticles on the germination of Chinese cabbage seeds. Environ Sci Pollut Res Int. 2015 Jul;22(14):10452–10462.

[24] Tumburu L, Andersen CP, Rygiewicz PT, Reichman JR. Phenotypic and genomic responses to titanium dioxide and cerium oxide nanoparticles in Arabidopsis germinants. Environ Toxicol Chem. 2015 Jan;34(1):70–83.

[25] Doshi R, Braida W, Christodoulatos C, Wazne M, O'Connor G. Nano-aluminum: transport through sand columns and environmental effects on plants and soil communities. Environ Res. 2008 Mar;106(3):296–303.

[26] Libralato G, Costa Devoti A, Zanella M, Sabbioni E, Mičetić I, Manodori L, Pigozzo A, Manenti S, Groppi F, Volpi Ghirardini A. Phytotoxicity of ionic, micro- and nano-sized iron in three plant species. Ecotoxicol Environ Saf. 2016 Jan;123:81–88.

[27] Bouguerra S, Gavina A, Ksibi M, Rasteiro Mda G, Rocha-Santos T, Pereira R. Ecotoxicity of titanium silicon oxide (TiSiO$_4$) nanomaterial for terrestrial plants and soil invertebrate species. Ecotoxicol Environ Saf. 2016 Jul;129:291–301.

[28] Ma Y, He X, Zhang P, Zhang Z, Guo Z, Tai R, Xu Z, Zhang L, Ding Y, Zhao Y, Chai Z. Phytotoxicity and biotransformation of La$_2$O$_3$ nanoparticles in a terrestrial plant cucumber (Cucumis sativus). Nanotoxicol. 2011 Dec;5(4):743–753.

[29] Kumar S, Bhanjana G, Sharma A, Sidhu MC, Dilbaghi N. Synthesis, characterization and on field evaluation of pesticide loaded sodium alginate nanoparticles. Carbohydr Polym. 2014 Jan 30;101:1061–1067.

[30] Zuverza-Mena N, Medina-Velo IA, Barrios AC, Tan W, Peralta-Videa JR, Gardea-Torresdey JL. Copper nanoparticles/compounds impact agronomic and physiological parameters in cilantro (Coriandrum sativum). Environ Sci Process Impacts. 2015 Oct;17(10):1783–1793.

[31] Wang Y, Hu J, Dai Z, Li J, Huang J. In vitro assessment of physiological changes of watermelon (Citrullus lanatus) upon iron oxide nanoparticles exposure. Plant Physiol Biochem. 2016 Nov;108:353–360.

[32] Lin D, Xing B. Phytotoxicity of nanoparticles: inhibition of seed germination and root growth. Environ Pollut. 2007 Nov;150(2):243–250.

[33] Ma X, Gurung A, Deng Y. Phytotoxicity and uptake of nanoscale zero-valent iron (nZVI) by two plant species. Sci Total Environ. 2013 Jan 15;443:844–849.

[34] Foltête AS, Masfaraud JF, Bigorgne E, Nahmani J, Chaurand P, Botta C, Labille J, Rose J, Férand JF, Cotelle S. Environmental impact of sunscreen nanomaterials: Ecotoxicity and genotoxicity of altered TiO$_2$ nanocomposites on Vicia faba. Environ Pollut. 2011 Oct;159(10):2515–2522.

[35] Andersen CP, King G, Plocher M, Storm M, Pokhrel LR, Johnson MG, Rygiewicz PT. Germination and early plant development of ten plant species exposed to titanium dioxide and cerium oxide nanoparticles. Environ Toxicol Chem. 2016 Sep;35(9):2223–2229.

[36] Martínez-Fernández D, Barroso D, Komárek M. Root water transport of Helianthus annuus L. under iron oxide nanoparticle exposure. Environ Sci Pollut Res Int. 2016 Jan;23(2):1732–1741.

[37] Grieger KD, Hansen SF, Sørensen PB, Baun A. Conceptual modeling for identification of worst case conditions in environmental risk assessment of nanomaterials using nZVI and C$_{60}$ as case studies. Sci Total Environ. 2011 Sep 1;409(19):4109–4124.

[38] Jośko I, Oleszczuk P, Skwarek E. Toxicity of combined mixtures of nanoparticles to plants. J Hazard Mater. 2017 Jun 5;331:200–209.

[39] Kumari M, Khan SS, Pakrashi S, Mukherjee A, Chandrasekaran N. Cytogenetic and genotoxic effects of zinc oxide nanoparticles on root cells of Allium cepa. J Hazard Mater. 2011 Jun 15; 190 (1–3): 613–621.

[40] Shen CX, Zhang QF, Li J, Bi FC, Yao N. Induction of programmed cell death in Arabidopsis and rice by single-wall carbon nanotubes. Am J Bot. 2010 Oct;97(10):1602–1609.

[41] Ochoa L, Medina-Velo IA, Barrios AC, Bonilla-Bird NJ, Hernandez-Viezcas JA, Peralta-Videa JR, Gardea-Torresdey JL. Modulation of CuO nanoparticles toxicity to green pea (Pisum sativum Fabaceae) by the phytohormone indole-3-acetic acid. Sci Total Environ. 2017 Nov 15;598:513–524.

[42] Ogunkunle CO, Jimoh MA, Asogwa NT, Viswanathan K, Vishwakarma V, Fatoba PO. Effects of manufactured nano-copper on copper uptake, bioaccumulation and enzyme activities in cowpea grown on soil substrate. Ecotoxicol Environ Saf. 2018 Jul 15;155:86–93.

[43] Tassi E, Giorgetti L, Morelli E, Peralta-Videa JR, Gardea-Torresdey JL, Barbafieri M. Physiological and biochemical responses of sunflower (Helianthus annuus L.) exposed to nano-CeO$_2$ and excess boron: Modulation of boron phytotoxicity. Plant Physiol Biochem. 2017 Jan;110:50–58.

[44] Chen X, O'Halloran J, Jansen MA. The toxicity of zinc oxide nanoparticles to Lemna minor (L.) is predominantly caused by dissolved Zn. Aquat Toxicol. 2016 May;174:46–53.

[45] Li X, Zhou S, Fan W, Effect of Nano-Al$_2$O$_3$ on the Toxicity and Oxidative Stress of Copper towards Scenedesmus obliquus. Int J Environ Res Public Health. 2016 Jun 9;13(6):575.

[46] Dolenc Koce J. Effects of exposure to nano and bulk sized TiO$_2$ and CuO in Lemna minor. Plant Physiol Biochem. 2017 Oct;119:43–49.

[47] Bergami E, Pugnalini S, Vannuccini ML, Manfra L, Faleri C, Savorelli F, Dawson KA, Corsi I. Long-term toxicity of surface-charged polystyrene nanoplastics to marine planktonic species Dunaliella tertiolecta and Artemia franciscana. Aquat Toxicol. 2017 Aug;189:159–169.

[48] Li J, Schiavo S, Rametta G, Miglietta ML, La Ferrara V, Wu C, Manzo S. Comparative toxicity of nano ZnO and bulk ZnO towards marine algae Tetraselmis suecica and Phaeodactylum tricornutum. Environ Sci Pollut Res Int. 2017 Mar;24(7):6543–6553.

[49] von Moos N, Maillard L, Slaveykova VI. Dynamics of sub-lethal effects of nano-CuO on the microalga Chlamydomonas reinhardtii during short-term exposure. Aquat Toxicol. 2015 Apr;161:267–275.

[50] Chen L, Zhou L, Liu Y, Deng S, Wu H, Wang G. Toxicological effects of nanometer titanium dioxide (nano-TiO$_2$) on Chlamydomonas reinhardtii. Ecotoxicol Environ Saf. 2012 Oct;84:155–162.

[51] Gunawan C, Sirimanoonphan A, Teoh WY, Marquis CP, Amal R. Submicron and nano formulations of titanium dioxide and zinc oxide stimulate unique cellular toxicological responses in the green microalga Chlamydomonas reinhardtii. J Hazard Mater. 2013 Sep 15;260:984–992.

[52] Simon DF, Domingos RF, Hauser C, Hutchins CM, Zerges W, Wilkinson KJ. Transcriptome sequencing (RNA-seq) analysis of the effects of metal nanoparticle exposure on the transcriptome of Chlamydomonas reinhardtii. Appl Environ Microbiol. 2013 Aug;79(16):4774–4785.

[53] Aruoja V, Dubourguier HC, Kasemets K, Kahru A. Toxicity of nanoparticles of CuO, ZnO and TiO$_2$ to microalgae Pseudokirchneriella subcapitata. Sci Total Environ. 2009 Feb 1;407(4):1461–1468.

[54]	Fujiwara K, Suematsu H, Kiyomiya E, Aoki M, Sato M, Moritoki N. Size-dependent toxicity of silica nano-particles to Chlorella kessleri. J Environ Sci Health A Tox Hazard Subst Environ Eng. 2008 Aug;43(10):1167–1173.

[55]	Manzo S, Miglietta ML, Rametta G, Buono S, Di Francia G. Toxic effects of ZnO nanoparticles towards marine algae Dunaliella tertiolecta. Sci Total Environ. 2013 Feb 15;445–446:371–376.

[56]	Golubev AA, Prilepskii AY, Dykman LA, Khlebtsov NG, Bogatyrev VA. Colorimetric Evaluation of the Viability of the Microalga Dunaliella Salina as a Test Tool for Nanomaterial Toxicity. Toxicol Sci. 2016 May;151(1):115–125.

[57]	Baun A, Sørensen SN, Rasmussen RF, Hartmann NB, Koch CB. Toxicity and bioaccumulation of xenobiotic organic compounds in the presence of aqueous suspensions of aggregates of nano-C(60). Aquat Toxicol. 2008 Feb 18;86(3):379–387.

[58]	Munk M, Brandão HM, Nowak S, Mouton L, Gern JC, Guimaraes AS, Yéprémian C, Couté A, Raposo NR, Marconcini JM, Brayner R. Direct and indirect toxic effects of cotton-derived cellulose nanofibres on filamentous green algae. Ecotoxicol Environ Saf. 2015 Dec;122:399–405.

[59]	Okupnik A, Pflugmacher S. Oxidative stress response of the aquatic macrophyte Hydrilla verticillata exposed to TiO$_2$ nanoparticles. Environ Toxicol Chem. 2016 Nov;35(11):2859–2866.

[60]	Chen P, Powell BA, Mortimer M, Ke PC. Adaptive interactions between zinc oxide nanoparticles and Chlorella sp. Environ Sci Technol. 2012 Nov 6;46(21):12178–12185.

[61]	Singh A, Singh NB, Hussain I, Singh H, Yadav V, Singh SC. Green synthesis of nano zinc oxide and evaluation of its impact on germination and metabolic activity of Solanum lycopersicum. J Biotechnol. 2016 Sep 10;233:84–94.

Glossary

Ab initio From the initiation; from first principles.

Absorption To intake a substance into the body.

Acetylcholinesterase An enzyme of the carboxylesterase family that catalyzes neurotransmitter acetylcholine breakdown to acetic acid and choline. Located in the neuromuscular junctions, they mediate synaptic transmission.

Acidophilic Acid loving.

Acquired immunodeficiency syndrome (AIDS) A life threatening disease of the immune system caused HIV infection that leads to destruction of CD4 T lymphocytes and CD4 count less than 200 cells/mm^3 and by breakdown of the body's immune defenses.

Action potential In neurophysiology, an action potential occurs when the membrane potential of an axon rapidly rises and falls.

Active immunity Exposure to pathogens and disease-causing organism or vaccine containing killed or inactivated organism can trigger the immune system to produce antibodies to that disease. Active immunity lasts long.

Acute exposure Exposure to a toxic substance or chemical once for a short duration, less than 14 days for humans.

Adenocarcinoma Cancer associated with mucus-secreting glands. Breast, lung, colon, colorectal, prostrate and esophageal cancers are adenocarcinomas.

Adverse Effects An abnormal, undesirable, or harmful effect caused by chemical, toxin, drugs, nanoparticles, etc. to an organism affecting its physiology, biochemistry and overall health and may bear lethal consequences.

Adrenergic Comprises the autonomic nervous system (ANS) where epinephrine or norepinephrine act as neurotransmitters.

Adsorption Is a surface phenomenon, by which atoms, ions or molecules adhere to an adsorbent surface forming a film of the adsorbate. Of the two types of adsorption, physical adsorption involves weak van der Waal's forces and chemical adsorption involves chemical bonds. Common adsorbents include colloids, clay, silica gel metals, etc.

Adult stem cells Undifferentiated cells that can both renew themselves and can differentiate into specialized lineages. They play a role in maintenance and repair of tissue.

Aerosol Suspension generated by natural or man-made sources, of fine solid particles or liquid droplets, in air or other gas.

Agglomerate Collection of particles, mass or aggregates.

Albumin A type of water soluble, globular simple proteins. Serum albumin is present in animals.

Algorithm In mathematics and computational sciences, a formula or set of rules to address the solution of a particular problem.

Allergen Any chemical substance, chemical from industry or environmental sources that enters the body through food or inhalation can trigger the immune system leading to allergic reaction.

Allergy diseases They are caused by hypersensitive reactions triggered by allergens. Diseases include hay fever, food allergies, atopic dermatitis, upper respiratory tract allergies, allergic asthma and anaphylaxis.

Alveoli Tiny sac at the terminal points of bronchioles in the lungs that are sites for gas exchange. Air sacs that allow rapid gaseous exchange of oxygen and carbon dioxide.

Amino acid Basic unit of a protein containing amine and a carboxylic acid group.

Anaphylactic shock An acute reaction due to allergic response and can lead to death.

Anaphylaxis Severe allergy response.

https://doi.org/10.1515/9783110579093-012

Angiogenesis It involves growth of blood vessels occurring throughout life in both health and disease, from within uterus till old age.

Anoxic Lack in oxygen.

Antibody Produced by B-lymphocyte cells in response to an antigen or foreign particle, leading to humoral immune response. This binding of the antigen causes its removal from the body.

Antibody-dependent cell-mediated cytotoxicity (ADCC) It is an immune reaction triggered by antibody, coating a target cell that is recognized by Fc receptor bearing effector cells that lyse the antibody-coated target cells.

Antigen Any molecule or foreign substance that can give rise to an immune response.

Antigen-presenting cells Cells that can present processed antigen to T cells. Examples: B cells, macrophages, dendritic cells.

Apoptosis Programmed cell death.

Arrhythmia Loss or variation of heartbeat.

Artery Blood vessels that carry oxygenated blood to the tissues.

Asphyxia Suffocating condition due to total lack of oxygen, can lead to death.

Asthma Chronic inflammatory disorder of the lung airways leading to airflow obstruction, mucous secretion bronchoconstriction and bronchospasm, edema of alveoli.

Astrocyte Star shaped supporting glial cell surrounding neurons in brain and spinal cord of the central nervous system.

Atomic Force Microscope (AFM) It is a microscopy technology for studying nanoscale materials, generating images at atomic resolution.

Autoantibody Present and contribute to autoimmunity. This antibody can react with host tissue or antigen.

Autoimmunity A disease state when the body fails to identify the host antigens and recognizing them as foreign generates immune response against them.

Autoimmune diseases Occurs due to autoimmune reactions in the body. Some diseases include type 1 diabetes, rheumatoid arthritis, lupus, coeliac disease.

Autonomic (nervous system) Controls involuntary functions of body.

Autotrophic algae Algae that photosynthesize and prepare their food from inorganic substances. They are the producers.

Auxotrophic algae These algae needs organic substances and vitamins together with inorganic nutrients for food through photosynthesis.

ACT Artemisinin based on combination therapies are the currently accepted treatments for falciparum malaria.

AD Alzheimer's disease is a progressive brain disorder that slowly destroys memory and thinking ability.

AF Aflatoxin includes family of poisonous carcinogenic toxins produced by certain fungi including *Aspergillus flavus* and *Aspergillus parasiticus* found on maize (corn), peanuts, cottonseed and tree nuts. They increase the risk of liver diseases and liver cancer.

AC Activated charcoal is a form of carbon composed of fine black powder made from bone char, petroleum coke, coal, or sawdust. The small, low-volume pores increases the surface area.

ALDH Aldehyde dehydrogenases are a group of enzymes that catalyze the aldehydes oxidation converting aldehydes to carboxylic acids.

AMI Acute myocardial infarction is a cardiovascular disorder resulting from acute obstruction of a coronary artery.

AMP Antimicrobial peptides are broad spectrum oligopeptides ranging from five to more than hundred amino acids.

APAF-1 Apoptotic protease-activating factor 1 is a major component of the apoptosome complex, and is known to undergo conformational change during mitochondrial apoptosis.

APP Acute Phase Proteins include a group of unrelated proteins, synthesized in the liver, whose plasma level concentrations reveal a marked increase (included under positive APP) or decrease (included under negative APP) in response to tissue injury, acute infections, burns or inflammation. The APPs include alpha 1-acid glycoprotein (AGP), C-reactive protein (CRP), complement components, fibrinogen, mannose binding proteins (MBP) and serum amyloid A (SAA).

Amyloid-β peptides Deposition of amyloid-beta peptide (Abeta) is associated with pathogenesis of Alzheimer's disease (AD) and is associated with neurodegeneration. Aβ is a cleavage product of transmembrane protein, APP (amyloid precursor protein) by the enzyme α-secretase prevents Aβ formation, but produces the neuroprotective sAPPα fragment. However, subsequent cleavage by β- and then γ-secretases leads to formation of Aβ.

B cells Immune cells, a type of white blood cells called lymphocytes that develop in the bone marrow. They produce antibodies that play a role in the humoral immune defenses.

Bacterium Prokaryotic microorganism, typically about 1000 nm in structure, single celled organism. They are categorized into two groups based on their cell wall architecture that includes gram negative and gram positive organisms. Some may cause diseases.

Basophil Occurs a frequency less than 1% granulocytes of circulating white blood cells, plays a role in allergy response, inflammation, release of histamine and other chemicals in blood.

Benthos. Organisms, both plants and animals, who prefer to stay in the ecological niche of sea or lake bottom. These animals include sea urchins, corals, sponges, acorn barnacles, etc.

Binding Association of two molecules could be receptor to ligand.

Biocompatible Compatible with living tissue, nontoxic or harmful.

Bioinformatics An interdisciplinary field of science involving concepts of molecular biology, genetics, biochemistry, computer science, mathematics and statistics.

Biomedical Nanotechnology Applications of nanotechnology toward biomedical science applications for human benefit.

BioMEMS Biomedical (or biological) microelectromechanical systems involving mechanical and microfabrication technologies used in biological applications in genomics, proteomics, molecular diagnostics, regenerative medicine, tissue engineering and implantable microdevices.

BioNEMS Nanoelectromechanical systems with biomedical applications.

Buckminster fullerene Chemical compound of carbon with the formula C_{60}. It has a truncated icosahedron structure with twenty hexagons and twelve pentagons, with a carbon atom at each vertex of polygon and a bond along each polygon edge. Its structure bears similarity with football. It is named after Buckminster Fuller.

Biosensor A device that can sense biological information like temperature, pressure, chemical presence, reaction product and convert it to electrical signals for detection.

Blastocyst A transient, preimplantation embryo composed of hollow ball of 150 to 200 cells formed after fertilization, contains inner cell mass, outer layer of cell called the trophoblast and fluid filled blastocoel.

Bone marrow stromal cells Bone marrow, including fibroblasts, adipocytes (fat cells) and bone- and cartilage-forming cells that provide support for blood cells formation.

Bone marrow Soft tissue of the bone cavities is the site of hematopoiesis, formation of blood cells.

Bronchial Related to lungs.

Bucky Balls Also called fullerenes, they were one of the first nanoparticles to be discovered in 1985 by Richard Smalley, Harry Kroto, and Robert Curl. They are composed 60 carbon atoms in interlocking hexagonal shapes.

BAL Bronchoalveolar lavage It is a diagnostic procedure to study cells and other components from bronchial and alveolar spaces used in diagnosis of infections and malignancies and lung diseases.

BALF bronchoalveolar lavage fluid It includes cells and other components from bronchial and alveolar spaces that could finds application in diagnosis of infection, lung disease and cancer.

BBB The blood–brain barrier (BBB) is a unique microvasculature of the central nervous system (CNS) acting as the interface between the blood circulation and the neural tissue.

BCR Membrane bound receptors on the B-cell surface is composed of membrane immunoglobulin (mIg) molecules that on binding to antigen, causes B-cell activation, triggering thereby a cascade of intracellular signaling reactions leading to the antigen processing and presentation to T cells.

BLI Bioluminescence imaging is a technology that applies light emitted by enzyme-catalyzed reactions for the study of molecular activity of molecules. Bioluminescent reporters utilize small chemical substrate for non-invasive imaging in cell biology and small animal studies.

BMS Bare-metal stent is an uncoated or uncovered stent composed of a mesh-like tube of thin wire made up of 316L stainless steel, cobalt chromium alloy. The fine application percutaneous coronary intervention (PCI) for indications including stable and unstable angina, acute myocardial infarction (MI), and CVS disease.

BPNs Enable therapeutic delivery of agents to the brain, and can readily reach disease cells with CNS disorders.

BSA Bovine serum albumin is a serum albumin protein derived from cows.

BUN A blood urea nitrogen (BUN) is a test that measures the amount of nitrogen in your blood that comes from the waste product urea.

Calmodulin Calcium binding protein in eukaryotic cells, plays role in immune response, inflammation, metabolism, apoptosis, muscle contraction, and nerve growth.

CAMP Cyclic adenosine monophosphate or 3′,5′-cyclic adenosine monophosphate acts as a second messenger in signal transduction in many biological responses.

Carbon Nanotubes These are carbon allotropes with a cylindrical nanostructure. Due to their structure they show strength, stiffness thermal conductivity, electrical properties with applications in field of electronics, optics, materials science and technology. They could be single walled (SWNTs), double walled (DWNTs) and multi walled (MWNTs).

Carbonyl A molecule with O in double bond arrangement to C (e.g. -CO-).

Carboxylic acid A molecule with C double bonded to O and single bonded to OH (e.g., -COOH).

Cardiac related to heart.

Cardiomyocytes Muscle cells that enable the heart to beat continuously and rhythmically.

Cartilage Type of connective tissue with cells embedded fibrous collagenous matrix.

Catacholamines Biologically active amine compounds that functions as neurotransmitter and hormones. For example, epinephrine, norepinephrine.

Cell It is the smallest unit of life consisting of cytoplasm enclosed by a membrane, containing biomacromolecules and cellular organelles that perform various functions.

Cell mediated immunity Cellular immunity involves cell mediated immune responses, activation of phagocytes, cytotoxic T cells, release of cytokines and chemokines on antigen exposure.

Chromosome Consists of DNA and regulatory proteins located in cell nucleus containing genetic information.

Clinical trial Experiments for clinical research on humans.

Clinical related to experiments performed in clinic.

Clone Cells, molecule or organism with identical genetic makeup or organisms with a single common ancestor.

Cluster of Differentiation It is used to identify cell surface molecules and has international nomenclature with CD number for cell surface molecules.

Colloid Composed of homogeneous non-crystalline substance dispersed in another substance. They include sols, aerosols, gels and emulsions.

Commensal Where one member gets benefitted from the other member in the association.

Complement Component of the immune system composed of a series of proteins involved in a stepwise biochemical reaction that leads to removal of infected cell, promotes inflammation.

Complementary Two molecules that fit or enhance qualities of each other or another.

Computerized Tomography (CT) It is a diagnostic imaging of internal organs, bones, soft tissue and blood vessels.

Conjugated It denotes double or triple bonds in a molecule that are separated by a single bond, across which some sharing of electrons occurs.

Conserved In genetics, it is the genetic code that is same in different species.

Cutaneous Related to the skin.

Cystic Structure or function or matter related to gallbladder or urinary bladder.

Cytocide Property of destroying cells.

Cytokine Biochemical molecules synthesized lymphocytes and monocytes that play important role in cell physiology and controlling cellular interactions in immunological reactions.

Cytology The science of study of cells.

Cytoplasm Includes all of the contents of the cell within the plasma membrane.

Cytotoxic Tcell T lymphocyte subset with CD8 marker and can kill infected cell or trans cancerous cell.

Cytotoxic Toxic to biological cells.

C4BP C4b-binding protein (C4BP) is a glycoprotein (500Kda) component in the complement cascade synthesized mainly in the liver.

CAPRA A non-IgE-mediated complement activation-related pseudoallergy pseudoallergy. Drugs like Ambisome and DaunoXome and agents like amphiphilic lipids can cause activation of complement by both alternative and classical pathways giving rise to C3a and C5a anphylatoxins that cause activation of basophil and mast cell leading to a type I hypersensitivity reaction.

CAT-Catalase Catalyzes the decomposition of hydrogen peroxide to water and oxygen. It protects cell from oxidative damage mediated by reactive oxygen species (ROS).

CB Carbon black is the product of material produced by the incomplete combustion of heavy petroleum products such as FCC tar, coal tar, diesel, etc. It is recently evaluated for carcinogenic potential causing discomfort to the upper respiratory tract, through mechanical irritation.

CBT Cognitive behavioral therapy (CBT) is a therapy for individuals suffering from psychological problems and disorders that helps to manage problems by changing the way of thinking and behavior.

CCBT Computerized cognitive behavioral therapy

CDK Cyclin-dependent kinases (CDKs) are protein kinases that needs a subunit – a cyclin. They play important roles in the control of cell division.

CF Cystic fibrosis is a progressive, genetic disorder causing persistent lung infections that lead to difficulty in breathing.

CGC Cerebellar granule cells are the smallest neurons of the brain that forms the thick granular layer of the cerebellar cortex.

CMA Chaperone-mediated autophagy (CMA), is a proteolytic systems that acts to degrade intracellular proteins in lysosomes.

DC– Direct current.

De novo– Created anew.

Demographic Referring to population size, structure and characteristic.

Denaturation A process by which proteins or nucleic acids lose their native form and structure.

Dendrimers Branched polymer molecules.

Dendrite A branched protoplasmic process of a neuron that conducts impulses toward the cell body. There are usually many to a cell, forming synaptic connections with other neurons.

Dendritic cells Type of white blood cells that can present processed antigens to T cells for their removal.

Deoxyribonucleic acid (DNA) Encodes genetic information, consists of deoxyribose (a sugar), phosphoric acid, and four bases (purines or pyrimidines), in two long chains that twist around each other to form a double helix.

Deposit feeder An organism that lives on and feeds on soft sediment.

Dermis Skin inner layer below the epidermis.

Desiccate Remove water.

Diagnosis To detect a cause and nature of a disease for treatment or prognosis.

Diatom Producers in the food chain.

Differentiation The process by which cells differentiate into cells with special function from plouripotent cells.

Diffusion Movement of substance from higher concentration zones to those at lower concentration.

Dimer A combination of two similar or identical molecules with the elimination of water.

Dinoflagellate Marine planktons.

Diploid Chromosome complete set that an individual inherits, one from either parents.

Dipole Equal and opposite charges separated by a distance.

Disease Disorder of structure or function in an organism.

Dissolved organic matter Dissolved molecules of organic materials derived from dead, decomposed organic matter.

Distal Away from the source or a point of attachment or origin.

Diurnal At the day time.

DNAase An enzyme that catalyzes the hydrolytic cleavage of phosphodiester linkages in the DNA backbone, thus degrading DNA.

Drug screening Screening of potential drug candidate after testing many potential candidates through assays.

DWNTs Double-walled carbon nanotubes.

DAF Decay-accelerating factor (DAF) or CD55 is a 70 kDa glycoprotein, encoded by gene on chromosome 1q32. It is a glycosyl phosphatidylinositol (GPI) anchored membrane protein expressed on RBC, lymphocytes, granulocytes, endothelium and epithelium. It functions by inhibiting classical and alternative pathway C3 converting enzymes and protects cells from complement mediated lysis.

DAMPs Damage-associated molecular pattern molecules (DAMPs) are endogenous molecules and released by damaged tissues that can activate innate immune system.

DFS Disease-free survival is defined as the period in which no signs and symptoms of the disease appears after a successful treatment.

DLS Dynamic Light Scattering (DLS) also known as Photon Correlation Spectroscopy (PCS) is used to detect size particles from below 5 nm to several microns.

DMSA The DMSA or dimercapto succinic acid is a short-lived nuclear medicine that is used to scan the kidneys and is a nuclear medicine test that enables detection of the kidney disorder.

DOX **Doxorubicin**, also known as **Adriamycin,** is a chemotherapy medication used for the treatment of cancer like breast cancer, bladder cancer, Kaposi's sarcoma, lymphoma, and leukemia.

Ectoderm The outer germ layer derived from the inner cell mass of the blastocyst. It forms the nervous system, sensory organs, skin, and related structures.

Edema Swelling, abnormal accumulation of fluid.

EEG Electroencephalogram is a monitoring method of brain activity by electrical signals.

Efferent conducting outwards or away from

Electrophoresis is used to separate macromolecules based on the mobility of ions in an electric field.

Electroporation Entry of DNA into cells by brief intense pulse of electricity to open cell pores.

Embryo An embryo develops after fertilization.

Embryonic stem cells Pluripotent stem cells in the inner cell mass of a blastocyst of an early-stage pre-implantation embryo.

Endocrine gland A gland that releases hormones into blood.

Endocytosis Cellular intake process by vesicles.

Endoderm The innermost layer of the cells of inner cell mass of the blastocyst. It leads to the development of lungs, respiratory structures and digestive organs.

Endogenous Growing inside an organ, part, or system.

Endoplasmic reticulum Cell organelles associated with protein and lipid synthesis.

Enzyme A biochemical catalyst that speeds up a chemical reaction.

Eosinophil Granulocytic white blood cell accounting for 1% to 4% of all leukocytes.

Epidermis The outer skin epithelial layer.

Epigenetic Factors other than genetic that cause heritable changes in an organism.

Epithelium Type of tissue that lines the outer surfaces of body, blood vessels.

Epitope Part of the antigen that is recognized by the antibody, B cells, T cells and generate an immune response.

Erythrocyte Red blood cells in humans containing hemoglobin and transports oxygen and carbon dioxide to and from cells.

Erythropoietin Also known as hematopoietin. It is a glycoprotein cytokine released by the kidney under hypoxia. It stimulates erythropoiesis or RBC production in the bone marrow.

Etiology The study to know the cause or origin of disease.

Eukaryote Cell with a true nucleus.

Eutrophic Nutrient rich aquatic environment.

Ex vivo outside the body.

Exocrine Glands that secretes chemicals and releases them through ducts.

Extravasation Movement of cells from capillaries to tissues observed during inflammation, tissue injury.

Exudate A fluid that flows from the circulatory system into injured areas or sites of inflammation.

EBC Exhaled breath condensate (EBC) is a condensate of exhaled gas and includes biomarkers of lung disease.

EMB Ethambutol is an antibiotic used with other medications to treat tuberculosis (TB).

EpCAM The epithelial cell adhesion molecule (EpCAM, CD326) is a glycoprotein and plays a role in different cellular physiological processes as cell adhesion signaling, cell migration, proliferation and differentiation. It is a well-known marker for carcinoma.

Fertilization The union of male and female gametes.

Fetus Developing human from within the mother.

Food chain Linear relationships of organisms through which nutrients and energy flows as one feeds on the other.

Functionalization of Nanotubes MWNTs or SWNTs are added with functional groups to the ends, sidewalls of nanotubes to add desired properties.

FADD Fas associated via death domain is an adaptor protein that plays major role in apoptosis

Fas ligand (FasL or CD95L) It is a type-II transmembrane protein that belongs to the tumor necrosis factor (TNF) family, which on binding to its receptor FasR induces apoptosis.

fC$_{60}$ Functionalized fullerene has found applications as biosensor with biomedical applications.

FcεRI The high-affinity IgE receptor, also known as FcεRI, or Fc epsilon RI, is the high-affinity receptor for the Fc region of immunoglobulin E (IgE), expressed on mast cells, basophils and eosinophils.

FUS Focused ultrasound (FUS) is a non-invasive, surgical technology based on imaging that uses ultrasound energy to target specific areas of the brain and body in the treatment, or diagnosis.

Ganglion Component of the nervous system. It is composed of nerve-cell bodies.

Gastro Related to stomach. Commonly used as gastroenteritis.

Gastrointestinal tract digestive tract.

GDP Full form is Guanosine diphosphate.

Gene A unit of genetic material or DNA that carries genetic information.

Genetic sequence Comprises of nucleotide sequences in DNA and RNA.

Genome The complete genetic material and information it contains in an organism.

Genomics Study of the genome.

Genotype The genetic make-up of an organism.

Genus A taxonomic hierarchy above species.

Germ layers Comprises of three layers including ectoderm, the mesoderm, and the endoderm.

Global warming Effects of undesired accumulation of carbon dioxide in the atmosphere, leading to greenhouse effect, global warming and climate change due to anthropogenic activities.

Glucose Simple monosaccharide with formula $C_6H_{12}O_6$.

Golgi complex/apparatus Cell organelles near the endoplasmic reticulum involved in post translational modification like glycosylation and transport of proteins

G-proteins Termed as guanine nucleotide-binding **proteins**, involved in signal transduction.

Granulocyte Type of leukocyte with granules in cytoplasm, including neutrophils, eosinophils and basophils.

Granuloma A lesion with tissue inflammation caused due to infection, inflammation or exposure to a chemical or agent from the environment like airborne agents.

Granulomatous reaction The process of formation of granuloma involves a chronic inflammatory response wherein macrophages lymphoid cells and epithelial cells.

Greenhouse effect It causes warming of earth's surface due to water vapor, carbon dioxide, methane and other gases in the air that forms a layer and traps the solar heat.

GTP Full form of guanosine triphosphate.

GALT Gut-associated lymphoid tissue is a component of the mucosa-associated lymphoid tissue (MALT) that provides immunity to the GI tract.

GCSF Granulocyte-colony stimulating factor (G-CSF or GCSF), is a glycoprotein cytokine that promotes survival, proliferation, differentiation and function of neutrophil precursors and mature neutrophils.

GDNF Glial cell line derived neurotrophic factor is involved in inflammation and pain.

GBM Glioblastoma multiforme is the primary malignant brain tumor in adults with poor treatment outcomes.

GO Gene ontology is a major bioinformatics application tool that understands gene and gene product attributes across all species.

Haploid number Chromosomes in male or female gametes in humans 23 numbers.

Haploid One set of autosome and one sex chromosome. It is half of diploid.

helper T cells A class of T cells with CD4 markers and plays a role in production of antibody, activating T cytotoxic cells and activating immune responses.

Hematopoietic stem cell Present in bone marrow. It is from where blood cells and platelets originates.

Hematopoietic Forming of blood cells from progenitor cells.

Hemolytic Lysis of red blood cells.

Hemorrhagic injury and bleeding.

Hemostasis Key step in healing of wound and stop of bleeding.

Hepatic Anything related to the liver.

Hepatocyte Liver cell.

Herbivore Primary consumer who draw their energy from consuming plants.

Heritable character A feature inherited by an individual.

Heterochromatin Tightly condensed DNA with no genetic expression.

Heterotrophic algae They draw their nourishment and energy from organic molecules.

Histamine Released by basophils and mast cells in response to allergen pollen, chemical compounds causing hypersensitive immune response.

Histology Study of animal, plant cells and tissues.

HLA complex Human Leukocyte Antigen complex codes for major histocompatibility complex (MHC) proteins and regulates immune system identification of self from foreign.

HIV (human immunodeficiency virus) Causes **HIV** infection and acquired immunodeficiency syndrome (AIDS) where the immune system fails leading to other infections and even can lead to cancer.

Holoplankton They live in the water column all through their life. For example, diatoms, dinoflagellates, etc.

Homeostasis A state of balance or equilibrium.

Hormone Acts as messengers, released from endocrine glands.

Humoral immunity Immunity conferred by the roles of antibodies released by B cells.

Hydrocarbon An organic compound consisting of Hydrogen and Carbon forming framework to many macromolecules.

Hydrophilic Water loving molecule or part that interacts with water or polar compounds.

Hydrophobic Water-hating molecules or those that repel water.

Hypoxia Oxygen deficiency in tissues such that normal functions are affected.

HARNS High aspect ratio nanoparticles (HARN) are nanoparticles that have length much more than their width. For example nanotubes, nanorods, etc.

Immune complex Complex of antigen and antibodies.

Immune response The physiological response of the immune system raised in against of contact with foreign substances.

Immunoassay Assay designed to detect proteins by the use of antigenic property or antibodies.

Immunocompetent Ability of the body to generate an immune response.

Immunoglobulin (Ig) Also termed as antibody.

Immunomodulatory Modulation of immune system or response.

Immunosuppression Reduction or inactivation of the immune responses.

Immunotoxin Proteins designed containing toxin together with an antibody or growth factor that can bind to target cells. They are used to target cancer cells.

Inflammatory response Immune response to injury, infections and shows symptoms of redness, heat and swelling.

In vitro In laboratory experimental set up.

In vivo Inside the body.

Incision A surgical cut made by sharp object like knife, sterile blades, etc.

Inner cell mass (ICM) Cells within the blastocyst

Intracellular fluid (ICF) Fluid within cells.

Intraperitoneal In the peritoneal cavity.

Ischemia Local and temporary deficiency of blood supply due to obstruction of the circulation into a body part.

Islets of Langerhans Location of beta cells that produces insulin in pancreas.

IV Injected into vein.

Im Intramuscular, injected into muscle

IFN Interferons (IFNs) are signaling proteins released by host cells in response to the viruses, bacteria, parasites and tumor cells.

INH Isoniazid or isonicotinylhydrazide is an antibiotic used for the treatment of tuberculosis.

IP3 Inositol trisphosphate or **inositol 1,4,5-trisphosphate** act as secondary messenger molecule in signal transduction and lipid signaling in biological cells.

IRT Immunoreactive trypsinogen is a diagnostic test screening applied in newborns to screen for cystic fibrosis (CF).

IT or intratracheal instillation involves the introduction of a substance directly into the trachea. It find application in detecting the toxicity of a substance to the respiratory system.

ITAM It is a conserved sequence of four amino acids that is repeated twice in the cytoplasmic tails of certain cell surface proteins of the immune system.

JAK The Janus kinase (Jak) are families of non-receptor tyrosine kinases including Jak1, Jak2, Jak3 and Tyrosine kinase 2 (Tyk2) in mammals. Cytokines bind to their receptors and activate Jak kinases, phosphorylate the receptors, creating docking sites for signaling molecules, signal transducer and activator of transcription (Stat) family.

JNK It includes a family of MAP-kinases involved in the regulation of cell proliferation, differentiation and apoptosis.

Keratin A structural protein that is fibrous in nature and occurs in horns and hair.

Keratinocyte Found in the epidermis, it is the outer layer of the skin and produces keratin.

Kupffer cells Macrophages found in the liver.

Kda Atomic mass unit equivalent to 1,000 Daltons used to describe molecular weight of large proteins.

KEGG Database containing molecular datasets and high throughput experimental genomics data used to understand the biological system, such as the cell, the organism and the ecosystem.

Larva Developmental stage.

LD50 The dose that causes 50% mortality.

Leukocytes White blood cells.

Ligand That can bind to the receptor.

Lipids With a hydrophilic polar heads and hydrophobic tail.

Lipophilic Loving lipids or fats.

Lipophobic Repulsion by lipids.

Liposomes Designed spherical entity with bilayer of lipid encasing a liquid. They are used in delivery of nanoparticles and drugs. Same as micelles.

LPS Lipopolysaccharide, layer of Gram-negative bacteria.

lymph nodes Part of the immune system where B cells, T cells are present.

Lymph An alkaline fluid that circulates in lymphatic system.

Lymphocyte Type of white blood cells that are of two types: T and B lymphocytes that has roles in immunodefenses.

Lymphoid Organs Organs of the immune system.

Lymphokines Type of cytokine produced by lymphocyte that play important role in immune responses.

Lysis Burst of cell wall of bacteria.

Lysosomes Cell organelle containing hydrolytic enzymes.

Lysozyme An enzyme that can break the walls of some bacteria.

Lactoferrin The protein present in milk and other secretions and on all mucosal surfaces throughout the body with bactericidal and iron-binding properties. In human milk, it provides iron absorption and protection against enteric infection in the newborns.

Laser induced fluorescence (LIF) It is a spectroscopic method in which an atom or molecule is excited to a higher energy level by the absorption of laser light followed by spontaneous emission of light.

Lymphotoxin (LT) It is a member of the tumor necrosis factor (TNF) superfamily.

Lactate dehydrogenase (LDH) It is an enzyme that catalyzes the conversion of lactate to pyruvic acid and back.

Macrobenthos Benthic organisms seen by the eye.

Macromolecule It includes biomolecules proteins, nucleic acids and polysaccharides.

Macrophage A phagocytic cell.

Macrophyte An alga or plant seen by the eye.

Magnetic Force Microscope (MFM) A special form of AFM that captures images of surfaces by using magnetic forces.

Major histocompatibility complex (MHC) Expressed on antigen presenting cells, they are molecules in the immune system that play major role in antigen processing and presentation to T cells. MHC genes are located in human chromosome 6. They include MHC I (HLA-A,B,C) and MHC II (HLA-D,DR).

mast cell A granulocyte found in tissue and are the major cells that play role in allergy reactions by the release of inflammatory mediators as histamine and leukotrienes.

Megaplankton Planktonic organisms that are greater than or equal to 2000 micrometers in size.

Meiosis The type of cell division in which a diploid germ cell produce gametes (sperm or eggs) that will carry half the normal chromosome number.

Membrane proteins Proteins in the plasma membrane.

Mesenchymal stem cells (MSCs) Multipotent in nature number and can differentiate into chondrocytes producing cartilage, osteoblasts develop into bone cells, myocytes producing muscle cells and adipocytes that give rise to adipose tissue.

Mesoderm A tissue layer in the embryo giving rise to connective tissues, including the muscular, skeletal, circulatory, lymphatic and urogenital systems, and the linings of the body cavities.

Messenger molecule A molecule which can convey information after it is received and sensed by chemical sensor.

Metamorphosis Developmental change from larva to an immature adult

Mitochondrion Cell organelle also called powerhouse of the cell

Moiety A portion of a molecular structure having some characteristic property of functional importance.

Molecular nanotechnology Involves a broad domain of engineering of all complex mechanical systems from the molecular level.

Monoclonal antibodies Antibodies produced by a single cell-binding to specific protein molecules. They find immense application in research, medicine and industry

Monocyte A large phagocytic white blood cell that develops into macrophage in tissues.

Monolayer A layer, film or coating that is only one atom or molecule thick.

Monomer Any molecule that can be bound to similar molecules to form a polymer.

Motile Capable of voluntary movement. Opposite of **Motif** – main element, theme,

MRI Magnetic resonance imaging.

Mucosa A mucous membrane; lining the hollow organ or body cavity.

Muller cells Neuroglial cells with fine fibers forming the supportive cells in the retina.

MWNTs Multi walled carbon nanotubes.

Myocardium Heart muscle.

MAC The membrane attack complex (MAC) is the final product of the complement activation pathways composed of subunits C5b, C6, C7, C8 and several C9 molecules by which a hole is formed on the cell leading to cell lysis and death.

MASP-1- The pattern-recognition molecules mannan-binding lectin (MBL) is association with serine proteases (MASPs) and the three ficolins play role in innate immune system.

MBL Mannose-binding lectin (MBL) is a lectin that can activate the lectin pathway of complement activation of innate immunity

Membrane cofactor protein MCP CD46 is a C3b/C4b-binding cell surface glycoprotein that inhibits complement activation on host cells.

MDCK cell line It is derived by S. H. Madin and N. B. Darby from the kidney tissue of an adult female cocker spaniel. In 1958 it was widely used in the study of **β-amyloid** peptide (Aβ) and other cellular processes.

MDR-Tb or Multidrug-resistant TB does not respond to isoniazid and rifampicin, the 2 most powerful anti-TB drugs.

MI Myocardial infarction occurs with decreased or stopped blood flow in any part of the heart causing damage to the heart muscle.

Macrophage inflammatory protein or MIP-1alpha and MIP-1beta are members of the CC chemokine subfamily.

Micronuclei Nuclear structure formed when a chromosome or a fragment of a chromosome is not incorporated into one of the daughter nuclei during cell division.

MPS Mononuclear phagocyte system is also termed as the reticuloendothelial system or macrophage system.

MS Multiple Sclerosis is a systemic autoimmune disorder disabling the brain and spinal cord.

Mtb *Mycobacterium tuberculosis* (MTB) is the causative agent of tuberculosis.

MRSA Bacterial strain resistant to many antibiotics.

MTT (4,5-dimethylthiazol-2-yl)-2,5-diphenyltetrazolium bromide) a colorimetric **assay** for detecting the metabolic activity of a cell.

Naked DNA DNA without any outer protein envelope.

Nano One billionth (1/1,000,000,000) or ~10^{-9} meters.

Nanocomposite Composed of two or more substances, of which at least one has a nanoscale dimension, such as nanoparticles dispersed throughout another solid material.

Nanocrystals Also known as nanoscale semiconductor crystals and around 10 nm in diameter.

Nanoelectronics Electronics on a nanometer scale, including molecular electronics and nanoscale devices.

Nanofabrication Tools and techniques and methods to create, assemble or form nanoscale structures.

Nanofluidics Science of flow of liquid or gas through nanoscale spaces.

Nanolithography The process of making nanometer-sized surface patterns.

Nanomachine Machine systems that functions in nanoscale.

Nanomachine Molecular machine created through molecular manufacturing.

Nanomanipulator A nanorobotic tool for manipulating nanoscale objects.

Nanomedicine Application of nanotechnology for detection, monitoring and targeting of diseases to improve human health.

Nanometer A billionth of meter.

Nanoparticle Particles of nanosize.

Nanoplankton Planktons of 2–20 micometer size.

Nanoscience- The science of matter at the nanoscale.

Nanosensor- A chemical or physical sensor made up of nanoscale components.

Nanotechnology Technology involving engineering and manufacturing of matter at nanometer scales.

Nanotoxicology Study of hazardous and potential of nanoparticles exposure on living organisms and environment.

Nasopharynx The nasal passages, mouth and upper throat.

Natriuretic Substance increasing the rate of excretion of sodium ion in the urine.

Natural killer (NK) cells Granulocutes that can target tumor cells and infected body cells.

Necrosis Cell death by lysis, enzymatic degradation together with inflammation.

Neonatal Infant during the first 4 weeks of postnatal life. In some study, the first 7 days is considered as neonatal life.

Neoplasisa New and abnormal formation of tumor.

Neoplastic Transformation of tissue toward tumor.

Nephritis Inflammation of the kidney that leads to its failure, accompanied by edema, hematuria, hypertension and proteinuria.

Nephropathy Any disorder or kidney abnormality.

Nephrosis Disease of the kidneys marked by degeneration of renal tubular epithelium.

Nephrotoxic Chemically toxic to kidney cells.

Neural stem cell An adult stem cell found in neural tissue that give rise to neurons and glial cells including astrocytes and oligodendrocytes.

Neural Related to nerve or nerves.

Neuron A nerve cell that transmits impulse or information to other neurons or cells by releasing neurotransmitters at synapses. It consists of a cell body and its processes—an axon and one or more dendrites. Neurons transmit information.

Neuropathy Any disease related to central or peripheral nervous system.

Neuropeptide Neurotransmitter peptides found in neural tissue including endorphins, enkephalins.

Neurotoxic Agent that is toxic to the nervous system causing adverse effects.

Neurotransmitter Released by an excited axon terminal of a presynaptic neuron. For example, acetylcholine.

Neutrophil Granulocytic white blood cell that play role in conferring protection against infection. OKT3: A monoclonal antibody that targets mature T cells.

Niche Ecological space occupied by a species

NIST National Institute of Standards and Technology.

NMR See Nuclear Magnetic Resonance.

Nociceptors Pain receptors.

Nodule Small node or boss that is solid and can be detected by touch.

non-ionizing radiation Electromagnetic radiation of low energy that do not ionize.

Nuclear Magnetic Resonance (NMR) An analytical technique involving absorption of certain electromagnetic frequencies by atomic nuclei.

Nucleic acids Polymers of nucleotidyl residues that are building blocks of nucleotides including DNA and RNA.

Nucleotide (nucleotidyl) Phosphorylated nucleosides; nucleotidyl residues are the monomeric units of RNA and DNA. They are composed of sugar (ribose or deoxyribose), phosphate and one of four nitrogen bases (purine or pyrimidine).

Nutrient cycling The pattern of transfer of nutrients between the components of a food web.

Nutrients Those constituents required by organisms for maintenance and growth.

NADPH Nicotinamide adenine dinucleotide phosphate is a coenzyme and serves as an electron carrier alternately oxidized as NADP+ and reduced NADPH.

NDV Newcastle disease is a disease in birds and poultry and can be transmitted to humans.

NLRs Nod like receptors that are components of innate immune system that can sense PAMPs on pathogen surface

Nanoparticle-rich diesel exhaust Nanoparticles in diesel exhaust potentially and have adverse effects profound on fetal development, pregnant mothers and overall health of the body.

Neuropilin-1 Neuropilin 1 (NRP1) is a transmembrane glycoprotein acting as a co-receptor for a extracellular ligands including class III/IV semaphorins, certain isoforms of vascular endothelial growth factor and transforming growth factor beta.

Obligate To make necessary or require, without alternative.

Occipital Pertaining to the back part of the head.

Olfaction (olfactory) Sense of smell.

Oligonucleotide A polymeric chain that consists of a small number of nucleotides.

Opportunistic infection An infection in an immunosuppressed person caused by an organism that does not usually trouble people with healthy immune systems.

Opsonin A biochemical substance that coats foreign antigens, increasing their susceptibility to macrophages and other leukocytes, thus increasing phagocytosis of the organism. Complement and antibodies are the two main opsonins in human blood.

Opsonization The process of coating of opsonins thus facilitating phagocytosis.

Organelle Subcellular component, located in the cytoplasm, surrounded by a membrane like Golgi body, lysosome, mitochondria, etc.

Osteoblast A bone-forming cell derived from mesenchyme to form the osseous matrix.

Osteocyte A mesodermal bone-forming cell entrapped within the bone matrix, helping to maintain bone as living tissue.

Occupational exposure limit (OEL) An OEL is an upper limit of acceptable concentration of a hazardous substance in air of workplace for a particular material. It is fixed by national authorities and enforced in industries and manufacturing units by legislation to protect occupational safety and health of workers.

Hydroxylated multiwalled carbon nanotubes Modified CNTs with hydroxyl group that promotes their applications in biomedical area.

Odorranalectin Small peptide lectin with potential for drug delivery and targeting the brain.

Palatine Related to the palate of the mouth.

Pancreatic beta cells Comprises the endocrine part of the pancreas and is involved in releasing insulin that regulates blood sugar levels. Type I diabetes is an autoimmune disorder when these cells are attacked and destroyed by antibodies.

Parasite An organism living on or in, and negatively affecting, another organism called the host.

Parenteral Medication route including intravenous, subcutaneous, intramuscular or mucosal.

Parietal Related to the walls of a cavity formed by the parietal bone, forming the roof and sides of the skull.

Particulate organic matter Particulate material derived from the decomposition of the nonmineral constituents of living organisms

Passive immunity Passive immunity can be naturally, mediated by transfer of maternal antibodies to the fetus through the placenta or artificial mediated by the transfer of antibodies or antiserum produced by another individual.

Pathogen A microorganism or agent capable of inducing disease.

Pathological Disease/adverse condition.

Pelagic Organisms living in the water column.

Peptide A short chain of amino acids joined by amide bonds.

Peptidoglycan layer Cross-linked polysaccharide chains with alternating sugars of β (1,4) linked *N*-acetylglucosamine (NAG) and *N*-acetylmuramic acid (NAM) in the cell wall of most bacteria.

Peyer's patches A collection of lymphoid tissues in the intestinal tract that are important component of gut-associated lymphoid tissue (GALT).

pH Measure of the acidity or basicity of water (-log10 of molar concentration of hydrogen ions in water)

Phagocyte A cell that can ingest and destroy particulate substances such as bacteria, protozoa, cells and cell debris, dust particles and colloids and contribute to innate immune defenses to the host body.

Phagocytosis Ingestion and digestion of bacteria and particles by phagocytes.

Pharmacodynamics The study involving drug action on living organisms.

Pharmacogenetics The study of the influence of genetical makeup of the organism on drug action.

Pharmacokinetics Study of drug metabolism emphasizing on aspects of absorption, duration of action, distribution in the body and method of excretion.

Pharynx The air passage from the nasal cavity to the larynx.

Phenotype The appearance or other characteristics of an organism.

Phytoplankton The plankton organism that can perform photosynthesis.

Phytotoxic A poisonous plant.

Plankton Organisms living suspended in the water column and cannot move against water currents.

Plasma cells Large antibody-producing cells that develop from B cells.

Plasmid An autonomous self-replicating extrachromosomal circular DNA present intracellularly and symbiotically in most bacteria, encoding a protein product that confers drug resistance or other bacterial phenotype.

Pluripotent stem cells Stem cells that can give rise to different types of cells. For example, embryonic stem cells or induced pluripotent (iPS) cells.

Polar liquid A liquid consisting of molecules that have an electric dipole moment like water, ethanol and liquefied ammonia.

Positron Emission Tomography (PET) Computerized Tomography.

Predation The consumption of one organism by another

Predator An organism that consumes another living organism.

Prognosis Based on correct diagnosis, deciding the future course of a disease or health and the patient's prospects for recovery or survival.

Prokaryote An organism with or without a true nucleus.

Proliferation Expansion of the number of cells by the continuous cell division.

Protein Polymer of amino acids linked by amide bonds.

Proteomics The study of proteins and their effect on the human body.

Protoplasm A thick, viscous colloidal substance that contains the contents of a living cell.

Protozoa The simplest unicellular animal, although some are colonial.

Pulmonary Pertaining to the lungs.

Peptide amphiphile Peptide-based molecules that self-assemble into structures including high aspect ratio nanofibers

Polycyclic aromatic hydrocarbons PAHs are environmental pollutants generated from incomplete combustion of coal, oil, petrol, diesel and wood.

Poly-amidoamine PAMAM is a class of dendrimer made up of repetitively branched subunits of amide and amine.

Pathogen-associated molecular patterns (PAMPs) Signatures of pathogen that is recognized by pattern recognition receptors (PRRs) as components of innate immune system.

Pediatric brain tumors Most common solid tumor in children in the brain leading to high rates of mortality and morbidity.

Proliferating cell nuclear antigen Protein for DNA polymerase that is maximally expressed during the S phase of the cell cycle. PCNA finds importance as a cancer biomarker.

Procalcitonin PCT is a 116 amino acid peptide belonging to the calcitonin (CT) superfamily of peptides

Pharmacodynamics Study of biochemical and physiologic effects of drugs.

Parkinson's disease Neurodegenerative disorder due to loss of brain cells and neurons that release dopamine.

Printer emitted particles The **particle** formed due to laser **printing** process by evaporation of VOCs (volatile organic compounds) and SVOCs (semi-volatile organic compounds).

Post-kala-azar dermal leishmaniasis PKDL is a complication after visceral leishmaniasis (VL) revealing rash in a patient who has recovered from VL.

Protein kinase C- PKC is a family of protein kinase enzymes that controls the signaling reactions by phosphorylating serine and threonine amino acid residues of other proteins by their hydroxyl groups.

Poly lactic-co-glycolic acid Synthesized by direct polycondensation of lactic acid and glycolic acid.

Poly-(l-lactide) It is a biodegradable and bioactive polyester

Polymethylmethacrylate Synthetic resin with transparent and plastic nature derived from polymerization of methyl methacrylate.

Peroxisome proliferator-activated receptor gamma PPAR-γ is a type II nuclear receptor, coded by the *PPARG* gene

Pattern recognition receptors PRRs are receptors expressed on macrophages, monocytes, neutrophils and epithelial cells that play important role in identifying the PAMS and DAMPS from pathogen or foreign or lysed cell surfaces and function in the innate immune system

Prostate-specific antigen Produced by the prostate gland and overexpression of PSA is an indication of prostate cancer.

Pulmonary surfactant Comprises of a mixture of lipids and proteins secreted by the epithelial type II alveolar cells into the alveolar space of the lungs.

Quantum dot A nanoscale crystal with a diameter ranging between 2–20 nm with unique electrical and optical properties.

Quantitative polymerase chain reaction Used for understanding gene transcription.

QSAR Used to predict the biological effect of chemical compounds on human body based on mathematical and statistical relations. It is used in lead optimization.

Radionuclide A radioactive isotope.

RBC Red blood cell (erythrocyte).

Reaction A process that transforms one or more chemical species into others.

Reagent A chemical that undergoes change as a result of a chemical reaction.

Receptor A structure that can identify and bind to a ligand.

Red blood cell (RBC) Erythrocyte.

Redox Oxidation-reduction reaction.

Regenerative Medicine Regenerative treatments including cellular therapy, gene therapy and tissue engineering approaches.

Renal Related to the kidney.

Reticuloendothelial system (RES) Comprises of phagocytes that engulf and remove foreign antigens and cell debris found in the human body.

Ribonucleic acid (RNA) The ribonucleotide polymer into which DNA is transcribed. that can synthesize proteins.

Rough ER Endoplasmic reticulum associated with ribosomes involved in protein synthesis.

Retinoblastoma Rare form of eye cancer that initiates from the immature retinal cells.

Randomised controlled trial Treatment protocol developed with reduced bias.

Registration, Evaluation, Authorization and Restriction of Chemicals (REACH) Regulation designed by the European Union in 2006.

Recommended exposure level (REL) It is an occupational exposure limit recommended by the United States National Institute for Occupational Safety and Health to the Occupational Safety and Health Administration (OSHA) in order to protect the health of workers associated with manufacturing or working with and exposed to substances hazardous to health.

Rifampicin Semisynthetic broad spectrum antibiotic produced from Streptomyces mediterrane, finds application in antitubercular medicine in treatment of tuberculosis and meningococcal meningitis.

Reactive oxygen species Causes oxidative stress.

Respiratory tract lining fluid RTLFs spreads over the respiratory tract epithelial cells (RTECs) from nasal mucosa to alveoli.

Scanning Electron Microscope (SEM); SEM uses a focused beam of high-energy electrons targeted to surface of solid specimens to generate signals.

Serum The straw colored clear liquid that separates from the blood when it is allowed to clot. It is from the serum that antibodies can be isolated.

Self-renewal A cell division in stem cells by which they can renew themselves.

Serotonin Chemically 5-hydroxytryptamine (5-HT) present in platelets, gastrointestinal mucosa and mast cells act as potent vasoconstrictor.

Sessile Incapable of voluntary movement and they remain attached to substratum.

Signaling molecules Molecules involved in signaling reactions.

Smooth ER Regions of endoplasmic reticulum devoid of ribosomes.

Species A population or groups that can reproduce within themselves and are reproductively isolated from all other populations.

Spleen A lymphoid organ in the abdominal cavity that plays important role in the immune system.

Sputum Substance expelled by coughing due to respiratory infection or other disease.

Stem cells Cells are capable to self-renew and to differentiate into different lineages.

Stratum corneum Outermost epidermal layer with many layers of flat, keratinized and enucleated cells.

Stromal cells Supporting cells capable of supporting growth of blood cells.

Subunit Vaccine A vaccine that uses merely one component of an infectious agent rather than the whole to stimulate an immune response.

Suppressor T cells A subset of T cells that reduces immune responses.

Surface markers Proteins expressed on the exterior surface of a cell that can be detected by antibodies.

SWCNTs Single-walled carbon nanotubes.

Synapse The junction between two neurons in a neural pathway, where one axon of one neuron terminates and comes into close proximity with the cell body or dendrites of another neuron.

small interfering RNA Double-stranded RNA molecules, 20–25 base pairs in length that prevents translation.

Systemic lupus erythematosus (SLE) Autoimmune disease that causes inflammation in the connective tissues and rashes in skin.

Superparamagnetic iron oxide nanoparticles (SPIONS) SPIONs consist of cores made of iron oxides targeted to the required area through external magnets.

Stage Specific Embryonic Antigen (SSEA) SSEA is a glycosphingolipid and a marker to stem cells.

Signal transducer and activator of transcription- STAT In response to cytokines and growth factors, STAT3 is phosphorylated by receptor-associated Janus kinases (JAK), forms homo- or heterodimers and acts as transcription activators.

T cells Small white blood cells that mature in the thymus and secrete lympho kines and contribute to T cell mediated immunity.

Thoracic In relation to the chest or thorax.

Thorax Body part within the neck and the diaphragm.

Thrombus Blood clot.

Thymus A primary lymphoid organ, near the chest, where T lymphocytes proliferate and mature.

Tomography A non-invasive imaging technique for detecting images of structures in a selected plane of tissue.

Toxicity Poisonous nature of chemical substance that can destroy living tissue and is a health hazard.

Transcutaneous (percutaneous) Through the skin.

Transdermal Through the skin.

Translational research Research focusing on the use of knowledge gathered from basic research to develop new drugs, treatments or therapies.

Transmembrane protein A membrane component, where a hydrophobic region of protein resides in the membrane and hydrophilic regions are exposed on either sides of the membrane.

Trophic level In a food chain, a level containing organisms of identical feeding habits.

Tb Tuberculosis is an infectious disease of the lungs caused by the bacteria *Mycobacterium tuberculosis*.

TCR Expressed by *T* cells that are able to identify the processed antigenic peptides expressed by major histocompatibility complex (MHC) molecule expressed on antigen presenting cells (APC).

Terminal deoxynucleotidyl transferase Tdt is a DNA polymerase expressed in immature, pre-B and pre-T cells.

Transforming growth factor beta 1 Member of transforming growth factor beta superfamily of cytokines that performs cellular functions, including cell growth, proliferation, differentiation, and apoptosis.

T helper cells T(H) Such cells are present in adaptive immunity that help to activate B cells to secrete antibodies and macrophages in order to destroy ingested microbes, and also help to activate cytotoxic T cells (CTLs) to kill infected target cells.

Toll like Receptors-TLRs Type I transmembrane pattern recognition receptors (PRRs) that can recognize the PAMPs and DAMPs from pathogens and foreign particles and initiate the innate and adaptive immune response.

Ultraplankton Planktonic organisms less than 2 micrometers in size.

Umbilical cord blood stem cells Stem cells collected from the umbilical cord at birth.

UV Ultraviolet Ultrafine Nanoparticles.

Vaccine A substance containing heat killed, inactivated and attenuated antigenic components from an infectious organism that can confer protection against infectious agents by stimulating immune response.

Vasoconstriction Reduction in the diameter of blood vessels.

Vasodilation Increase in the diameter of blood vessels.

vascular cell adhesion molecule-1

vascular endothelial growth factor

WBC White blood cell.

Zooplankton Animals with planktonic existence.

Zooxanthellae Dinoflagellates living endosymbiotically in association with other invertebrate groups (e.g., corals).

Abbreviations

$^1O_2$	Singlet oxygen
8-nitroG	8-nitroguanine
AC	Activated charcoal
ACGIH	American Conference of Governmental Industrial Hygienists
AChE	Acetylcholinesterase
AD	Alzheimer's disease
ADCC	Antibody-dependent cellular cytotoxicity
AF	Aflatoxin
AgNPs	Silver nanoparticles
ALDH	Aldehyde dehydrogenase
ALP	Alkaline phosphatase
AMP	Antimicrobial peptide
APU	Amphiphilic polyurethane
APC	Antigen presenting cells
APP	Acute phase proteins
AuNPs	Gold nanoparticles
$A\beta$	Amyloid-β peptides
BALF	Bronchoalveolar lavage fluid
BBB	Blood–brain barrier
BBBD	Blood–brain barrier disruption
BCSFB	Blood cerebrospinal fluid barrier
BM	Bone marrow
BMEC	Brain microvessel endothelial cells
BP	Bulk particles
Br_2	Bromine
BSA	Bovine serum albumin
BUN	Blood–urea nitrogen
CARPA	Complement activation-related pseudo allergy
C. carpio	Cyprinus carpio
C. catla	Catla catla
C. dilutus	Chironomus dilutus
C. dubia	Ceriodaphnia dubia
C. elegans	Caenorhabditis elegans
C. riparius	Cophixalus riparius
C4-BP	C4-binding protein
Ca^{2+}	Calcium
cAMP	Cyclic adenosine monophosphate
CaO	Calcium oxide
CAPRA	Complement activation related pseudo allergy
CAT	Catalase
CB	Carbon black
CCBT	Computerized cognitive behavioral therapy
CCl_4	Carbon tetrachloride
CD	Cluster of differentiation
CD19	B-cell marker
CD22	B-cell marker
CD25	T-cell marker
CD3	T-cell surface marker associated with T cell receptor (TCR) complex
CD4	T cell surface marker
CD8	T cytotoxic cell marker

https://doi.org/10.1515/9783110579093-013

CDK	Cyclin-dependent kinase
CdS	Cadmium sulfide
CdSe	Silica-coated cadmium selenide
CdTe	cadmium telluride
CeO2NPs	Ceria nanoparticles
CF	Cystic fibrosis
CGC	Cerebellar granule cells
CHCl$_3$	Chloroform
CHO	Chinese hamster ovary
Cl$_2$	Chlorine
cLSM	Confocal laser scanning microscopy
CMA	Chaperone-mediated autophagy
CMC	Cellulose microcrystal
CMI	Cell-mediated immunity
CN	Cellulose nanomaterial(s)
CNH	Carbon nanohorns
CNP	2′,3′-cyclic nucleotide 3′-phosphodiesterase
CNS	Central nervous system
CNT	Carbon nanotubes
Co	Cobalt
CoO	Cobalt monoxide
COPD	Chronic obstructive pulmonary disease
CPDs	Phosphorus containing dendrimers
CREB	cAMP response element binding
CRP	C-reactive protein
CSC	*Cancer stem cell*
CSNP	Chitosan nanoparticle
CT	Computed tomography
CTL	Cytotoxic T lymphocyte
CuO	Copper oxide
CuONP	Copper oxide nanoparticle
CVS	Cardiovascular system
D. subspicatus	*Desmodesmus subspicatus*
D. magna	*Daphnia magna*
D. rerio	*Danio rerio*
D. subspicatus	*Desmodesmus subspicatus*
DA	Dopamine
DAF	Decay accelerating factor
DAMP	Damage-associated molecular pattern
DFS	Disease free survival
DG	Dentate gyrus
DISC	Death-inducing signaling complex
DLS	Dynamic light scattering
DMSA	Dimercaptosuccinic acid
DNAPLs	Dense non aqueous phase liquids
DNTT	DNA nucleotidyl terminal transferase
Dox	Docetaxal
DOX	Doxorubicin
DWCNT	Double-walled carbon nanotubes
E. heterophylla	*Euphorbia heterophylla*
E. affinis	*Eurytemora affinis*
E. coli	*Escherichia coli*
EBC	exhaled breath condensate

EC	Endothelial cells
ECHA	European Chemical Agency
EGR	Early growth response
EHS	Environmental Health and Safety
EMB	Ethambutol
EPA	US Environmental Protection Agency
EpCAM	Epithelial cell adhesion molecule
ECM	Extracellular matrix
F. heteroclitus	*Fundulus heteroclitus*
FADD	Fas-associated death domain
FAR	Farnesol
Fas ligand	Fas L
fC_{60}	Functionalized fullerene
FcεRI	Fcε receptor
FDA	Food and Drug Administration
Fe-bLf	Iron-saturated bovine lactoferrin
FeNP	Iron nanoparticle
Fe_3O_4	Magnetite
FQDs	Fluorescent Quantum Dots
FSA	UKs Food Standard Agency
f-SWCNT	Functionalized SWCNT
FUS	Focused ultrasound
GALT	Gut-associated lymphoid tissue
GBM	Glioblastoma multiforme
GCSF	Granulocyte colony stimulating factor
GDNF	Glial cell-line derived neurotrophic factor
GF	Growth factors
GFAP	Glial fibrillary acidic protein
GI	Gastrointestinal
GMO	Genetically modified food
GO	Graphene oxide
GO	Gene ontology
GPx	Glutathione peroxidase
GSNP	Gelatin-siloxane nanoparticle
H. azteca	*Hyalella azteca*
HA	Hyaluronic acid
HAND	HIV-associated neurocognitive disorders
H_2O_2	Hydrogen peroxide
HARNS	High aspect ratio nanoparticle
HD	High Definition
HD	Heart Disease
HEK293	Human embryonic kidney cells 293
hES cells	Human embryonic stem cells
HI	Humoral immunity
HIV	Human immunodeficiency virus
Hg	Mercury
hMSC	Human mesenchymal stem cell
hNPCs	Human neural progenitor cells
HSC	Hematopoietic stem cell
HupA	Huperzine A
IAP	Inhibit apoptosis
i.m.	Intramuscular
i.p.	Intraperitoneal

i.v.	Intravenous
ICP-OES	Inductively coupled plasma-optical emission spectroscopy
IFN	Interferon
Ig	Immunoglogulin
IHD	Ischemic heart disease
IL	Interleukin
IL-2R	IL-2 receptor
INH	Isoniazid
IONP	Iron oxide nanoparticle
IP3	Inositol trisphosphate
IRT	Immunoreactive trypsin
ISR	In-stent restenosis
IT	Intratracheal instillation
ITAM	Immunoreceptor tyrosine-based activation motifs
IWGN	Interagency Working Group on Nanotechnology
JAK	Janus tyrosine kinase
JE	Japanese encephalitis
JNK	c-Jun N-terminal kinase
kDa	kiloDalton
KEGG	Kyoto Encyclopedia of Genes and Genomes
L	Ligand
L. luteola	*Lymnaea luteola*
L. plumulosus	*Leptocheirus plumulosus*
L. plumulosus	*Leptocheirus plumulosus*
L. stagnalis	*Lymnaea stagnalis*
L. rohita	*Labeo rohita*
La_2O_3	Lanthanum oxide
$LaPO_4$	Lanthanum(III) Phosphate
LDH	lactate dehydrogenase
Lf	Lactoferrin
Li_2TiO_3	Lithium titanate
LIF	Laser induced fluorescence
LNPs	Lipid-based nanoparticles
LPO	Lipid peroxidation
LPS	Lipopolysaccharide
LT	Lymphotoxin
LTP	Long-term potentiation
MAC	Membrane attack comples
MASP-1	Mannan-binding lectin serine protease 1
MBL	Mannose binding lectin
MBP	Mannan-binding protien
MCP	Membrane cofactor protein
MDA	Malondialdehyde
MDCK	Madin-Darby canine kidney
MDR-Tb	Multidrug-resistant Tb
mESC	Mouse embryonic stem cell
MgO	Magnesium oxide
MHC	Major histocompatibility
MI	Myocardial infarction
MIP	Macrophage inflammatory protein
MNi	Micronuclei
MNP	Metal nanoparticle
MoNP	Molybdenum nanoparticle

MoO(3)	Molybdenum trioxide
MPS	Mononuclear phagocyte system
MRAM	Magnetic Random Access Memory
MRI	Magnetic resonance imaging
MRSA	Methicillin-resistant *Staphylococcus aureus*
MS	Multiple sclerosis
MS	Mass spectrometry
MSC	Mesenchymal stem cells
MSDS	Material safety data sheet
MSI	Mass spectrometry imaging
MSN	Mesoporous silica nanoparticle
Mtb	*Mycobacterium tuberculosis*
MTT	-(4,5-dimethylthiazol-2-yl)-2,5-diphenyltetrazolium bromide
MWCNT	Multi-walled carbon nanotube
MWM	Morris water maze
NADPH	Nicotinamide adenine dinucleotide phosphate
NCCMH	National Collaborating Centre for Mental Health
NCI	National Cancer Institute
NCRPM	National Council on Radiation Protection and Measurements
NDV	Newcastle disease virus
NEDO	New Energy and Industrial Technology Development Organization
NGC	Nerve guidance conduit
NGF	nerve growth factor
NHS	National Health Service
Ni	Nickel
NiO	Nickel oxide
NiONP	Nickel oxide nanoparticle
NIOSH	The United States National Institute for Occupational Safety and Health
NK	Natural killer
NLC	Nanostructured lipid carrier
NLRP-3	NOD-like receptor family
NLR	Nod-like Receptor
NM	Nanomaterial
nm	Nanometer
NO	Nitric oxide
NRDE	Nanoparticle-rich diesel exhaust
NRP-1	Neuropilin-1
NS	Nanosphere
NSC	Neural stem cell
NSCLC	Non-small cell lung cancer
NSE	Nanoscale science and engineering
NW	Nanoworm
nZVI	Zerovalent iron
O. latipes	*Oryzias latipes*
O. melastigma	*Oryzias melastigma*
O. mossambicus	*Oreochromis mossambicus*
O. mykiss	*Oncorhynchus mykiss*
O_2^-	Superoxide
$OC\text{-}Fe_3O_4$	Oleic acid coated iron oxide nanoparticles
OCT4	Octamer 4
OECD	Organization for Economic Cooperation and Development
OECD	Organization for Economic Co-operation and Development
OEL	Occupational exposure limit

OH	Hydroxyl radical
OH-MWCNTs	Hydroxylated multiwalled carbon nanotubes
OL	Odorranalectin
OPN	Osteopontin
OS	Osteocalcin
OSTP	Office of Science and Technology Policy
P. subcapitata	*Pseudokirchneriella subcapitata*
P. subcapitata	*Pseudokirchneriella subcapitata*
P. dumerilii	*Platynereis dumerilii*
PECAM	Platelet endothelial cell adhesion molecule
P. falciparum	*Plasmodium falciparum*
PA	Peptide amphiphile
PAH	Polycyclic aromatic hydrocarbons
PAMAM	Polyamidoamine
PAMPs	Pathogen-associated molecular patterns
PBAE	Poly(beta-amino ester)
PBCA	Poly(n-butylcyano-acrylate)
PBTs	Pediatric brain tumors
PCNA	Proliferating cell nuclear antigen
PCT	Procalcitonin
PD	Pharmacodynamics
PD	Parkinson's disease
PEG	Polyethylene glycol
PEPs	Printer emitted particles
Peroxynitrite	$ONOO^-$
PET	Positron emission tomography
PFOB	Perfluorocarbon nanoparticle
PGA	Poly gamma glutamic acid
PGN	Peptidoglycan
PK	Pharmacokinetics
PKC	Protein kinase C
PKDL	Post-kala-azar dermal leishmaniasis
PLGA	Poly lactic-co-glycolic acid
PLLA	Poly-(l-lactide)
PM	Particulate matter
PMMA	Polymethylmethacrylate
PMN	Phagocytic mononuclear
PNC	Particle number concentration
PP	Peyer's patches
PPARγ	Peroxisome proliferator-activated receptor gamma
PRRs	Pattern recognition receptors
PS	Polystyrene
PSA	Prostrate specific antigen
PSC	Pluripotent stem cells
PSf	Pulmonary surfactant
Pt	Platinum
PTPRC	Protein tyrosine phosphatase receptor type C
PVAc	Poly(vinyl) acetate
PVM	Perivascular macrophage
PVP	Polyvinyl pyrrolidone
PZA	Pyrazinamide
QD	Quantum Dot
qPCR	Quantitative polymerase chain reaction

QSAR	Adverse outcome quantitative structure–activity relationship
R. philippinarum	*Ruditapes philippinarum*
Rb	Retinoblastoma
RCT	Randomized controlled trial
RE	Reticuloendothelial
REACH	Registration, Evaluation, Authorization and Restriction of Chemicals
REL	Recommended exposure level
RGD	Cyclic arginylglycylaspartic acid
RIF	Rifampicin
ROS	Reactive oxygen species
RTLF	Respiratory tract lining fluid
RT-PCR	Reverse transcription-polymerase chain reaction
S. droebachiensis	*Strongylocentrotus droebachiensis*
SAEC	Small airway epithelial cells
SAS	Synthetic amorphous SiO_2
SC	Stem Cell
SC	Stratum corneum
SCI	Spinal cord injury
SD	Sprague Dawley
SEERS	Surface-enhanced resonance Raman Scattering
SEM	Scanning electron microscope
SeNPs	Selenium nanoparticles
SERS	Surface Enhanced Raman Scattering
SHRA	Suwannee River humic acid
shRNA	Short hairpin RNA
SiO_2	Silicon dioxide
SiO_2NP	Silica nanoparticle
siRNA	Small interfering RNA
SLE	Systemic lupus erythematosus
SLN	Solid lipid nanoparticle
SMAC	Second mitochondria-derived activator of caspases
SMR	Standardized mortality ratio
SOD	Superoxide dismutase
SOM	Self organizing map
SPIO	Superparamagnetic iron oxide
SPION	Superparamagnetic iron oxide nanoparticle
SPMN	Superparamagnetic nanoparticle
SSEA	Stage Specific Embryonic Antigen
STAT	Signal transducer and activator of transcription
SHp	Stroke homing peptide
SUF	Serum ultrafiltrate
SWCNT	Single-walled carbon nanotube
T. pyriformis	*Tetrahymena pyriformis*
Tb	Tuberculosis
TC	T cytotoxic cells
TCC	Triclocarban
TCR	T cell Receptor
TdT	Terminal deoxynucleotidyl transferase
TE	Transposable element
TEOS	Tetraethyl orthosilicate
TGF	Transforming growth factor
TGF-β1	Transforming growth factor beta 1

Th	Helper T cells
TiO$_2$	Titanium dioxide
TJ	Tight junction
TLRs	Toll-like receptors
TLV	Threshold Limit Values
TMC	N-trimethylated chitosan
TNF	Tumor necrosis factor
TRAcP	Tartrate-resistant acid phosphatase
TRADD	TNF receptor-associated death domain
TRAIL	TNF-related apoptosis-inducing ligand
TRAP	Tartrate-resistant acid phosphatase
TTA	Tumor-targeting aptamer
TWIST-1	Twist-related protein 1
UFP	Ultrafine particles
UNPs	Ultrafine nanoparticles
usGO-AuNPs	Ultra-small graphene oxide-supported gold nanoparticles
USPION	Ultrafine superparamagnetic iron oxide nanoparticles
V. fischeri	*Vibrio fischeri*
VCAM-1	Vascular cell adhesion molecule-1
VEGF	Vascular endothelial growth factor
VL	Visceral leishmaniasis
WBC	White blood cells
WHO	World health organization
WO$_3$ NPs	Tungsten (IV) oxide nanoparticles
Xe	Xenon
Y$_2$O$_3$	Yttrium oxide
ZnO	Zinc oxide
ZrO2	Zirconium dioxide
α-Fe$_2$O$_3$	Hermatite
γ-Fe$_2$O$_3$	Maghemite
µg	Microgram
R. leguminosarum	*Rhizobium leguminosarum*
U. elongata	*Ulmus elongata*
T. aestivum	*Triticum aestivum*
O. sativa	*Oryza sativa*
V. radiata	*Vigna radiata*
H. verticillata	*Hydrilla verticillata*
P. vulgaris	*Phaseolus vulgaris*
L. sativum	*Lepidium sativum*
S. alba	*Sinapis alba*
S. saccharatum	*Sorghum saccharatum*
nano-TiSiO4	Titanium silicon oxide nanomaterial
L. lycopersicum	*Lycopersicon lycopersicum*
C. sativum	*Coriandrum sativum*
A. cepa	*Allium cepa*
L. sativum	*Lepidium sativum*
L. utisassimmum	*Linum utisassimmum*
C. sativus	*Cucumis sativus*
T. aestivum	*Triticum aestivum*
E. angustifolia	*Elaeagnus angustifolia*
S. lycopersicum	*Solanum lycopersicum*

Index

https://doi.org/10.1515/9783110579093-014